Oxidative Stress in Human Toxicology

Oxidative Stress in Human Toxicology

Editor

Tim Hofer

MDPI • Basel • Beijing • Wuhan • Barcelona • Belgrade • Manchester • Tokyo • Cluj • Tianjin

Editor
Tim Hofer
Norwegian Institute of Public Health
Oslo, Norway

Editorial Office
MDPI
St. Alban-Anlage 66
4052 Basel, Switzerland

This is a reprint of articles from the Special Issue published online in the open access journal *Antioxidants* (ISSN 2076-3921) (available at: https://www.mdpi.com/journal/antioxidants/special_issues/Oxidative_Toxicology).

For citation purposes, cite each article independently as indicated on the article page online and as indicated below:

LastName, A.A.; LastName, B.B.; LastName, C.C. Article Title. *Journal Name* **Year**, *Volume Number*, Page Range.

ISBN 978-3-0365-8054-8 (Hbk)
ISBN 978-3-0365-8055-5 (PDF)

Contents

antioxidants

Editorial

Oxidative Stress in Human Toxicology

Tim Hofer

Department of Environmental Health, Norwegian Institute of Public Health, P.O. Box 222 Skøyen, N-0213 Oslo, Norway; tim.hofer@fhi.no; Tel.: +47-21076671

This Special Issue (same name as title) focuses on human exposure to foreign chemicals (xenobiotics) that cause oxidative stress. Certain xenobiotics and/or their metabolites (e.g., quinones) can directly mediate the formation of reactive oxygen species (ROS; e.g., superoxide and peroxides) [1] that can inflict damage to biomolecules [2,3] and/or affect signaling pathways. Xenobiotics can also indirectly cause oxidative stress by affecting protective proteins (i.e., antioxidant enzymes and metal transporters/chelators) and their expression (epigenetics). Increased ROS levels can cause antioxidant depletion. Some xenobiotics, however, are antioxidants [4,5]. Describing adverse outcome pathways (AOPs) related to toxicants and oxidative stress is presently a hot topic as well as suggesting relevant experimental models (e.g., what are suitable animals for studying human conditions, and, how can new 3D human cell culture systems be used) for detecting xenobiotics' eventual toxic effects.

Drug-induced liver injury (DILI) is a serious challenge for pharmaceutical companies as DILI often appears post-approval in a minority of patients when the drug is administered to thousands of patients [6]. Oxidative damage can be a major mechanism in DILI. In their new review, M.T. Donato and L. Tolosa describe liver cell-based assays (e.g., using ROS-sensitive fluorescent probes) suitable for high-content screening to detect drug-induced oxidative stress in vitro [6]. The suitability of using traditional 2D liver cell monolayers [4] versus new 3D culture systems is also discussed.

Unborn (embryos and fetuses) and new-born babies often represent the most sensitive groups, but the reason for this is not always clear. T.B. Jeong et al. found that two-week-old weaning mice were more vulnerable towards carbon tetrachloride (CCl_4)-induced liver injury than eight-week-old young adult mice [7]. Measured differences in basal hepatic glutathione (GSH (reduced form)) levels (higher in weaning mice) and the expression of hepatic enzymes such as cytochrome (CYP) P450s, GSH transferases, and GSH reductases may explain this age-related (developmental) susceptibility [7]. Adult mice had lower levels of the hepatic phase I enzymes CYP2E1 and CYP3A but had higher levels of phase II GSH S-transferase enzymes. CCl_4 markedly lowered the hepatic GSH level only in weaning mice that suffered from various types of damages including oxidative stress-inflicted ones, e.g., from ROS and lipid peroxidation.

Vancomycin is an antibiotic used to treat bacterial infections. It is often administrated intravenously. A side effect is that it can cause kidney injury, i.e., Vancomycin-associated acute kidney injury (VAKI) for which oxidative stress is thought to be a main mechanism. In a clinical metabolomic study by H.-S. Lee et al. [8], serum concentrations of amino acids and amino acid derivatives/metabolites were compared between human patients who met VAKI criteria vs. non-VAKI subjects (divided into three subgroups, two of which did not receive Vancomycin). Significant differences among the groups were found for some measured parameters, particularly for the tryptophan (metabolized via the kynurenine pathway) metabolite serotonin (5-HT) and its metabolite 5-hydroxyindoleacetic acid (5-HIAA), as well as for the ratio 5-HIAA/5-HT. The authors conclude that their study demonstrates that an increased 5-HIAA/5-HT ratio has the potential to act as a novel biomarker for VAKI detection and that this ratio can be a useful indicator for oxidative stress and inflammation [8].

Citation: Hofer, T. Oxidative Stress in Human Toxicology. *Antioxidants* **2021**, *10*, 1159. https://doi.org/10.3390/antiox10081159

Received: 14 July 2021
Accepted: 16 July 2021
Published: 21 July 2021

Silver nanoparticle (AgNP) interaction with cultured human adherent (lung epithelial A549) and suspension (monocytic U937) cells under static and dynamic flow conditions was investigated by K.E. Burns et al. [9]. The measured parameters included AgNP cell association, oxidative stress-related (intracellular ROS and GSH/GSSG ratio (GSSG is oxidized glutathione)) levels, activation of oxidative stress sensitive signaling pathways (phosphorylation of tumor suppressor protein p53 and nuclear factor kappa B (NFκB)), and production of inflammatory cytokines (interleukin-1β (IL-1β) and tumor necrosis factor-α (TNF-α)). AgNP cell association was higher for A549 than for U937 cells. For adherent cells, dynamic flow reduced AgNP contact due to the disruption of sedimentation. For suspension cells, however, dynamic flow increased AgNP cell association. Oxidative stress levels were found to be a function of cell type and flow conditions. U937 monocytes secreted considerably more cytokines (also when untreated) than alveolar epithelial A549 cells [9].

Arabic gum (*Acacia senegal*) is a plant extract known to have antioxidant and cytoprotective properties [10]. Using young adult rats, I. El-Garawani et al. investigated if Arabic gum, dissolved and administrated daily through drinking water, protected against oxidative stress and damage from a single intravenous high dose of Ioxitalamate (Telebrix-35®), an iodinated contrast medium used for X-ray imaging. Ioxitalamate caused various toxicities. Apoptosis in kidneys and blood leucocyte cytotoxicity were more severe after 14 days than after one day of Ioxitalamate injection. Arabic gum restored the oxidative stress status in kidneys, exerted antigenotoxic potential against DNA damage in peripheral blood leucocytes and bone marrow, and protected against bone marrow mitotic index alterations, results that can be attributed to metabolites of Arabic gum [10].

The emission of mercury (Hg) into the atmosphere, e.g., during power plant coal burning, leads to widespread global Hg contamination. After settlement, microbes convert Hg into methylmercury (MeHg) that possesses an increased lipophilicity (it readily crosses the placenta and concentrates in the fetal brain [11]) and neurotoxicity with distinct fetal and adult clinical forms. In their new review, M. Fujimura and F. Usuki describe several mechanisms of how MeHg mediates cellular toxicity [12]. Not only does MeHg disturb antioxidant proteins and enzymes, causing oxidative stress due to its high affinity for selenohydryl (SeH) and sulfhydryl (SH) groups as well as selenides (Se-Se), but MeHg also prevents the synthesis (Se-dependent) of selenoenzymes due to mRNA degradation, affects cellular signaling pathways such as those involving redox-regulated Kelch-like ECH-associated protein 1 (Keap1), and, induces neuronal hyperactivity and apoptosis [12].

A majority (85–90%) of Alzheimer's and Parkinson's disease cases are thought to result from nongenetic causes. Thus, environmental factors including chemical exposures may play strong roles. Post-mortem analyses of affected brain regions often reveal oxidative damage. These diseases develop with age and bioaccumulation of foreign chemicals (xenobiotics) [13] is suspected. A dark brown pigment present in catecholamine neurons called neuromelanin binds xenobiotics (both organic substances and metals) as well as endogenously formed potentially toxic metabolites (e.g., dopamine oxidation products) and excess metals (e.g., iron) that are stored away inside membrane-encapsulated tiny balls. The neuromelanin content increases almost linearly with age inside dopaminergic and noradrenergic brain neurons. Old neurons are sensitive to toxic insults and are lost in disease. In their new review, A. Cappucciati et al. [14] hypothesize that xenobiotics somehow disturb the build-up of neuromelanin and/or its protective function, which causes toxicity (including ROS formation) as xenobiotics and endogenous metabolites and excess metals are no longer stored away. The authors review the latest knowledge on neuromelanin biosynthesis, list neuromelanin-interacting xenobiotics, and suggest suitable experimental models for studying neuromelanin-related toxicity [14]. The authors question the use of rodents such as mice and rats (devoid or low in neuromelanin) as models when resembling human (high in neuromelanin) degenerative diseases.

In conclusion, oxidative stress from xenobiotic exposure is relevant for various human conditions and their induction. Identifying, developing, and using increasingly more suitable experimental models may be key to success in human toxicology research.

Funding: This research received no external funding.

Conflicts of Interest: The author declares no conflict of interest.

References

1. Steffensen, I.L.; Dirven, H.; Couderq, S.; David, A.; D'Cruz, S.C.; Fernandez, M.F.; Mustieles, V.; Rodriguez-Carrillo, A.; Hofer, T. Bisphenols and Oxidative Stress Biomarkers-Associations Found in Human Studies, Evaluation of Methods Used, and Strengths and Weaknesses of the Biomarkers. *Int. J. Environ. Res. Public Health* **2020**, *17*, 3609. [CrossRef] [PubMed]
2. Hofer, T. Oxidation of 2′-deoxyguanosine by H_2O_2-ascorbate: Evidence against free $OH^•$ and thermodynamic support for two-electron reduction of H_2O_2. *J. Chem. Soc. Perkin Trans. 2* **2001**, *2*, 210–213. [CrossRef]
3. Moller, L.; Hofer, T.; Zeisig, M. Methodological considerations and factors affecting 8-hydroxy-2′-deoxyguanosine analysis. *Free. Radic. Res.* **1998**, *29*, 511–524. [CrossRef] [PubMed]
4. Hofer, T.; Jorgensen, T.O.; Olsen, R.L. Comparison of food antioxidants and iron chelators in two cellular free radical assays: Strong protection by luteolin. *J. Agric. Food Chem.* **2014**, *62*, 8402–8410. [CrossRef]
5. Mohamed, Y.M.A.; Vik, A.; Hofer, T.; Andersen, J.H.; Hansen, T.V. Polyunsaturated fatty acid-derived chromones exhibiting potent antioxidant activity. *Chem. Phys. Lipids* **2013**, *170*, 41–45. [CrossRef] [PubMed]
6. Donato, M.; Tolosa, L. High-Content Screening for the Detection of Drug-Induced Oxidative Stress in Liver Cells. *Antioxidants (Basel)* **2021**, *10*, 106. [CrossRef] [PubMed]
7. Jeong, T.B.; Kwon, D.; Son, S.W.; Kim, S.H.; Lee, Y.H.; Seo, M.S.; Kim, K.S.; Jung, Y.S. Weaning Mice and Adult Mice Exhibit Differential Carbon Tetrachloride-Induced Acute Hepatotoxicity. *Antioxidants (Basel)* **2020**, *9*, 201. [CrossRef] [PubMed]
8. Lee, H.-S.; Kim, S.-M.; Jang, J.-H.; Park, H.-D.; Lee, S.-Y. Serum 5-Hydroxyindoleacetic Acid and Ratio of 5-Hydroxyindoleacetic Acid to Serotonin as Metabolomics Indicators for Acute Oxidative Stress and Inflammation in Vancomycin-Associated Acute Kidney Injury. *Antioidants (Basel)* **2021**, *10*, 895. [CrossRef]
9. Burns, K.E.; Uhrig, R.F.; Jewett, M.E.; Bourbon, M.F.; Krupa, K.A. Characterizing the Role of Biologically Relevant Fluid Dynamics on Silver Nanoparticle Dependent Oxidative Stress in Adherent and Suspension In Vitro Models. *Antioxidants (Basel)* **2021**, *10*, 832. [CrossRef] [PubMed]
10. El-Garawani, I.; Hassab El-Nabi, S.; El Kattan, A.; Sallam, A.; Elballat, S.; Abou-Ghanima, S.; El Azab, I.H.; El-Seedi, H.R.; Khalifa, S.A.M.; El-Shamy, S. The Ameliorative Role of Acacia senegal Gum against the Oxidative Stress and Genotoxicity Induced by the Radiographic Contrast Medium (Ioxitalamate) in Albino Rats. *Antioxidants (Basel)* **2021**, *10*, 221. [CrossRef] [PubMed]
11. Duale, N.; Gutzkow, K.B.; Hofer, T.; Lindeman, B. Impact of Environmental Pollutants on Placentation. In *Human Placental Trophoblasts. Impact on Maternal Nutrition*; Duttaroy, A.K., Basak, S., Eds.; CRC Press: Boca Raton, FL, USA, 2016; pp. 345–359.
12. Fujimura, M.; Usuki, F. Methylmercury-Mediated Oxidative Stress and Activation of the Cellular Protective System. *Antioxidants (Basel)* **2020**, *9*, 1004. [CrossRef] [PubMed]
13. Hofer, T.; Myhre, O.; Peltola-Thies, J.; Hirmann, D. Analysis of elimination half-lives in MamTKDB 1.0 related to bioaccumulation: Requirement of repeated administration and blood plasma values underrepresent tissues. *Environ. Int.* **2021**, *155*, 106592. [CrossRef] [PubMed]
14. Capucciati, A.; Zucca, F.A.; Monzani, E.; Zecca, L.; Casella, L.; Hofer, T. Interaction of Neuromelanin with Xenobiotics and Consequences for Neurodegeneration; Promising Experimental Models. *Antioxidants (Basel)* **2021**, *10*, 824. [CrossRef] [PubMed]

antioxidants

Article

Sidestream Smoke Extracts from Harm-Reduction and Conventional Camel Cigarettes Inhibit Osteogenic Differentiation via Oxidative Stress and Differential Activation of intrinsic Apoptotic Pathways

Nicole R. L. Sparks [1,2], Lauren M. Walker [1,2], Steven R. Sera [1,3], Joseph V. Madrid [1], Michael Hanna [1], Edward C. Dominguez [2] and Nicole I. zur Nieden [1,2,3,*]

1 Department of Molecular, Cell & Systems Biology and Stem Cell Center, College of Natural and Agricultural Sciences, University of California Riverside, Riverside, CA 92521, USA
2 Environmental Toxicology Graduate Program, University of California Riverside, Riverside, CA 92521, USA
3 Cell, Molecular and Developmental Biology Graduate Program, University of California Riverside, Riverside, CA 92521, USA
* Correspondence: nicolezn@ucr.edu

Citation: Sparks, N.R.L.; Walker, L.M.; Sera, S.R.; Madrid, J.V.; Hanna, M.; Dominguez, E.C.; zur Nieden, N.I. Sidestream Smoke Extracts from Harm-Reduction and Conventional Camel Cigarettes Inhibit Osteogenic Differentiation via Oxidative Stress and Differential Activation of intrinsic Apoptotic Pathways. *Antioxidants* **2022**, *11*, 2474. https://doi.org/10.3390/antiox11122474

Academic Editor: Tim Hofer

Received: 10 November 2022
Accepted: 12 December 2022
Published: 15 December 2022

Abstract: Epidemiological studies suggest cigarette smoking as a probable environmental factor for a variety of congenital anomalies, including low bone mass, increased fracture risk and poor skeletal health. Human and animal in vitro models have confirmed hypomineralization of differentiating cell lines with sidestream smoke being more harmful to developing cells than mainstream smoke. Furthermore, first reports are emerging to suggest a differential impact of conventional versus harm-reduction tobacco products on bone tissue as it develops in the embryo or in vitro. To gather first insight into the molecular mechanism of such differences, we assessed the effect of sidestream smoke solutions from Camel (conventional) and Camel Blue (harm-reduction) cigarettes using a human embryonic stem cell osteogenic differentiation model. Sidestream smoke from the conventional Camel cigarettes concentration-dependently inhibited in vitro calcification triggered by high levels of mitochondrially generated oxidative stress, loss of mitochondrial membrane potential, and reduced ATP production. Camel sidestream smoke also induced DNA damage and caspase 9-dependent apoptosis. Camel Blue-exposed cells, in contrast, invoked only intermediate levels of reactive oxygen species insufficient to activate caspase 3/7. Despite the absence of apoptotic gene activation, damage to the mitochondrial phenotype was still noted concomitant with activation of an anti-inflammatory gene signature and inhibited mineralization. Collectively, the presented findings in differentiating pluripotent stem cells imply that embryos may exhibit low bone mineral density if exposed to environmental smoke during development.

Keywords: developmental toxicity; embryonic stem cells; osteoblasts; tobacco smoke solution; oxidative stress; hypomineralization

1. Introduction

Women make up a staggering 20% of global smokers [1], 10–20% of which continue to smoke during pregnancy [2]. This is despite intensive public health initiatives that have publicized the detrimental effects of tobacco use during pregnancy. Among these are a plethora of neurological and physical developmental defects, including a decrease in overall body length, as well as head and chest circumference [3]. Smoking during any point in pregnancy also associates with derailed bone measurements in the offspring, such as bone age, bone mineral content and density, as well as neonate bone mineral quality [3,4]. Epidemiological studies have shown an increased risk of bone-development malformations, such as cleft lip and palate, and risk of fractures following minor injuries [3,5,6].

Despite most of this research surfacing from conventional cigarettes, "harm-reduction" products have become increasingly popular to consumers. As an alternative to conventional cigarettes, pregnant smokers will opt to use cigarettes with lower tar and nicotine content, previously advertised as "light" cigarettes with the assumption that such products are safer to use. However, this perception is concerning as users of harm-reduction products often engage in compensatory smoking to offset reduced nicotine delivery. Consequently, the frequency of smoking-associated cancer deaths seems equally high in those who use harm-reduction products [7].

To evaluate the comparative developmental toxicity of harm-reduction and conventional cigarette smoke as it pertains to skeletal malformations that arise upon exposure while in utero our group has recently turned to a well-established osteogenic differentiation model that is based on human embryonic stem cells (hESCs) [8]. Using this model, we have previously evaluated the skeletal embryotoxicity of mainstream (MS) and sidestream (SS) smoke solutions generated from multiple conventional and harm-reduction cigarettes [9]. Comparing MS and SS smoke solutions, our study supported the notion that tobacco has detrimental effects on human osteogenic differentiation with SS smoke, which burns off the tip of the cigarette [10], being more developmentally osteotoxic than MS smoke, which is the smoke actively inhaled by the smoker. Generally, developmental toxicity may be elicited by toxicants due to the cytotoxic nature of the exposure, because the toxicant alters cell fate trajectories in the absence of cytotoxicity [11], or by a combination of both. Intriguingly, sidestream smoke solutions from harm-reduction products elicited differentiation defects at sub-toxic concentrations, while developmental effects for conventional tobacco product solutions followed their general cytotoxicity. This differential effect was strongest for cigarettes of the Camel brand.

We then confirmed the adverse skeletal outcome associated with sidestream smoke encounters in a whole-animal exposure regimen using the zebrafish. At high concentrations, sidestream cigarette smoke from several brands, including Camel, reduced hatching success and overall survival in the developing larvae [12]. At lower, non-lethal concentrations, craniofacial defects were noted that were grounded in impacted cartilage formation and/or bone mineralization independent of whether the skeletal structures were mesoderm or neural crest-derived. Numerous epidemiological, animal and in vitro studies, including ours, therefore now demonstrate the adverse effects of tobacco-related products towards developing embryos and human skeletal health, yet, the molecular underpinnings of the differential adverse skeletal effects caused by harm-reduction products in relation to conventional products remain unclear.

As one possible molecular initiation event, oxidative stress has recently been discussed to contribute to the embryotoxicity of a variety of different chemicals [13–15], including tobacco smoke and its extracts [16]. Arising as a consequence of chemical or environmental insult, oxidative stress is classically defined as the "imbalance of reducing and oxidizing equivalents where the latter predominates" [17]. In such cases, increased production of reactive oxygen species (ROS) contributes to a loss of tissue function [18–21]. Due to this existing relationship between developmental inhibition and oxidative stress, the purpose of the current study was to evaluate the contribution of oxidative stress to the differential developmental osteotoxicity of Camel smoke solutions observed in vitro.

2. Materials and Methods

2.1. Cell Culture

Human H9 embryonic stem cells (ESCs), acquired from WiCell (WiCell Research Institute, Madison, WI, USA), were maintained in mTeSR® medium (Stem Cell Technologies, Vancouver, BC, Canada) and kept in the undifferentiated state at 37 °C in a humid 5% CO_2 environment. Pluripotent colonies were passaged every 5 days by dissociating cells with Accutase® (Innovative Cell Technologies, Inc., San Diego, CA, USA) and a cell scraper. Cells were replated on Matrigel (BD Biosciences, Franklin Lakes, NJ, USA) coated culture plates.

2.2. *Osteogenic Differentiation*

At confluency, pluripotent colonies were induced to undergo osteogenesis with control differentiation medium consisting of Dulbecco's modified Eagle's medium (DMEM; Gibco, Waltham, MA, USA) containing 15% FBS (Atlanta Biologicals, Flowery Branch, GA, USA), 1% non-essential amino acids (NEAA; Gibco), 1:200 penicillin/streptomycin (Gibco), and 0.1 mM β-mercaptoethanol (Sigma-Aldrich, Waltham, MA, USA) for 5 days as described [8]. Starting from the fifth day of culture, control differentiation medium was supplemented for the remaining differentiation duration with osteogenic factors: 0.1 mM β-glycerophosphate (βGP; Sigma-Aldrich), 50 μg/mL ascorbic acid (AA; Sigma-Aldrich), and 1.2×10^{-7} M $1.25(OH)_2$ Vitamin D_3 (VD_3; Calbiochem, San Diego, CA, USA).

2.3. *Production of Smoke Solution*

Commercially available conventional and harm-reduction Camel cigarettes were purchased from a local retailer and used to make MS and SS smoke solutions with a method described previously in detail [22,23]. Smoke solutions were generated using a University of Kentucky smoking machine that took a 2.2 s puff of MS every minute. MS smoke solution was generated by pulling 30 puffs of MS smoke through 10 mL of DMEM culture medium. During MS smoke production, SS smoke solution was produced by collecting the smoke that burned off the end of the cigarette and pulling it through 10 mL of DMEM. SS smoke was collected continuously, while MS smoke was collected during each puff. Both MS and SS solutions were made at concentrations of 3 puff equivalents (PE). Immediately after preparation, smoke solutions were filtered through a 0.2 micron Acrodisc® PSF Syringe Filter (Pall Corporation, Port Washington, NY, USA), aliquoted into sterile Eppendorf tubes, and stored in a −80 °C freezer until used. Experiments were performed using either MS or SS at previously identified concentrations alongside an untreated control [9]. Specifically, both Camel and Camel Blue MS effective doses were at 1 PE, Camel SS at 0.2 PE and Camel Blue SS at 0.05 PE. All selected effective doses were found to be in the physiological range and thus very relevant to actual human exposure [9]. All non-effective doses were used at 0.01 PE. Osteogenic differentiation of hESCs was induced as described above, and cultures were treated with smoke solution throughout the experiment. Smoke solutions were replenished with each media change.

2.4. *Calcium Assay*

To determine osteogenic differentiation yield, matrix calcium was quantified by reacting cells lysed in modified radioimmunoprecipitation (RIPA) buffer with Arsenazo III (Genzyme) and subsequent measurement at 655 nm [24]. The concentration of total calcium in the sample was calculated based on a $CaCl_2$ standard and normalized to the total protein content of the sample measured with the Lowry method (Bio-Rad, Hercules, CA, USA) [24].

2.5. *Antioxidant and Caspase Inhibitor Treatment*

To counteract tobacco-induced oxidative stress, three antioxidants were used concomitantly with tobacco treatment during days 0–7 of differentiation: ascorbic acid (AA; Sigma-Aldrich) [10 μM], dl-α-tocopherol acetate (vitamin E; Supelco, Sigma-Aldrich) [10 μM], and glutathione reduced ethyl ester (GSH-OEt; Sigma-Aldrich) [500 μM]. The antioxidants were replenished with every medium change.

To explore the involvement of caspases 4, 8 and 9 in tobacco-related inhibition of osteogenic differentiation, tobacco-treated cultures were simultaneously dosed with caspase 4 inhibitor (4i; Z-LEVD-FMK, PromoCell GmbH, Heidelberg, Germany), caspase-8 inhibitor (8i; Z-IETD-FMK, R&D Systems, Minneapolis, MN, USA) or caspase 9 inhibitor (9i; Z-LEHD-FMK, R&D Systems), all at 3 μM, during days 0–7 of differentiation. Inhibitor-supplemented medium was replaced with every medium change.

2.6. Superoxide Anion Detection

Generation of superoxide anion was determined using a Lumimax Superoxide Anion Detection Kit (Agilent Technologies, Santa Clara, CA, USA). H9 cells were trypsinized, washed with phosphate-buffered saline (PBS), and resuspended in fresh medium to incubate for 30 min at 37 °C. A total of 5 × 10^5 cells were incubated in superoxide anion assay medium including 0.1 mM luminol solution and 125 µM enhancer at room temperature for 30 min. The chemiluminescent light emissions of superoxide anion were measured with a Lucetta™ luminometer (Lonza, Basel, Switzerland).

2.7. MitoSOX Assay

Superoxide formation specially produced by mitochondria was assessed using the commercially available MitoSOX Red Mitochondrial Superoxide Indicator dye (ThermoFisher M36008, ThermoFisher, Waltham, MA, USA). Adherent cells were washed with PBS and incubated with 2.5 µM MitoSOX in PBS for 10 min in the dark. Cells were then immediately imaged on a Nikon Ti fluorescent microscope. MitoSOX positive cells were identified using NIH ImageJ analysis software as outlined by Jensen [25].

2.8. MitoTracker Staining and Mitochondrial Analysis

Stress-related changes in mitochondrial morphology were visualized and quantified using the MitoTracker Deep Red FM fluorescent dye (ThermoFisher). H9 cells were trypsinized, washed with PBS, and resuspended in 200 nM MitoTracker dye prepared in fresh medium. Cells were incubated in darkness for 20 min at 25 °C, washed with PBS, and fixed in 4% paraformaldehyde for 15 min at room temperature. Fixed cells were then washed three times with PBS and permeabilized with 0.1% Triton X-100 in PBS for 15 min at room temperature. Cells were washed again and counterstained with 1 µg/mL 4′, 6-diamidino-2-phenylindole (DAPI) in PBS for 30 min. Cells were washed three more times with PBS and resuspended in PBS supplemented with 2% FBS and 1 mM ethylenediaminetetraacetic acid (EDTA). Cell number was quantified and adjusted to a concentration of 5 × 10^5 cells/mL. Cells were spun onto pre-coated Shandon Single Cytoslides (ThermoFisher) using a Shandon Cytospin 3 (Shandon). For each treatment, a 100 µL volume of fixed and stained cellular suspension was loaded into a cytospin funnel and centrifuged at 200 rpm for 5 min at low acceleration/deceleration settings. Slides were allowed to air-dry overnight before mounting with Fluoro-Gel (Electron Microscopy Sciences, Hatfield, PA, USA) for imaging.

Z-stack images were taken using a Leica DMi8 fluorescent confocal microscope and max projected to flatten out each image. Resultant images were pre-processed in ImageJ 1.48v (NIH) to prepare for mitochondrial morphological analysis. The MitoTracker Deep Red channel was first separated from the nuclear DAPI channel to allow for specific analysis of the mitochondria. Images were further processed using the Mitochondrial Network Analysis (MiNA) ImageJ plug-in (https://github.com/ScienceToolkit/MiNA, accessed on 16 November 2021) to prepare images for evaluation of mitochondrial networks within individual cells. Using the default MiNA settings, images were subjected to a 2-pixel gaussian blur rendering through the Enhance Local Contrast median filter, and a final processing through an unsharp mask tool to yield a "skeleton" or tracing of the mitochondrial networks in a given cell. Mitochondrial networks were evaluated using the MiNA analysis method as outlined by Valente et al. [26]. The mitochondrial network length, number of branches, and area of mitochondrial footprint were then automatically measured from the program-generated skeletons to quantify tobacco-related changes to mitochondrial networks.

2.9. Caspase 3/7 Stain

For determination of activated caspases 3/7, cells were incubated for 1 h in a 1X caspase 3/7 reagent conjugated to carboxyfluorescein fluorochrome (Guava Technologies, Hayward, CA, USA). The fluorescent signal was detected in cells where the reagent is covalently bound to the activated caspases, any unbound reagent was washed away with 1X apoptosis buffer provided by the manufacturer. Cells were observed and imaged on a Nikon Ti fluorescent microscope.

2.10. Apoptosis RT^2 Profiler qPCR Array

The correlation between tobacco exposure and apoptosis was examined based on expression changes of 84 apoptosis-associated genes using a Qiagen human Apoptosis RT^2 Profiler Array (complete gene list accessible at https://geneglobe.qiagen.com/us/product-groups/rt2-profiler-pcr-arrays, accessed on 5 May 2021). For this, hESCs were differentiated into osteoblasts as described above with concomitant exposure to either solvent, non-effective, or effective doses (50% inhibition of calcification) of tobacco smoke solutions as determined from concentration response curves [9]. RNA was isolated using the NucleoSpin RNA kit (Macherey-Nagel, Düren, Germany) and examined for RNA integrity using the Agilent 2100 Bioanalyzer. Only samples with an RNA integrity number of >8 were used for further processing. Five hundred nanograms of RNA were input into a cDNA reaction as described before [8]. qPCR reactions were set up using iQ SYBR Green Supermix (Bio-Rad) and 12.5 ng of cDNA per array well and cycled in a Bio-Rad iQ5 qPCR machine. Data were uploaded to the Qiagen Data Analysis Center at www.SABiosciences.com/pcrarraydataanalysis.php accessed on 5 May 2021, which uses the $\Delta\Delta C_T$ method [27] and a two-tailed *t*-test with equal variance for analysis. Normalization was performed to all five housekeeping genes (*ACTB*, *B2M*, *GAPDH*, *HPRT*, *RPLP0*). Heatmaps and volcano plots were made with GraphPad Prism (version 9.4.1, GraphPad Software Inc., San Diego, CA, USA) and VENN diagrams were generated with Venny 2.0 (https://bioinfogp.cnb.csic.es/tools/venny/index.html, accessed on 6 June 2022) [28].

2.11. Real-Time Quantitative PCR

Changes in cellular stress-related gene expression related to DNA damage, growth arrest, and apoptosis were assessed using real-time quantitative PCR (qPCR) measurements of *GADD45α*, *GADD45β*, and *GADD45γ* isoform expression. RNA was extracted from cells and subsequently purified using the NucleoSpin RNA kit (Macherey-Nagel) protocol. Isolated RNA was quantified using a NanoDrop® 1000 spectrophotometer (ThermoFisher) at 260 nm. Synthesis of cDNA was performed using 25 ng of total RNA as a template and a cDNA mastermix as described before [8]. Quantitative PCR analysis utilized resultant 25 ng of cDNA transcripts and iQ SYBR Green Supermix (Bio-Rad) on the CFX Connect thermocycler (Bio-Rad). Reactions were programmed for 5 min of initial denaturing at 94 °C, followed by 40 cycles of denaturing at 94 °C for 45 s and annealing at 60 °C for 45 s. The $\Delta\Delta C_T$ method [27] was used to calculate n-fold expression in target gene expression by normalizing target C_T values to their respective *GAPDH* expression values. Primer sequences for human *GADD45α* were 5′-TTACTCAAGCAGTTACTCCCTACA-3′ and 5′-CCTTCTTCATTTTCACCTCTTTCCA-3′, for *GADD45β* they were 5′-ATGACATCGCCCTGCAAATC-3′ and 5′-GTGACCAGGAGAC AATGCAG-3′, and for *GADD45γ* they were 5′-CGCGCTGCAGATCCATTTTA-3′ and 5′-GGGGTTCGAAATGAGGATGC-3′. Primer sequences for human *GAPDH* and *RUNX2* were as described [8].

2.12. Comet Assay and Analysis

Comet assay slides were prepared prior to cell collection by coating clean frosted microscope slides (ThermoFisher) with 1% normal melting agarose (NMA, Sigma) in PBS. A volume of 75 µL of melted NMA was pipetted directly to the surface of each slide and immediately covered with a coverslip. The NMA layer was allowed to solidify at 25 °C for 10 min prior to coverslip removal. Slides were stored at −20 °C until use. Cells were trypsinized, washed with PBS, counted and resuspended in PBS supplemented with 2% FBS. A cell suspension of 200,000 cells per ml in 0.1% low melting point agarose (LMA, Fisher Bioreagents, Pittsburgh, PA, USA) was prepared immediately prior to distributing cells onto slides for the assay. NMA-coated slides were allowed to come to room temperature before coating with the LMA-cell mixture. The cell-LMA agarose solution was pipetted directly onto each slide in a volume of 75 μL and immediately covered with a coverslip. The LMA-cell layer was allowed to solidify at 25 °C for 10 min prior to coverslip removal. An additional 5 min solidification period at 25 °C was observed after the coverslip was removed. The slides were then placed in a slide tray and the cells were lysed in comet assay lysis buffer (1.2 M NaCl, 100 mM Na2EDTA, 0.1% sodium lauryl sarcosinate, 0.26 M NaOH, pH > 13) for 1 h at 25 °C. Next, lysis buffer was aspirated off the slides and replaced with electrophoresis solution (0.03 M NaOH, 2 mM Na2EDTA, pH~12.3) for 20 min at 4 °C. Slides were transferred to an electrophoresis box and subjected to electrophoresis for 20 min at 1 V/cm and 4 °C. Slides were rinsed in ddH$_2$O for 1 min, then immersed in 70% ethanol for 5 min. Slides were air-dried overnight and stained with 1 μg/mL DAPI in PBS for 30 min.

Slides were rinsed with PBS and air dried prior to imaging on a Nikon Ti Eclipse fluorescent microscope. Three slides per treatment group were prepared and scored visually. Fifty cells per slide were scored. Fluorouracil (5-FU; Sigma-Aldrich) treated cells were evaluated as a positive control. To ensure accuracy of comet parameter measurements, DAPI signal intensity was subjected to a color threshold in ImageJ (commands used: Image > Adjust > Color Threshold).

Cells were assessed visually according to comet tail length and Olive tail moment. Comet tail length was manually quantified in ImageJ by measuring the distance spanned by comet tails (if present) from the edge of the cell nucleus to the furthest end of the comet tail. Olive tail moment (OTM) was determined by multiplying the distance between the center of the comet head and tail and the percentage of total DNA in the comet tail as previously described.

Cells were designated as class 0 (undamaged cell; 0–10% of DNA in comet tail) while cells with tails were ranked according to increasing tail size: class 1 (minimally damaged; 10–20% of DNA in comet tail), class 2 (moderately damaged; 20–40% of DNA in comet tail), class 3 (severely damaged; 40–60% of DNA in comet tail), and class 4 (maximally damaged; 70–90% of DNA in comet tail).

2.13. Western Blotting

Prior to lysing, cells were pretreated for 30 min with 1 mM sodium orthovanadate to inhibit protein tyrosine phosphatases. Cells were then lysed with RIPA buffer (pH 7.4 150 mM NaCl, 2 mM EDTA, 50 mM Tris-HCl pH 7.4, 1% NP-40, 0.5% sodium deoxycholate, 0.1% sodium dodecyl sulfate (SDS), 1 mM sodium orthovanadate, 1 mM sodium fluoride, 1 mM phenylmethylsulfonyl fluoride (PMSF), and 1:100 Halt Protease Inhibitor Cocktail (ThermoFisher) to collect whole-cell protein lysates. A modified Lowry protein assay (Bio-Rad DC™ protein assay) was used to determine protein concentration in fresh lysates. For Western blot analysis, equal amounts of protein per treatment group were loaded into a 6–10% SDS/polyacrylamide gel and separated by electrophoresis prior to electrophoretic transfer to a polyvinylidene difluoride (PDVF) membrane. Membranes were immediately blocked in 5% bovine serum albumin (BSA) in tris-buffered saline with tween 20 (TBS-T) for 30 min at room temperature on an orbital shaker. Membranes were incubated with one of the following primary antibodies (1:1000) for 2 h at room temperature with shaking: mouse anti-caspase 8 (CST 9746S), mouse anti-caspase 9 (CST 9508S), rabbit anti-phospho-caspase 9 (Tyr153) (abcam ab79202), rabbit anti-c-Abl (CST 2862S), rabbit anti-phospho-c-Abl (Tyr245) (CST 2868S), rabbit anti-phospho-c-Abl (Tyr412) (CST 247C7S), rabbit anti-caspase 4 (ab22687) mouse anti-actin (CST 3700S). Membranes were subsequently incubated for 1 h at room temperature with horseradish peroxidase-conjugated anti-rabbit (CST 7074S) or anti-mouse (CST 7076S) secondary antibody (1:2000). Bands were visualized using chemiluminescence substrate (SuperSignal West Pico PLUS Chemiluminescent Substrate, ThermoFisher) and the Bio-Rad ChemiDoc MP System imager.

2.14. Live/Dead Assay

Healthy live and apoptotic cell populations were quantified using the LIVE/DEAD Viability/Cytotoxicity Kit (ThermoFisher L3224). H9 cells were trypsinized, washed with PBS, and resuspended in fresh medium containing 0.1 µM calcein AM and 8 µM ethidium homodimer-1 (EthD-1) as described [29]. Cells were incubated away from light for 20 min at 25 °C, washed with PBS, and centrifuge-strained to encourage a single-cell suspension (Fisher Scientific, 08-771-23). Cells were resuspended in ice-cold PBS supplemented with 2% FBS and immediately analyzed on a FACSCalibur Flow Cytometer (BD Biosciences). Fluorescence was detected at excitation/emission at λ = 494/517 nm and 517/617 nm. Cytometer gating was set using unstained untreated samples and adjusting forward scatter and side-scatter light. For each sample, 10,000 events were collected.

2.15. Mitochondrial Membrane Potential

Changes in mitochondrial membrane potential were assessed using the commercially available JC-1 Dye (ThermoFisher T3168). Cells were trypsinized, washed with PBS, and resuspended in fresh medium containing 5 µM JC-1. Cells were incubated for 20 min at 25 °C, washed with PBS, and centrifuge-strained to break up cells into a single-cell suspension (Fisher Scientific, 08-771-23). Cells were subsequently resuspended in ice-cold PBS supplemented with 2% FBS and immediately analyzed on a FACSCalibur Flow Cytometer (BD Biosciences). Fluorescence was detected at excitation/emission at λ = 488/530 nm and 488/585 nm. Cytometer gating was set using unstained untreated samples and adjusting forward scatter and side-scatter light. For each sample, 10,000 events were collected.

2.16. ATP:AMP Assays

ATP and AMP levels were quantified via an ATP Determination Kit (ThermoFisher A22066) and an AMP ELISA (Kamiya Biomedical Company KT-52769, Seattle, WA, USA), respectively. For both assessments, cells were washed with PBS, trypsinized and resuspended in PBS.

In the ATP assay, a cell suspension for each treatment group was counted and readjusted to a final concentration of 12.5 × 10^6 cells/mL. For each reading, 10 µL of ATP standard or sample was combined with 90 µL of the ATP reaction solution provided by the ATP Determination Kit immediately before to recording the reaction luminescence output using a Lucetta™ luminometer (Lonza). Relative ng amounts of ATP in each treatment group were determined from an ATP standard curve.

For AMP determination, each cell suspension was adjusted to a final concentration of 1.25 × 10^6 cells/mL. PBS-suspended cells were prepared for ELISA analysis by freeze–thawing three times with gentle mixing between freezing followed by centrifugation at 1000× *g* for 15 min at 4 °C. Cell lysates were assessed according to the manufacturer's ELISA protocol and final optical density was measured at 450 nm using an iMark microplate reader (Bio-Rad). Relative ng amounts of AMP were determined using a standard curve constructed from an AMP standard.

2.17. Statistical Analysis

Assay results were statistically assessed with a paired student's *t*-test or one-way Analysis of Variance (ANOVA) as appropriate (GraphPad Prism 9.4.1, GraphPad Software Inc.). Normality and equal variance were examined with a Shapiro–Wilk test and Bartlett's test, respectively. Non-normal data were log transformed before proceeding to a one-way ANOVA to compare the mean of each treatment with the mean of the control group. A Dunnett's test was used to correct for multiple comparisons. For all conducted tests, p-values below 0.05 were considered significant.

3. Results

3.1. Embryotoxicity of SS Smoke Extract Is Associated with Oxidative Stress

We have previously determined that Camel and Camel Blue smoke solutions impede osteogenic differentiation of hESCs when added throughout the differentiation period from d0 to d20 [9]. While this exposure regimen is commonplace in developmental toxicity assays that rely on ESCs [30–35], it does not illuminate the critical window of susceptibility to certain chemicals. Yet, knowing the window of susceptibility may allow a first insight into the molecular mechanism of toxicity and the potential identity of the target cell or tissue. Therefore, in the current study we performed an initial experiment in which we dosed cells with an effective dose, determined from a concentration-response curve as the puff equivalent concentration that reduced calcification to 50% [9], and a non-effective dose (no effect), separating the differentiation process into two exposure windows (Figure 1A).

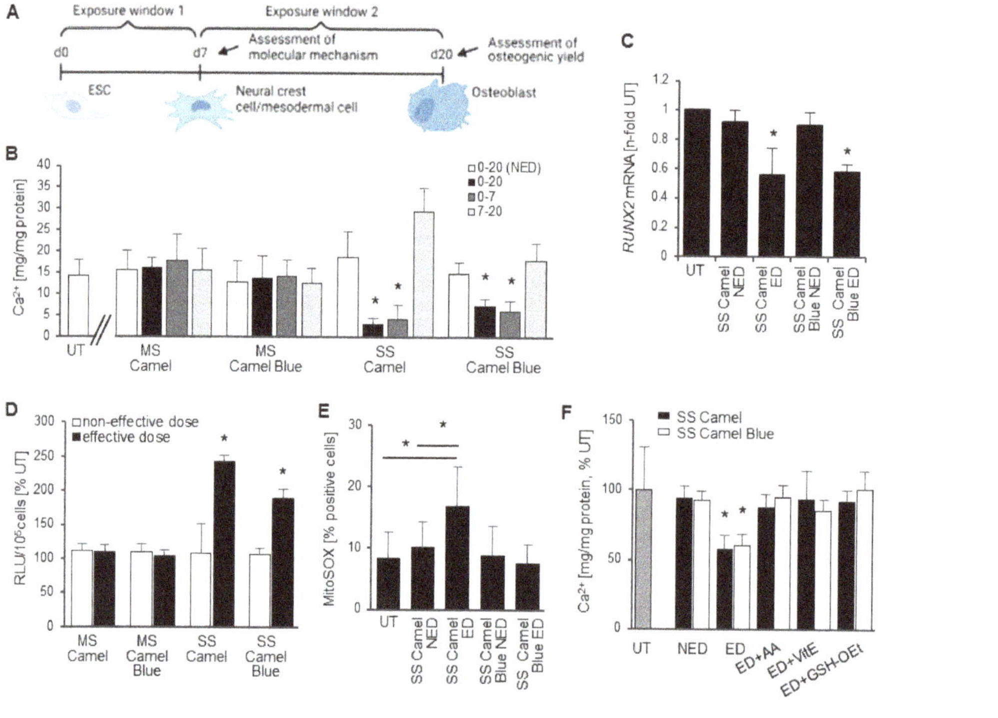

Figure 1. Differentiation inhibition caused by harm-reduction tobacco exposure occurred through intermediate levels of reactive oxygen species. (**A**) Schematic of experimental design and exposure scheme. Created with BioRender.com. (**B**) Mineralization assay using Arsenazo III to determine osteogenic differentiation efficiency after exposure to previously identified effective doses of Camel cigarette smoke extracts [9] within different time windows. (**C**) *RUNX2* mRNA expression measured with qRT-PCR, normalized to *GAPDH* and compared to the untreated control, set at 1; $n = 3 \pm$ SD. (**D**) Superoxide anion content measured upon reaction of the cells with luminol and charted as percent of the untreated cultures; d7, $n = 3 \pm$ SD. (**E**) Cells were exposed for seven days, incubated with MitoSOX, photographed and positive cells counted. Only Camel exposure elicited a significant increase specifically in mitochondrial oxidative stress. (**F**) Calcium deposit was quantified from cultures exposed for 20 days with and without concomitant addition of antioxidants. Effective doses of tobacco smoke solutions and extracts reduced calcification, which was rescued with antioxidant treatment; $n = 3 \pm$ SD. * $p < 0.05$, One-Way ANOVA versus untreated cultures. AA, ascorbic acid; ED, effective dose; GSHOEt, glutathione reduced ethyl ester; MS, mainstream; NED, non-effective dose; RLU, relative light unit; SS, sidestream; UT, untreated solvent control; VitE, Vitamin E.

Camel and Camel Blue SS smoke solutions elicited a differentiation defect in the first exposure window (d0–7) that encompasses exit from pluripotency and the specification of cells with osteogenic potential (Figure 1B), but not in the second exposure window (d7–20), which represents the maturation of the osteoprogenitors. By day seven, cells have differentiated into neural crest and mesodermal progenitors that both have the potential to generate mesenchymal cells that continue to differentiate and mature into osteoblasts [8]. Based on this result, it could be speculated that the differentiation defect observed after exposure to tobacco smoke might stem from an error in the early commitment of cells with osteogenic potential rather than from interference with the maturation process. This notion was exemplified by a reduction in the mRNA of *RUNX2*, a master regulatory transcription factor of the osteogenic lineage, in tobacco treated cells measured on d7 (Figure 1C).

To determine the contribution of oxidative stress to the differential developmental osteotoxicity observed between harm-reduction and conventional Camel cigarette smoke solutions, we next investigated the level of superoxide anion ($O_2^{\bullet-}$) that had been generated seven days into exposure. MS Camel and MS Camel Blue showed no statistical difference in $O_2^{\bullet-}$ content, while the Camel SS smoke extract revealed a 2.4-fold increase in the effective dose over the non-effective dose (Figure 1D). In contrast, the Camel Blue SS effective dose evoked elevated $O_2^{\bullet-}$ content in the range between 1.6- and 1.9-fold. Further evaluation of the potential source of the $O_2^{\bullet-}$ uncovered elevated mitochondrial oxidative stress in the Camel SS, but not in Camel Blue SS extract also on day 7 (Figure 1E). Treatment with antioxidants during smoke extract exposure equalized $O_2^{\bullet-}$ content to solvent controls (Supplemental Figure S1) and rescued calcification in both effective doses (Figure 1F), causally implicating oxidative stress in the observed osteogenic defect.

3.2. *Camel SS and Camel Blue SS Smoke Differentially Alter Oxidative Stress Associated Transcripts*

In an effort to link oxidative stress-mediated embryotoxicity to altered gene regulation, we next performed a Qiagen RT^2 Apoptosis Profiler qPCR array (Figure 2). A heatmap generated from all transcripts across all treatment groups indicated a close relationship between the untreated control, Camel SS and Camel Blue SS non-effective doses (Figure 2A,B), with only minimal changes in gene expression patterns between these groups. In contrast, global apoptotic gene regulation was significantly different in the effective doses of Camel SS and Camel Blue SS smoke extract (Figure 2C) with both sharing 31 differentially regulated mRNAs. In line with the elevated $O_2^{\bullet-}$ levels measured in the effective doses, mRNAs involved in ROS signaling were highly expressed in the effective dose of Camel SS smoke extract and to a lesser extent in the effective dose of Camel Blue SS. These same mRNAs were mainly unaltered in the non-effective doses (Figure 2C). Volcano plots further underscored the differential regulation of additional mRNAs not shared between the two smoke solutions (Figure 2D,E) pointing to alternate mechanisms of toxicity.

3.3. *Conventional Camel SS, but Not the Harm-Reduction Camel Blue SS Smoke Extract Elicits Apoptotic Gene Expression and Activates Executioner Caspases*

The high $O_2^{\bullet-}$ levels found in Camel SS effective doses cultures occurred in the presence of up-regulated *CASP9* mRNA (Figure 3A), a transcript identified to be uniquely regulated in Camel SS ED (Figure 2D). Respective Western blots with an antibody against caspase 9 specifically phosphorylated at Y153, a well-established activation mark [36], detected increased levels in the Camel SS effective dose (Figure 3B). Concomitantly, and in line with the unique regulation of *CASP3* mRNA (Figure 2D), downstream executioner caspases 3 and 7 were found to be highly activated in Camel SS (Figure 3C). In contrast, the intermediate intracellular $O2^{\bullet-}$ levels released upon exposure to Camel Blue SS were associated with marginal activation signals observed in the caspase3/7 stain (Figure 3C). Indeed, *CASP9* mRNA was not significantly elevated (Figure 3A). Conversely, however, caspase 4 showed higher mRNA and protein expression unique to Camel Blue SS (Figures 2D and 3A,C).

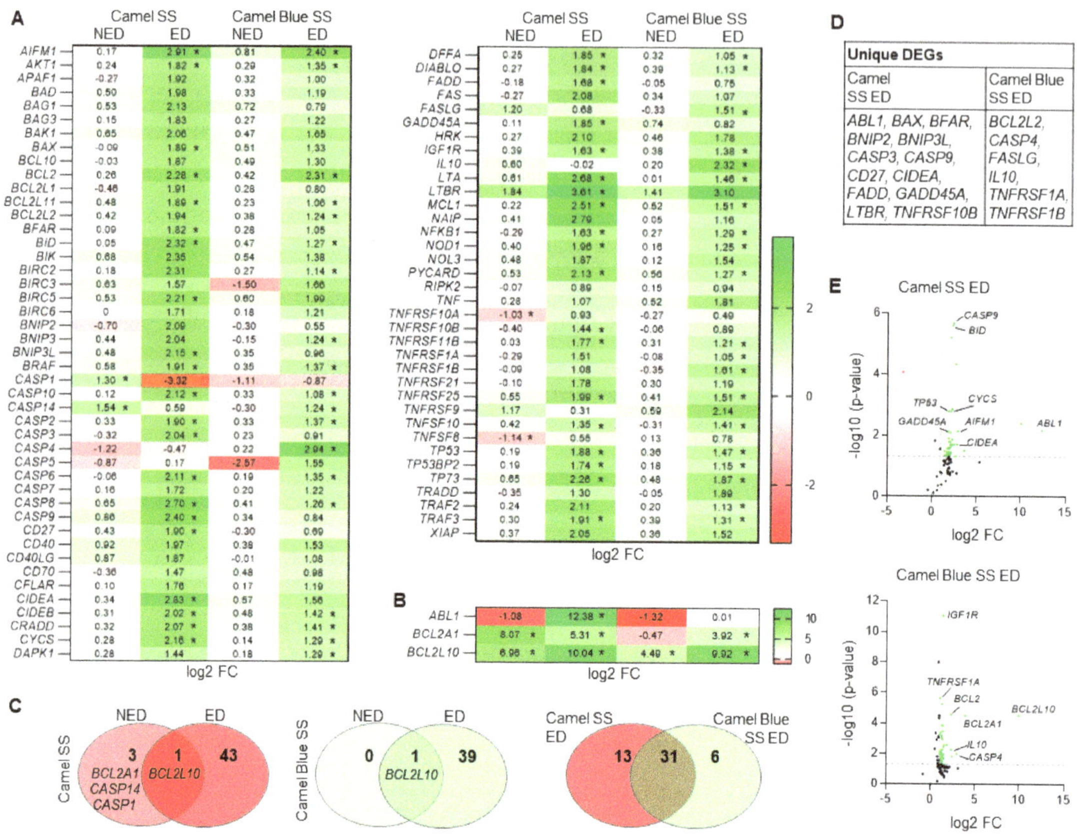

Figure 2. Medium throughput transcript analysis reveals common and differentially expressed apoptosis-related mRNAs in hESCs exposed to conventional and harm-reduction Camel smoke extracts. (**A**,**B**) Heat maps of apoptotic mRNAs de-regulated in tobacco exposed hESCs as measured with the RT^2 qPCR array for apoptosis. * $p < 0.05$ vs UT. (**C**) VENN diagrams showing relations between differentially expressed transcripts in dependence of cigarette smoke extract type ($\log_2$FC > 1; Adj. p-value < 0.05). Venn diagrams were generated with Venny 2.0 [28]. (**D**) Table identifying the transcripts uniquely regulated by the two cigarette smoke extracts. (**E**) Volcano Plot for Camel SS ED and Camel Blue SS ED exposed cells. ED, effective dose; FC, fold change; NED, non-effective dose; SS, sidestream; UT, untreated solvent control.

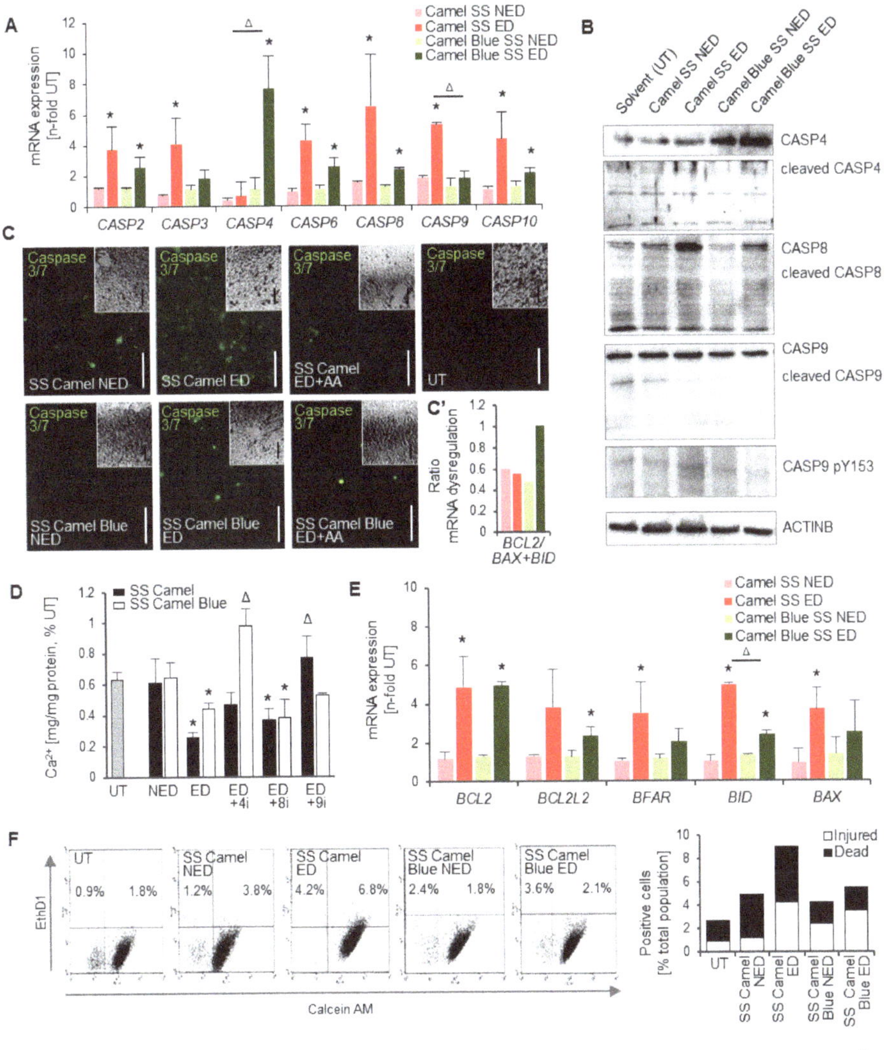

Figure 3. Camel Blue SS elicits a weaker apoptosis response than Camel SS. (**A**) RT^2 qPCR array for apoptosis identified distinct expression patterns of various caspase isoforms between Camel and Camel Blue SS smoke exposed cells. $n = 3 \pm$ SD. * $p < 0.05$, two-tailed *t*-test. $^{\Delta}$ $p < 0.05$, two-tailed *t*-test between Camel SS ED and Camel Blue SS ED. (**B**) Western blots revealed the differential activation of caspases associated with extrinsic and intrinsic apoptotic pathways. (**C**) Accordingly, the executioner caspases 3/7 were highly activated in cells exposed to Camel, but only mildly when exposed to Camel Blue. Antioxidant treatment inhibited this activation. Insets show brightfield images of the same field of view. Bar = 100 µM. (**C'**) The ratio of *BCL2* to *BAX* mRNA expression suggested an anti-apoptotic response in Camel Blue SS cultures. (**D**) Inhibition of these caspases rescued calcification in cells treated with effective doses of tobacco products; $n = 5 \pm$ SD. * $p < 0.05$, One-Way ANOVA versus untreated cultures. $^{\Delta}$ $p < 0.05$, One-Way ANOVA versus ED. (**E**) Some proapoptotic genes were found upregulated in both Camel and Camel Blue SS cultures. $n = 5 \pm$ SD. * $p < 0.05$, two-tailed *t*-test. $^{\Delta}$ $p < 0.05$, two-tailed *t*-test between Camel SS ED and Camel Blue SS ED. (**F**) LIVE/DEAD assay revealed cell death in cells exposed to conventional smoke extracts only. 4i, caspase 4 inhibitor; 8i, caspase 8 inhibitor; 9i, caspase 9 inhibitor; AA, ascorbic acid; ED, effective dose; NED, non-effective dose; SS, sidestream; UT, untreated solvent control.

Inhibition of caspase 9 in Camel SS effective doses and inhibition of caspase 4 in Camel Blue SS effective doses rescued calcification, providing an isoform-specific causal link between caspase activation and differentiation inhibition for both Camel products (Figure 3D). Of note, *CASP8* mRNA (Figure 3A) as well as higher levels of total and cleaved caspase 8 protein expression (Figure 3B) were also noted for Camel SS and Camel Blue SS, and, although higher in Camel SS than Camel Blue SS, were not followed further due to the dual activation by both tobacco products.

Changes in mRNA expression were also observed for other apoptosis-related genes (Figure 3E). Significant upregulation of mRNA expression for the pro-apoptotic factor *BID* was observed in both effective doses, while *BAX* up-regulation was specific to the Camel SS effective dose. Anti-apoptotic genes *BCL2L10*, *BCL2A1*, and *BCL2* were also upregulated in both Camel SS and Camel Blue SS effective doses (Figures 2B and 3E). However, the anti-apoptotic gene *BCL2A1* was highest in the SS Camel non-effective dose—potentially explaining the survival noted in those cultures. Similarly, despite the activation of multiple apoptotic genes, the ratio between *BCL2* and *BAX/BID* was the most beneficial for survival in the effective dose of SS Camel Blue (Figure 3C′).

Indeed, an increased Bcl2/Bax ratio has been suggested to act as an early protective response to apoptotic stimuli, since the enhanced expression of the anti-apoptotic protein Bcl2 can counteract the rising Bax expression [37]. In line with this notion, a Live/Dead assay revealed Camel SS effective dose cultures to possess the highest percentage of dead cells, while the Camel Blue SS exposed cultures had a lower percentage of fully dead cells (Figure 3F). Both, however, had a similar number of injured cells (Figure 3F).

3.4. *Conventional Camel, but Not the Harm-Reduction Camel Blue Smoke Solution Elicits a DNA Damage Response*

The noted severity of apoptosis in the Camel SS effective dose may also be caused by an upregulation in genes associated with genotoxicity response. Indeed, the cell cycle arrest and DNA damage genes *GADD45α*, *GADD45β*, *GADD45γ*, *BNIP2*, and *DFFA* mRNA were significantly upregulated exclusively in cells treated with a Camel SS effective dose (Figure 4A).

Furthermore, mRNA for the DNA-damage response kinase *ABL1* was only upregulated in the Camel SS effective dose and was notably the highest regulated gene observed (Figures 2B and 4A). Western blot analysis confirmed greater ABL1 phosphorylation at Y412 and Y245 (Figure 4B), residues that contribute to full kinase activation in the event of DNA damage and subsequent DNA repair response [38,39].

Additional mRNA markers related to cellular stress, *CIDEA*, *CIDEB*, and *TP53* were also examined. *CIDEB* mRNA was upregulated in both Camel SS and Camel Blue SS effective doses while significance could only be established for *CIDEA* mRNA in Camel SS treated cultures. *TP53* mRNA levels were conspicuously upregulated for Camel SS and Camel Blue SS effective doses, suggesting adverse molecular misregulation by both reatments.

Confirmatory assessment for DNA lesions was then performed using a Comet Assay. Camel SS effective dose exposed cells demonstrated a significantly higher proportion of the severely damaged comet phenotype and larger DNA lesion tails (Figure 4C,D), an effect not observed in the Camel Blue SS effective dose. Co-treatment with a caspase 9 inhibitor reduced observed comet severity in Camel SS smoke extract-treated cells. As apoptosis-driven DNA fragmentation can be visualized as comets in the comet assay, it is possible that caspase 9 inhibitor treatment reduced comet severity by preventing caspase 9 mediated apoptotic signaling. Ascorbic acid supplement of Camel SS cultures did not reduce comet severity, suggesting that antioxidant treatment was not efficient to prevent DNA damage during the exposure period despite lowering $O_2^{\bullet-}$ content and rescuing osteogenesis (compare Figure 1F and Supplemental Figure S1).

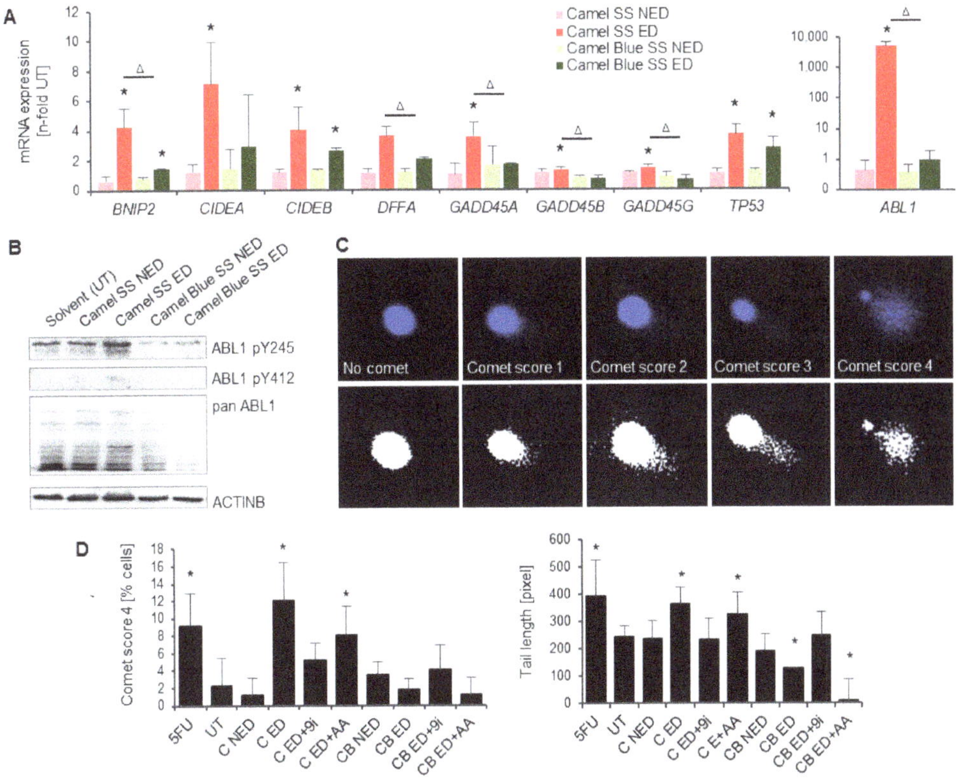

Figure 4. Reduced viability in hESCs exposed to conventional Camel extract is due to DNA damage. (**A**) RT^2 qPCR array for apoptosis found upregulation of genes associated with DNA damage response in Camel SS smoke exposed cells. $n = 3 \pm$ SD. * $p < 0.05$, two-tailed t-test versus UT, $^{\Delta}$ $p < 0.05$, two-tailed t-test between Camel SS ED and Camel Blue SS ED. (**B**) Western blots confirmed ABL1 activation in Camel SS effective doses at the protein level. (**C**,**D**) Comet assays confirm DNA damage in response to Camel exposure, which was absent in Camel Blue exposed cells and cells treated with antioxidant. $n = 3 \pm$ SD. * $p < 0.05$, One-Way ANOVA versus untreated cultures. Scale bar = 63×. 9i, caspase 9 inhibitor; AA, ascorbic acid; ED, effective dose; NED, non-effective dose; SS, sidestream; UT, untreated solvent control.

3.5. Camel and Camel Blue Damage Mitochondria with Differential Severity

Since high levels of mitochondria-specific $O_2^{\bullet -}$ were observed in Camel SS effective dose cultures, mitochondrial health and morphological parameters were next evaluated to determine if mitochondrial dysfunction was also a factor in observed osteogenic inhibition. Mitochondrial membrane potential was significantly reduced in the Camel SS effective dose, which suggested a disturbance to mitochondrial function (Figure 5A). In contrast, mitochondrial membrane potential in Camel Blue SS effective doses remained unchanged from the untreated group. Further, comparative analysis of cellular AMP and ATP levels revealed that the AMP/ATP ratio was increased in the Camel SS effective dose (Figure 5B). This outcome signifies a potential underproduction of ATP and further evidence of mitochondrial dysfunction in Camel SS treated cells.

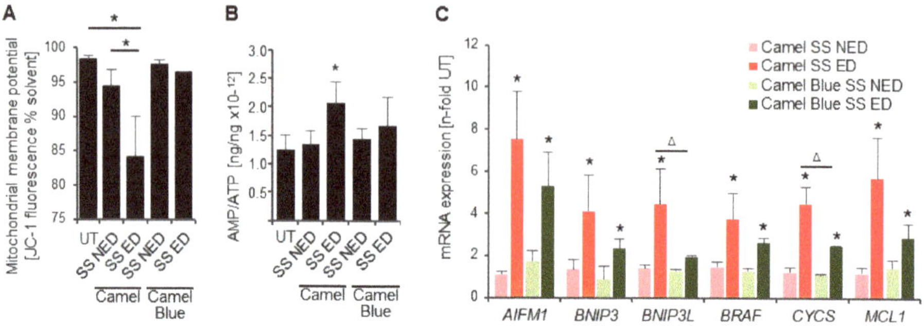

Figure 5. Deterioration of mitochondrial health in exposed hESCs. (**A**) Mitochondrial membrane potential measurements revealed a reduced membrane potential in Camel exposed cells as a sign for execution of the intrinsic apoptotic pathway. * $p < 0.05$, One-Way ANOVA versus untreated or NED cultures (**B**) AMP-to-ATP ratio was increased in Camel SS ED, suggesting mitochondrial dysfunction. * $p < 0.05$, One-Way ANOVA versus untreated cultures. (**C**) qPCR array analysis revealed upregulation of mRNAs associated with integral mitochondrial apoptosis in both Camel SS and Camel Blue SS ED. * $p < 0.05$, two-tailed *t*-test versus untreated control, $^{\Delta}$ $p < 0.05$, two-tailed *t*-test between Camel SS ED and Camel Blue SS ED. ED, effective dose; NED, non-effective dose; SS, sidestream; UT, untreated solvent control.

Additional evidence to support this notion came from the qPCR array. *BRAF*, *BNIP3* and *CYCS* were upregulated by the Camel SS effective dose, while only a mild upregulation was observed in the Camel Blue SS effective dose (Figure 5C). In addition, *BNIP3L* was exclusively upregulated in Camel SS effective dose only (Figure 5C, compare Figure 2D). Given that the latter two proteins are involved in permeabilization of the mitochondrial outer membrane to prepare for cytochrome C release [40–44], it follows that they could have caused the mitochondrial membrane potential depolarization in Camel SS cultures. Collectively, these results suggest a strong mitochondrial-driven apoptotic response in Camel SS cultures. However, the similar *AIFM1* expression patterns between both Camel and Camel Blue SS indicates a potential issue with mitochondrial function was not only occurring in Camel SS smoke extract, but in Camel Blue SS smoke extract also.

Quantification of mitochondrial number from confocal images of Mitotracker stained cells found that the total mitochondrial population in the Camel SS effective dose did not deviate from the untreated (Figure 6A′), despite the altered mitochondrial function that was measured in this group. Cells treated with Camel Blue SS, conversely, featured a significant increase in total mitochondrial "bright spots". Collectively, these outcomes infer that Camel SS and Camel Blue SS elicited differential responses in mitochondrial behavior, which validated investigating these outcomes in further detail.

Mitochondrial networks were characterized by measuring mitochondrial network branch lengths, branch numbers per network, and overall mitochondrial footprint. Cells exposed to Camel SS showed reduced network interconnection (Figure 6A) as well as a significant increase in mitochondrial branch length and a decrease in the overall mitochondrial footprint (Figure 6B). Concurrent treatment of Camel SS effective doses with an inhibitor to caspase 9, the caspase activated to a higher extent in this condition, but not to caspase 4, restored network morphology (Figure 6A), and rescued mitochondrial footprint (Figure 6B). Camel Blue SS exposed cells, in contrast, demonstrated network branch lengths that were not significantly different from the untreated cultures (Figure 6B). Trends towards decreased branch lengths were detectable, however. Cells exposed to Camel Blue SS also demonstrated more highly branched networks and a significantly reduced mitochondrial footprint.

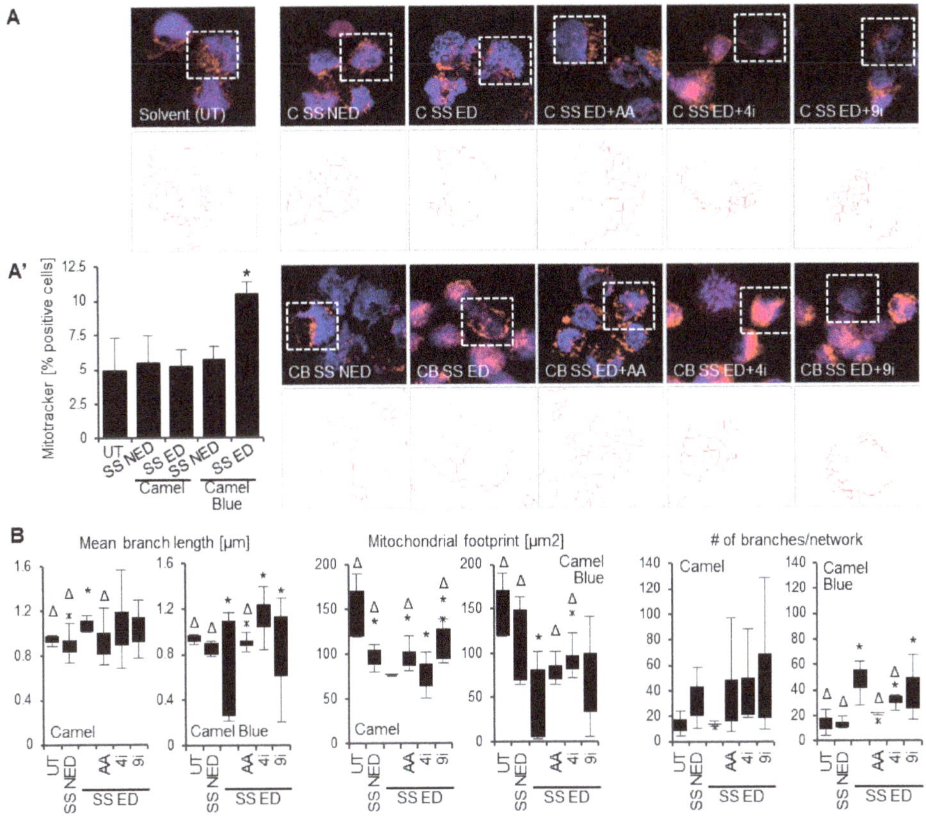

Figure 6. Tobacco smoke exposure elicits changes in mitochondrial networks. (**A**) MitoTracker and MiNA visualization of mitochondrial networks, magnification 63×. (**A'**) MitoTracker dye analysis revealed increased mitochondrial signal in the Camel Blue SS effective dose. * $p < 0.05$, One-Way ANOVA versus untreated cultures. (**B**) Changes to mitochondrial networks were assessed via mean branch length, mitochondrial footprint, and branches per network. * $p < 0.05$, One-Way ANOVA versus untreated cultures, $^{\Delta}$ $p < 0.05$, One-Way ANOVA versus ED, x denotes max or min outliers. 4i, caspase 4 inhibitor; 9i, caspase 9 inhibitor; C, Camel; CB, Camel Blue; ED, effective dose; NED, non-effective dose; SS, sidestream; UT, untreated solvent control.

Simultaneous treatment of Camel Blue SS effective doses with caspase 9 inhibitor yielded abnormal mitochondrial network morphology that was comprised of shorter fragmented networks mixed with some filamentous, interconnected networks (Figure 6A). Caspase 9 inhibitor treatment did not significantly rescue mean branch length or mitochondrial footprint (Figure 6B), suggesting a lack of intrinsic mitochondrial apoptosis response. Treatment with caspase 4 inhibitor did, however, restore interconnected mitochondrial network morphology (Figure 6A), rescued mitochondrial footprint (Figure 6B) and decreased the number of branches per network (Figure 6B) suggesting caspase 4 involvement in aberrant mitochondrial behaviors under Camel Blue SS treatment.

Cellular redox status also appeared to partly influence mitochondrial network morphology in both Camel SS and Camel Blue SS effective doses. Concurrent treatment with antioxidant ascorbic acid rescued branch length and slightly restored mitochondrial footprint in Camel SS cultures (Figure 6B). Camel Blue SS effective doses with ascorbic acid reestablished interconnected mitochondrial network morphology (Figure 6A) and modestly restored the mean branch length range by reducing the incidence of short branch fragments (Figure 6B) implicating a causal relationship between oxidative stress and mitochondrial health.

4. Discussion

Tobacco smoke has been shown to reduce bone density and increase osteopenia and osteoporosis rates in adults [45]. Due to the shortage of information on toxic effects of tobacco smoke on the developing skeleton, we previously used hESCs directed through an osteogenic lineage to assess the potency of tobacco products to inhibit calcification in differentiating osteogenic cultures [9]. Based on this endpoint, differential embryotoxicity was observed in cultures treated with cigarette smoke solutions from MS and SS smoke of Camel and Camel Blue cigarettes. Indeed, other studies found that SS smoke from conventional cigarettes was more potent than MS smoke, in very diverse endpoints such as free radical species levels, angiogenesis, oviductal function, adverse IVF outcome, sperm motility, and attachment ability of peri-implantation embryonic cells [46–53]. Regarding differentiating osteoblasts, our previous study found that conventional MS smoke did not hinder the viability of developing osteoblasts or their differentiation [9]. However, SS smoke always showed detrimental effects at lower concentrations than MS smoke and inhibited both calcification and the survival of the osteogenic cultures. We extend this finding here to a specific susceptibility window that does not include maturation events, but rather represents an early window of development in which cells commit and specify. During this early time window, only SS smoke elicited oxidative stress, but not MS smoke.

Since it has not been unequivocally answered whether harm-reduction products confer reduced harm on differentiating osteogenic cells, in the current study, smoke extract from a harm-reduction cigarette was compared to a conventional cigarette. Our data show exposure to Camel Blue SS smoke demonstrated differentiation inhibition in developing osteoblasts at sub-cytotoxic concentrations. This enhanced potency of the harm-reduction SS smoke may be explained by the alteration in chemical composition associated with the process of lowering tar and nicotine content during manufacturing. During this process, other constituents found in the complex chemical blend of cigarette smoke such as nitrate, nitrogen oxide, and tobacco-specific nitrosamines may be enriched [54]. Individually, these chemicals can cause adverse health effects in mammalian cells [55–58]. Not only can concentrations of such chemicals be higher in harm-reduction cigarettes because of the processing required to reduce content of other carcinogens, but smokers also overcompensate for delivered nicotine by smoking more cigarettes or by inhaling deeper while smoking [7]. For these reasons it is likely that concentrations of such harmful chemicals are even higher in mothers who have difficulty quitting smoking while pregnant and that the harmful effects of those chemicals are compounded in their unborn fetuses. Of note in this context is the limitation of our study to assess aqueous smoke extracts only. At the same time, however, the detection of such noteworthy skeletal malformations despite the omission of solvent-dilutable organic compounds, which are in themselves highly toxic, speaks for the problematic association of tobacco smoke exposure with skeletal defects.

We further provide here a potential first insight into why embryotoxicity elicited by the conventional Camel SS smoke is carried by its general (unspecific) cytotoxicity and why Camel Blue SS smoke exposure may instead confer embryotoxicity through authentic chemical impact on developmental processes. While it may be assumed that the chemicals in tobacco smoke solutions trigger signaling cascades that are detrimental to development, another potential mechanism of action is that they induce mild oxidative stress. As we show here, the developmental toxicity of the harm-reduction tobacco product seems defined by its ability to create ROS at levels that are insufficient to kill the cells. As recently put forth by Hansen and Harris [59], teratogenesis is defined by dysmorphogenetic events that may precede excessive cell death. While cytotoxicity focuses on the accumulation of ROS, the impediment of antioxidant capacities, and consequent cell death, developmental toxicity may be the result of untimely regulation of critical cellular signaling rather than the result of early cell death. Our results outlined here offer support for this notion in that while

Camel SS and Camel Blue SS effective doses both demonstrated reduced osteogenesis and upregulation of pro-apoptotic gene expression, Camel Blue SS effective cultures specifically failed to achieve complete caspase cascade activation and cell death as seen in Camel SS effective doses. It follows, then, that additional molecular or cellular players may be involved in the differential outcomes mediated by different tobacco products.

Along this argumentation, disruption of normal tissue redox balances has also been reported to interfere with normal biological processes that can lead to pathological outcomes including DNA damage [60–62]. Given that evidence of DNA damage was exclusively observed in cultures exposed to Camel SS effective doses, our results suggest that elevated ROS could cause a DNA damage-mediated mode of action for embryotoxicity outcomes following Camel but not Camel Blue exposure. However, this notion was not supported by experimental evidence, as antioxidant treatment did not reduce DNA damage cell populations in Camel SS effective doses. Nonetheless, given that DNA damage is well-reported to activate caspase 9-mediated apoptosis [63–65], it follows that concurrent *CASP9* mRNA upregulation and posttranslational activation of caspase 9 protein in Camel SS effective dose cultures support a biochemical basis for intrinsic apoptotic responses observed exclusively with Camel exposure.

During the intrinsic apoptosis process, caspase 9 has been reported to regulate the collapse of mitochondrial membrane potential, causing further mitochondrial disruption [66,67]. Aberrant mitochondrial behavior is well-associated with stress responses and a variety of pathologies [68,69], and also reported to follow mitochondrial membrane depolarization and precede mitophagy and apoptosis [60,70]. In accordance with elevated caspase 9 activity, our study found concurrent functional and morphological mitochondrial changes exclusively in Camel effective dose cultures that were reversed when caspase 9 was inhibited. Indeed, BAX, transcripts for which were exclusively up-regulated in Camel SS ED cells, may be activated by DNA damage [71] and can form a channel in the mitochondrial membrane that eventually mediates cytochrome C release [72,73], a notable outcome of Camel SS exposure. Through cytochrome C release BAX helps to assemble the apoptosome from APAF1 and CASP9, which activates and cleaves CASP3 downstream, ultimately executing cell death [74], another noted outcome of Camel SS exposure. The reverse relationship has also been described where BAX activation is CASP9 dependent [75]. This context provides further support for the notion that DNA damage likely elicited the caspase 9-driven cytotoxic responses that may be responsible for osteogenic inhibition following SS Camel exposure.

Mitochondrial morphological changes observed in the absence of altered mitochondrial function or DNA damage in Camel Blue SS smoke effective dose cultures hint at divergent modes of embryotoxic action for the conventional versus the harm-reduction tobacco product of this brand. In point of fact, the mitochondria and the endoplasmic reticulum (ER) have an important crosstalk relationship that may be of further relevance in the context of development and this study [76–80]. In contrast with Camel SS ED cultures, Camel Blue SS ED cells demonstrated upregulation of caspase 4 mRNA and protein with only little caspase 9 activation. Notably, caspase 4 has been proposed to function as an ER-stress specific caspase [37,81–83], not least because it mainly localizes to the ER [81]. Given that treatment with caspase 4 rescued mitochondrial network morphology, it is possible that ER stress was also involved in the differential molecular response of cells exposed to the Camel Blue effective dose. Notably, disruption to normal crosstalk between the mitochondria and ER has been implicated in developmental, metabolic and degenerative diseases [77–80]. Moreover, ER stress can interfere with protein folding processes [84], which may be to the detriment of osteogenic differentiation as new sets of proteins required for differentiation progression may be made incorrectly or not at all. Secretion of critical regulatory proteins by osteoblasts and their precursors could also be adversely impacted in this context. Follow up studies, however, are required to confirm ER stress and dysfunction in Camel Blue exposed cultures.

In COS-7 cells, overexpression of caspase-4 induces activation of the death protease CASP3 [85]. Conversely, biochemical and genetic evidence suggests that inhibition of CASP4 ablates ER stress-induced apoptosis [86–88]. Why, in our study then, CASP4 activation does not lead to CASP3 activation in CB SS cells remains nebulous, but it may be speculated that CASP4 plays additional roles in processes other than apoptosis. In support of this notion, there is evidence in the literature to suggest a link between CASP4, IL10, one of the six unique transcripts found to be regulated in Camel Blue SS ED exposed cells, and inflammation. For instance, CASP4 mediates non-canonical activation of the NLRP3 inflammasome in human myeloid cells [89]. Further, it has been proposed that CASP4 may function mainly via the NFκB signal pathway in inflammatory responses, leading to NFκB-dependent transcriptional up-regulation and secretion of important cytokines and chemokines [90]. While many of the NFκB targets are classical pro-inflammatory cytokines, the NF-κB family member p50 is also known to promote transcription of the anti-inflammatory IL-10 [91,92].

Aside from its anti-inflammatory properties, IL-10 exhibits osteoblastogenic or osteolytic attributes, depending on dosage. While IL-10-deficient mice exhibit reduced bone formation [93] and low physiological concentrations of IL-10 induce osteoblastogenesis by activating the p38 MAPK signaling pathway in human MSCs, higher pathological doses of IL-10 inhibit osteoblastogenesis by activating NFκB signaling [94]. These data corroborate the notion that the Camel Blue SS mediated increase in IL-10 may underlie the noted phenotypic alterations in our cultures.

Aside from the CASP4/IL-10 axis, a second potential mechanism for the dampened osteogenic differentiation efficiency can be construed from the known relationship between IL-10 and Fas ligand, the transcript for which was also uniquely regulated upon Camel Blue SS exposure. IL-10 production downstream of Fas ligand-mediated engagement of the Fas receptor represents an important anti-inflammatory mechanism in monocytes [95] that occurs in the absence of executioner caspase activation and thus apoptosis. Again, it may be through IL-10 that osteogenic differentiation is inhibited. In addition, IL-10 may protect cells from Fas-induced apoptosis by dampening CASP8 activation, as is the case in intestinal epithelial cells [96], which may be the reason for the lower CASP8 response in Camel Blue SS ED we observe. The enhanced expression of *TNFSRF1B* in Camel Blue SS may additionally tip the balance towards survival, not apoptosis, by activating AKT and NFκB [97].

Supplementarily, the involvement of Fas may also have direct consequences for osteogenic cell fate specification. Osteoblasts constitutively express FAS and FASLG [98,99] and gld mice, which have a defect in the FASLG-mediated apoptotic pathway, have increased bone trabecular volume in vivo [100]. Mice without functional Fas-ligand also have an increased recruitment of cells of mesenchymal origin and an abnormal pattern of differentiation and maturation of the newly formed mesenchymal tissues [101]. In reverse, this would imply that overactivation of FASLG may decrease proliferation of bone progenitors and total volume of newly formed bone, which is what we find here, and a failure to recruit mesenchymal progenitor cells. Taken together, independently of how IL-10, FASLG and CASP4 specifically interact, the increase in IL-10 may therefore underlie the phenotypic alterations noted in the Camel Blue SS mediated osteogenic defect.

5. Conclusions

The results from this study imply a causal link between the tobacco-induced osteogenic differentiation effect and oxidative stress. However, observed differential mRNA upregulation patterns between Camel SS and Camel Blue SS suggested that a precise molecular interplay may be responsible for ultimate phenotypic outcomes following exposure to a specific tobacco product. As follows, this result also implies that while Camel SS and Camel Blue SS actively harm differentiating osteoblasts, both products may act distinctly upon cellular regulatory mechanisms. Conceptually, our finding remains important despite global tobacco education programs that generally caused smoking to decline. This is because approximately 10–20% of women in the United States still smoke during pregnancy and home exposure to secondhand smoke is prevalent amongst pregnant women in low- and middle-income households [2].

Supplementary Materials: The following supporting information can be downloaded at: https://www.mdpi.com/article/10.3390/antiox11122474/s1, Figure S1: Superoxide anion content measured upon reaction of the cells with luminol in response to antioxidant treatment.

Author Contributions: Conceptualization, N.I.z.N.; methodology, N.R.L.S., L.M.W., S.R.S., M.H., J.V.M. and E.C.D.; formal analysis, N.R.L.S., L.M.W., S.R.S. and J.V.M., writing—original draft N.R.L.S. and L.M.W.; writing—review and editing, N.I.z.N.; visualization, N.R.L.S., L.M.W., S.R.S., and J.V.M.; supervision, N.R.L.S., L.M.W., and N.I.z.N.; project administration, N.I.z.N.; funding acquisition, N.I.z.N. All authors have read and agreed to the published version of the manuscript.

Funding: This study was supported by Tobacco-Related Disease Research Program (TRDRP) grant numbers 19KT-0017H and 25IP-0018 and National Institutes of Health R01 DE025330. N.R.L.S. acknowledges the support from a TRDRP Cornelius Hopper Diversity Award, a TRDRP pre-doctoral fellowship [24DT-0002], a fellowship from the International Foundation for Ethical Research and a University of California Riverside Chancellor's Postdoctoral Fellowship. L.M.W. is a recipient of a Graduate Research Fellowship from the National Science Foundation [DGE-1326120] and a National Institute of Environmental Health Sciences T32 traineeship [ES018827].

Institutional Review Board Statement: Not applicable.

Informed Consent Statement: Not applicable.

Data Availability Statement: The data presented in this study are available in the article and supplementary material.

Acknowledgments: The authors would like to thank Prue Talbot and Monique Williams (University of California Riverside) for helping with the occasional production of tobacco smoke solutions.

Conflicts of Interest: The authors declare no conflict of interest. The funders had no role in the design of the study; in the collection, analyses, or interpretation of data; in the writing of the manuscript; or in the decision to publish the results.

References

1. Warner, K.E. The role of research in international tobacco control. *Am. J. Public Health* **2005**, *95*, 976–984. [CrossRef] [PubMed]
2. Tong, V.T.; Dietz, P.M.; Farr, S.L.; D'Angelo, D.V.; England, L.J. Estimates of smoking before and during pregnancy, and smoking cessation during pregnancy: Comparing two population-based data sources. *Public Health Rep.* **2013**, *128*, 179–188. [CrossRef] [PubMed]
3. Macdonald-Wallis, C.; Tobias, J.H.; Davey Smith, G.; Lawlor, D.A. Parental smoking during pregnancy and offspring bone mass at age 10 years: Findings from a prospective birth cohort. *Osteoporos. Int.* **2011**, *22*, 1809–1819. [CrossRef]
4. Godfrey, K.; Walker-Bone, K.; Robinson, S.; Taylor, P.; Shore, S.; Wheeler, T.; Cooper, C. Neonatal bone mass: Influence of parental birthweight, maternal smoking, body composition, and activity during pregnancy. *J Bone Miner Res.* **2001**, *16*, 1694–1703. [CrossRef]
5. Centers for Disease Control and Prevention (CDC). Smoking prevalence among women of reproductive age—United States, 2006. *MMWR Morb. Mortal. Wkly. Rep.* **2008**, *57*, 849–852.
6. Parviainen, R.; Auvinen, J.; Pokka, T.; Serlo, W.; Sinikumpu, J.J. Maternal smoking during pregnancy is associated with childhood bone fractures in offspring—A birth-cohort study of 6718 children. *Bone* **2017**, *101*, 202–205. [CrossRef] [PubMed]
7. Burns, D.M.; Benowitz, N.L. Public health implications of changes in cigarette design and marketing. In *National Cancer Institute. Risks Associated with Smoking Cigarettes with Low Machine-Measured Yields of Tar and Nicotine*; Smoking and Tobacco Control Monograph No. 13; NIH Publication No. 02-5074; US Department of Health and Human Services, National Institutes of Health, National Cancer Institute: Bethesda, MD, USA, 2001; pp. 1–12.
8. Sparks, N.R.L.; Martinez, I.K.C.; Soto, C.H.; zur Nieden, N.I. Low osteogenic yield in human pluripotent stem cells associates with differential neural crest promoter methylation. *Stem Cells* **2018**, *36*, 349–362. [CrossRef]
9. Martinez, I.K.C.; Sparks, N.R.L.; Madrid, J.V.; Talbot, P.; zur Nieden, N.I. Exposure to cigarette smoke impedes human in vitro osteoblast differentiation independently of nicotine. *Nicotine Tob. Res.* **2022**, *24*, 1921–1926. [CrossRef]

10. U.S. Department of Health and Human Services. *The Health Consequences of Involuntary Exposure to Tobacco Smoke: A Report of the Surgeon General. Secondhand Smoke What It Means to You*; U.S. Department of Health and Human Services, Centers for Disease Control and Prevention, Coordinating Center for Health Promotion, National Center for Chronic Disease Prevention and Health Promotion, Office on Smoking and Health: Washington, DC, USA, 2006.
11. Buck, K.K.; zur Nieden, N.I. Risk Assessment Using Human Pluripotent Stem Cells: Recent Advances in Developmental Toxicity Screens. In *Stem Cells in Birth Defects Research and Developmental Toxicology*; Rasmussen, T.P., Ed.; Wiley: Hoboken, NJ, USA, 2018; pp. 91–117.
12. Karmach, O.; Madrid, J.V.; Dasgupta, S.; Volz, D.C.; zur Nieden, N.I. Embryonic Exposure to Cigarette Smoke Extract Impedes Skeletal Development and Evokes Craniofacial Defects in Zebrafish. *Int. J. Mol. Sci.* **2022**, *23*, 9904. [CrossRef]
13. Hansen, J.M. Oxidative stress as a mechanism of teratogenesis. *Birth Defects Res. C Embryo Today* **2006**, *78*, 293–307. [CrossRef]
14. Dennery, P.A. Effects of oxidative stress on embryonic development. *Birth Defects Res. C Embryo Today* **2007**, *81*, 155–162. [CrossRef] [PubMed]
15. Kovacic, P.; Somanathan, R. Mechanism of teratogenesis: Electron transfer, reactive oxygen species, and antioxidants. *Birth Defects Res. C Embryo Today* **2006**, *78*, 308–325. [CrossRef] [PubMed]
16. Carnevali, S.; Luppi, F.; D'Arca, D.; Caporali, A.; Ruggieri, M.P.; Vettori, M.V.; Caglieri, A.; Astancolle, S.; Panico, F.; Davalli, P.; et al. Clusterin decreases oxidative stress in lung fibroblasts exposed to cigarette smoke. *Am. J. Respir. Crit. Care Med.* **2006**, *174*, 393–399. [CrossRef] [PubMed]
17. Sies, H. *Oxidative Stress*; Academic Press: London, UK, 1985.
18. Weinbrenner, T.; Cladellas, M.; Isabel Covas, M.; Fito, M.; Tomas, M.; Senti, M.; Bruguera, J.; Marrugat, J. High oxidative stress in patients with stable coronary heart disease. *Atherosclerosis* **2003**, *168*, 99–106. [CrossRef] [PubMed]
19. Abramson, J.L.; Hooper, W.C.; Jones, D.P.; Ashfaq, S.; Rhodes, S.D.; Weintraub, W.S.; Harrison, D.G.; Quyyumi, A.A.; Vaccarino, V. Association between novel oxidative stress markers and C-reactive protein among adults without clinical coronary heart disease. *Atherosclerosis* **2005**, *178*, 115–121. [CrossRef]
20. Kaneto, H.; Katakami, N.; Kawamori, D.; Miyatsuka, T.; Sakamoto, K.; Matsuoka, T.A.; Matsuhisa, M.; Yamasaki, Y. Involvement of oxidative stress in the pathogenesis of diabetes. *Antioxid. Redox. Signal.* **2007**, *9*, 355–366. [CrossRef]
21. Mena, S.; Ortega, A.; Estrela, J.M. Oxidative stress in environmental-induced carcinogenesis. *Mutat. Res.* **2009**, *674*, 36–44. [CrossRef]
22. Knoll, M.; Shaoulian, R.; Magers, T.; Talbot, P. Ciliary beat frequency of hamster oviducts is decreased in vitro by exposure to solutions of mainstream and sidestream cigarette smoke. *Biol. Reprod.* **1995**, *53*, 29–37. [CrossRef]
23. Knoll, M.; Talbot, P. Cigarette smoke inhibits oocyte cumulus complex pick-up by the oviduct in vitro independent of ciliary beat frequency. *Reprod. Toxicol.* **1998**, *12*, 57–68. [CrossRef]
24. Davis, L.A.; Dienelt, A.; zur Nieden, N.I. Absorption-based assays for the analysis of osteogenic and chondrogenic yield. *Methods Mol. Biol.* **2011**, *690*, 255–272. [CrossRef]
25. Jensen, E.C. Quantitative analysis of histological staining and fluorescence using ImageJ. *Anat. Rec.* **2013**, *296*, 378–381. [CrossRef] [PubMed]
26. Valente, A.J.; Maddalena, L.A.; Robb, E.L.; Moradi, F.; Stuart, A.J. A simple ImageJ macro tool for analyzing mitochondrial network morphology in mammalian cell culture. *Acta Histochem.* **2017**, *119*, 315–326. [CrossRef] [PubMed]
27. Livak, K.J.; Schmittgen, T.D. Analysis of relative gene expression data using real-time quantitative PCR and the 2(-Delta Delta C(T)) Method. *Methods* **2001**, *25*, 402–408. [CrossRef] [PubMed]
28. Oliveros, J.C. (2007–2015) Venny. An Interactive Tool for Comparing Lists with Venn's Diagrams. Available online: https://bioinfogp.cnb.csic.es/tools/venny/index.h (accessed on 6 June 2022).
29. Puig-Sanvicens, V.A.; Semino, C.E.; zur Nieden, N.I. Cardiac differentiation potential of human induced pluripotent stem cells in a 3D self-assembling peptide scaffold. *Differentiation* **2015**, *90*, 101–110. [CrossRef]
30. zur Nieden, N.I.; Kempka, G.; Ahr, H.J. Molecular multiple endpoint embryonic stem cell test–a possible approach to test for the teratogenic potential of compounds. *Toxicol. Appl. Pharmacol.* **2004**, *194*, 257–269. [CrossRef]
31. zur Nieden, N.I.; Davis, L.A.; Rancourt, D.E. Comparing three novel endpoints for developmental osteotoxicity in the embryonic stem cell test. *Toxicol. Appl. Pharmacol.* **2010**, *247*, 91–97. [CrossRef]
32. Seiler, A.E.; Spielmann, H. The validated embryonic stem cell test to predict embryotoxicity in vitro. *Nat. Protoc.* **2011**, *6*, 961–978. [CrossRef]
33. Madrid, J.V.; Sera, S.R.; Sparks, N.R.L.; zur Nieden, N.I. Human Pluripotent Stem Cells to Assess Developmental Toxicity in the Osteogenic Lineage. *Methods Mol. Biol.* **2018**, *1797*, 125–145. [CrossRef]
34. Walker, L.; Baumgartner, L.; Keller, K.C.; Ast, J.; Trettner, S.; zur Nieden, N.I. Non-human primate and rodent embryonic stem cells are differentially sensitive to embryotoxic compounds. *Toxicol. Rep.* **2014**, *2*, 165–174. [CrossRef]
35. Walker, L.M.; Sparks, N.R.L.; Puig-Sanvicens, V.; Rodrigues, B.; zur Nieden, N.I. An Evaluation of Human Induced Pluripotent Stem Cells to Test for Cardiac Developmental Toxicity. *Int. J. Mol. Sci.* **2021**, *22*, 8114. [CrossRef]
36. Raina, D.; Pandey, P.; Ahmad, R.; Bharti, A.; Ren, J.; Kharbanda, S.; Weichselbaum, R.; Kufe, D. c-Abl tyrosine kinase regulates caspase-9 autocleavage in the apoptotic response to DNA damage. *J. Biol. Chem.* **2005**, *280*, 11147–11151. [CrossRef] [PubMed]
37. Bian, Z.M.; Elner, S.G.; Elner, V.M. Dual involvement of caspase-4 in inflammatory and ER stress-induced apoptotic responses in human retinal pigment epithelial cells. *Investig. Ophthalmol. Vis. Sci.* **2009**, *50*, 6006–6014. [CrossRef] [PubMed]

38. Brasher, B.B.; Van Etten, R.A. c-Abl has high intrinsic tyrosine kinase activity that is stimulated by mutation of the Src homology 3 domain and by autophosphorylation at two distinct regulatory tyrosines. *J. Biol. Chem.* **2000**, *275*, 35631–35637. [CrossRef] [PubMed]
39. Pluk, H.; Dorey, K.; Superti-Furga, G. Autoinhibition of c-Abl. *Cell* **2002**, *108*, 247–259. [CrossRef] [PubMed]
40. Imazu, T.; Shimizu, S.; Tagami, S.; Matsushima, M.; Nakamura, Y.; Miki, T.; Okuyama, A.; Tsujimoto, Y. Bcl-2/E1B 19 kDa-interacting protein 3-like protein (Bnip3L) interacts with bcl-2/Bcl-xL and induces apoptosis by altering mitochondrial membrane permeability. *Oncogene* **1999**, *18*, 4523–4529. [CrossRef] [PubMed]
41. D'Alessio, M.; De Nicola, M.; Coppola, S.; Gualandi, G.; Pugliese, L.; Cerella, C.; Cristofanon, S.; Civitareale, P.; Ciriolo, M.R.; Bergamaschi, A.; et al. Oxidative Bax dimerization promotes its translocation to mitochondria independently of apoptosis. *FASEB J.* **2005**, *19*, 1504–1506. [CrossRef]
42. Kubli, D.A.; Ycaza, J.E.; Gustafsson, A.B. Bnip3 mediates mitochondrial dysfunction and cell death through Bax and Bak. *Biochem. J.* **2007**, *405*, 407–415. [CrossRef]
43. Shelton, S.N.; Shawgo, M.E.; Robertson, J.D. Cleavage of Bid by executioner caspases mediates feed forward amplification of mitochondrial outer membrane permeabilization during genotoxic stress-induced apoptosis in Jurkat cells. *J. Biol. Chem.* **2009**, *284*, 11247–11255. [CrossRef]
44. Shamas-Din, A.; Kale, J.; Leber, B.; Andrews, D.W. Mechanisms of Action of Bcl-2 Family Proteins. *Cold Spring Harb. Perspect. Biol.* **2013**, *5*, a008714. [CrossRef]
45. Al-Bashaireh, A.M.; Haddad, L.G.; Weaver, M.; Chengguo, X.; Kelly, D.L.; Yoon, S. The Effect of Tobacco Smoking on Bone Mass: An Overview of Pathophysiologic Mechanisms. *J. Osteoporos.* **2018**, *2018*, 1206235. [CrossRef]
46. Melkonian, G.; Le, C.; Zhang, W.; Talbot, P.; Martins-Green, M. Normal patterns of angiogenesis and extracellular matrix deposition in chick chorioallantoic membranes are disrupted by mainstream and sidestream cigarette smoke. *Toxicol. Appl. Pharmacol.* **2000**, *163*, 26–37. [CrossRef] [PubMed]
47. Melkonian, G.; Chung, L.; Marr, R.; Tong, C.; Talbot, P. Mainstream and sidestream cigarette smoke inhibit growth and angiogenesis in the day 5 chick chorioallantoic membrane. *Toxicol. Sci.* **2002**, *68*, 237–248. [CrossRef] [PubMed]
48. Valavanidis, A.; Haralambous, E. A comparative study by electron paramagnetic resonance of free radical species in the mainstream and sidestream smoke of cigarettes with conventional acetate filters and 'bio-filters'. *Redox. Rep.* **2001**, *6*, 161–171. [CrossRef] [PubMed]
49. Gieseke, C.; Talbot, P. Cigarette smoke inhibits hamster oocyte pickup by increasing adhesion between the oocyte cumulus complex and oviductal cilia. *Biol. Reprod.* **2005**, *73*, 443–451. [CrossRef] [PubMed]
50. Neal, M.S.; Hughes, E.G.; Holloway, A.C.; Foster, W.G. Sidestream smoking is equally as damaging as mainstream smoking on IVF outcomes. *Hum. Reprod.* **2005**, *20*, 2531–2535. [CrossRef] [PubMed]
51. Lin, S.; Tran, V.; Talbot, P. Comparison of toxicity of smoke from traditional and harm-reduction cigarettes using mouse embryonic stem cells as a novel model for preimplantation development. *Hum. Reprod.* **2009**, *24*, 386–397. [CrossRef]
52. Polyzos, A.; Schmid, T.E.; Piña-Guzmán, B.; Quintanilla-Vega, B.; Marchetti, F. Differential sensitivity of male germ cells to mainstream and sidestream tobacco smoke in the mouse. *Toxicol. Appl. Pharmacol.* **2009**, *237*, 298–305. [CrossRef]
53. Riveles, K.; Tran, V.; Roza, R.; Kwan, D.; Talbot, P. Smoke from traditional commercial, harm-reduction, and research brand cigarettes impairs oviductal functioning in hamsters (*Merocricetus auratus*) in vitro. *Human Reprod.* **2007**, *22*, 346–355. [CrossRef]
54. Hoffmann, D.; Hoffmann, I. The changing cigarette, 1950–1995. *J. Toxicol. Environ. Health* **1997**, *50*, 307–364. [CrossRef]
55. Fujinaga, M.; Baden, J.M.; Mazze, R.I. Susceptible period of nitrous oxide teratogenicity in Sprague-Dawley rats. *Teratology* **1989**, *40*, 439–444. [CrossRef]
56. Nayak, B.N.; Ray, M.; Persaud, T.V. Maternal and fetal chromosomal aberrations in mice following prenatal exposure to subembryotoxic doses of lead nitrate. *Acta Anat.* **1995**, *135*, 185–188. [CrossRef] [PubMed]
57. Sasaki, S.; Sata, F.; Katoh, S.; Saijo, Y.; Nakajima, S.; Washino, N.; Konishi, K.; Ban, S.; Ishizuka, M.; Kishi, R. Adverse birth outcomes associated with maternal smoking and polymorphisms in the N-Nitrosamine-metabolizing enzyme genes NQO1 and CYP2E1. *Am. J. Epidemiol.* **2008**, *67*, 719–726. [CrossRef] [PubMed]
58. Zeman, C.; Beltz, L.; Linda, M.; Maddux, J.; Depken, D.; Orr, J.; Theran, P. New questions and insights into nitrate/nitrite and human health effects: A retrospective cohort study of private well users' immunological and wellness status. *J. Environ. Health* **2011**, *74*, 8–18. [PubMed]
59. Hansen, J.M.; Harris, C. Redox control of teratogenesis. *Reprod. Toxicol.* **2013**, *35*, 165–179. [CrossRef]
60. Elmore, S.P.; Qian, T.; Grissom, S.F.; Lemasters, J.J. The mitochondrial permeability transition initiates autophagy in rat hepatocytes. *FASEB J.* **2001**, *15*, 2286–2287. [CrossRef]
61. Burgoyne, J.R.; Mongue-Din, H.; Eaton, P.; Shah, A.M. Redox signaling in cardiac physiology and pathology. *Circ. Res.* **2012**, *111*, 1091–1106. [CrossRef]
62. Lepka, K.; Berndt, C.; Hartung, H.P.; Aktas, O. Redox Events as Modulators of Pathology and Therapy of Neuroinflammatory Diseases. *Front. Cell Dev. Biol.* **2016**, *4*, 63. [CrossRef]
63. D'Sa-Eipper, C.; Leonard, J.R.; Putcha, G.; Zheng, T.S.; Flavell, R.A.; Rakic, P.; Kuida, K.; Roth, K.A. DNA damage-induced neural precursor cell apoptosis requires p53 and caspase 9 but neither Bax nor caspase 3. *Development* **2001**, *128*, 137–146. [CrossRef]
64. Ochs, K.; Kaina, B. Apoptosis induced by DNA damage O6-methylguanine is Bcl-2 and caspase-9/3 regulated and Fas/caspase-8 independent. *Cancer Res.* **2000**, *60*, 5815–5824.

65. Sharma, P.; Bhusan Jha, A.; Shanker Dubey, R.; Pessarakli, M. Reactive Oxygen Species, Oxidative Damage, and Antioxidative Defense Mechanism in Plants under Stressful Conditions. *Am. J. Bot.* **2012**, *2012*, 217037. [CrossRef]
66. Chen, M.; Guerrero, A.D.; Huang, L.; Shabier, Z.; Pan, M.; Tan, T.H.; Wang, J. Caspase-9-induced mitochondrial disruption through cleavage of anti-apoptotic BCL-2 family members. *J. Biol. Chem.* **2007**, *282*, 33888–33895. [CrossRef] [PubMed]
67. Eeva, J.; Nuutinen, U.; Ropponen, A.; Mättö, M.; Eray, M.; Pellinen, R.; Wahlfors, J.; Pelkonen, J. Feedback regulation of mitochondria by caspase-9 in the B cell receptor-mediated apoptosis. *Scand. J. Immunol.* **2009**, *70*, 574–583. [CrossRef] [PubMed]
68. Yu, T.; Sheu, S.S.; Robotham, J.L.; Yoon, Y. Mitochondrial fission mediates high glucose-induced cell death through elevated production of reactive oxygen species. *Cardiovasc. Res.* **2008**, *79*, 341–351. [CrossRef] [PubMed]
69. Makino, A.; Scott, B.T.; Dillmann, W.H. Mitochondrial fragmentation and superoxide anion production in coronary endothelial cells from a mouse model of type 1 diabetes. *Diabetologia* **2010**, *53*, 1783–1794. [CrossRef]
70. Edwards, J.L.; Quattrini, A.; Lentz, S.I.; Figueroa-Romero, C.; Cerri, F.; Backus, C.; Hong, Y.; Feldman, E.L. Diabetes regulates mitochondrial biogenesis and fission in mouse neurons. *Diabetologia* **2010**, *53*, 160–169. [CrossRef]
71. Sawada, O.; Perusek, L.; Kohno, H.; Howell, S.J.; Maeda, A.; Matsuyama, S.; Maeda, T. All-trans-retinal induces Bax activation via DNA damage to mediate retinal cell apoptosis. *Exp. Eye Res.* **2014**, *123*, 27–36. [CrossRef]
72. Korsmeyer, S.J.; Wei, M.C.; Saito, M.; Weiler, S.; Oh, K.J.; Schlesinger, P.H. Pro-apoptotic cascade activates BID, which oligomerizes BAK or BAX into pores that result in the release of cytochrome c. *Cell Death Differ.* **2000**, *7*, 1166–1173. [CrossRef]
73. Große, L.; Wurm, C.A.; Brüser, C.; Neumann, D.; Jans, D.C.; Jakobs, S. Bax assembles into large ring-like structures remodeling the mitochondrial outer membrane in apoptosis. *EMBO J.* **2016**, *35*, 402–413. [CrossRef]
74. Zou, H.; Li, Y.; Liu, X.; Wang, X. An APAF-1.cytochrome c multimeric complex is a functional apoptosome that activates procaspase-9. *J. Biol. Chem.* **1999**, *274*, 11549–11556. [CrossRef]
75. Keep, O.; Rajalingam, K.; Kimmig, S.; Rudel, T. Bak and Bax are non-redundant during infection- and DNA damage-induced apoptosis. *EMBO J.* **2007**, *26*, 825–834. [CrossRef]
76. Marchi, S.; Patergnani, S.; Pinton, P. The endoplasmic reticulum-mitochondria connection: One touch, multiple functions. *Biochim. Biophys. Acta* **2014**, *1837*, 461–469. [CrossRef] [PubMed]
77. Cagalinec, M.; Liiv, M.; Hodurova, Z.; Hickey, M.A.; Vaarmann, A.; Mandel, M.; Zeb, A.; Choubey, V.; Kuum, M.; Safiulina, D.; et al. Role of Mitochondrial Dynamics in Neuronal Development: Mechanism for Wolfram Syndrome. *PLoS Biol.* **2016**, *14*, e1002511. [CrossRef] [PubMed]
78. Theurey, P.; Tubbs, E.; Vial, G.; Jacquemetton, J.; Bendridi, N.; Chauvin, M.A.; Alam, M.R.; Le Romancer, M.; Vidal, H.; Rieusset, J. Mitochondria-associate endoplasmic reticulum membranes allow adaptation of mitochondrial metabolism to glucose availability in the liver. *J. Mol. Cell Biol.* **2016**, *8*, 129–143. [CrossRef]
79. Filadi, R.; Theurey, P.; Pizzo, P. The endoplasmic reticulum-mitochondria coupling in health and disease: Molecules, functions and significance. *Cell Calcium* **2017**, *62*, 1–15. [CrossRef] [PubMed]
80. Angebault, C.; Fauconnier, J.; Patergnani, S.; Rieusset, J.; Danese, A.; Affortit, C.A.; Jagodzinska, J.; Mégy, C.; Quiles, M.; Cazevieille, C.; et al. ER-mitochondria cross-talk is regulated by the Ca2+ sensor NCS1 and is impaired in Wolfram syndrome. *Sci. Signal.* **2018**, *11*, eaaq1380. [CrossRef]
81. Hitomi, J.; Katayama, T.; Eguchi, Y.; Kudo, T.; Taniguchi, M.; Koyama, Y.; Manabe, T.; Yamagishi, S.; Bando, Y.; Imaizumi, K.; et al. Involvement of caspase-4 in endoplasmic reticulum stress-induced apoptosis and Abeta-induced cell death. *J. Cell Biol.* **2004**, *165*, 347–356. [CrossRef] [PubMed]
82. Yamamuro, A.; Kishino, T.; Ohshima, Y.; Yoshioka, Y.; Kimura, T.; Kasai, A.; Maeda, S. Caspase-4 Directly Activates Caspase-9 in Endoplasmic Reticulum Stress-Induced Apoptosis in SH-SY5Y Cells. *J. Pharmacol. Sci.* **2011**, *115*, 239–243. [CrossRef]
83. Li, C.; Wei, J.; Li, Y.; He, X.; Zhou, Q.; Yan, J.; Zhang, J.; Liu, Y.; Liu, Y.; Shu, H.B. Transmembrane Protein 214 (TMEM214) mediates endoplasmic reticulum stress-induced caspase 4 enzyme activation and apoptosis. *J. Biol. Chem.* **2013**, *288*, 17908–17917. [CrossRef]
84. Oslowski, C.M.; Urano, F. Measuring ER stress and the unfolded protein response using mammalian tissue culture system. *Methods Enzymol.* **2011**, *490*, 71–92. [CrossRef] [PubMed]
85. Yukioka, F.; Matsuzaki, S.; Kawamoto, K.; Koyama, Y.; Hitomi, J.; Katayama, T.; Tohyama, M. Presenilin-1 mutation activates the signaling pathway of caspase-4 in endoplasmic reticulum stress-induced apoptosis. *Neurochem. Int.* **2008**, *52*, 683–687. [CrossRef]
86. Kamada, S.; Washida, M.; Hasegawa, J.; Kusano, H.; Funahashi, Y.; Tsujimoto, Y. Involvement of caspase-4(-like) protease in Fas-mediated apoptotic pathway. *Oncogene* **1997**, *15*, 285–290. [CrossRef] [PubMed]
87. Lin, R.; Sun, Y.; Li, C.; Xie, C.; Wang, S. Identification of differentially expressed genes in human lymphoblastoid cells exposed to irradiation and suppression of radiation-induced apoptosis with antisense oligonucleotides against caspase-4. *Oligonucleotides* **2007**, *17*, 314–326. [CrossRef] [PubMed]
88. López-Antón, N.; Rudy, A.; Barth, N.; Schmitz, M.L.; Pettit, G.R.; Schulze-Osthoff, K.; Dirsch, V.M.; Vollmar, A.M. The marine product cephalostatin 1 activates an endoplasmic reticulum stress-specific and apoptosome-independent apoptotic signaling pathway. *J. Biol. Chem.* **2006**, *281*, 33078–33086. [CrossRef] [PubMed]
89. Schmid-Burgk, J.L.; Gaidt, M.M.; Schmidt, T.; Ebert, T.S.; Bartok, E.; Hornung, V. Caspase-4 mediates non-canonical activation of the NLRP3 inflammasome in human myeloid cells. *Eur. J. Immunol.* **2015**, *45*, 2911–2917. [CrossRef]

90. Lakshmanan, U.; Porter, A.G. Caspase-4 interacts with TNF receptor-associated factor 6 and mediates lipopolysaccharide-induced NF-kappaB-dependent production of IL-8 and CC chemokine ligand 4 (macrophage-inflammatory protein-1). *J. Immunol.* **2007**, *179*, 8480–8490. [CrossRef]
91. Cao, S.; Zhang, X.; Edwards, J.P.; Mosser, D.M. NF-kappaB1 (p50) homodimers differentially regulate pro- and anti-inflammatory cytokines in macrophages. *J. Biol. Chem.* **2006**, *281*, 26041–26050. [CrossRef]
92. Porta, C.; Rimoldi, M.; Raes, G.; Brys, L.; Ghezzi, P.; Di Liberto, D.; Dieli, F.; Ghisletti, S.; Natoli, G.; De Baetselier, P.; et al. Tolerance and M2 (alternative) macrophage polarization are related processes orchestrated by p50 nuclear factor kappaB. *Proc. Natl. Acad. Sci. USA* **2009**, *106*, 14978–14983. [CrossRef]
93. Dresner-Pollak, R.; Gelb, N.; Rachmilewitz, D.; Karmeli, F.; Weinreb, M. Interleukin 10-deficient mice develop osteopenia, decreased bone formation, and mechanical fragility of long bones. *Gastroenterology* **2004**, *127*, 792–801. [CrossRef]
94. Chen, E.; Liu, G.; Zhou, X.; Zhang, W.; Wang, C.; Hu, D.; Xue, D.; Pan, Z. Concentration-dependent, dual roles of IL-10 in the osteogenesis of human BMSCs via P38/MAPK and NF-kappaB signaling pathways. *FASEB J.* **2018**, *32*, 4917–4929. [CrossRef]
95. Daigle, I.; Rückert, B.; Schnetzler, G.; Simon, H.U. Induction of the IL-10 gene via the fas receptor in monocytes–an anti-inflammatory mechanism in the absence of apoptosis. *Eur. J. Immunol.* **2000**, *30*, 2991–2997. [CrossRef]
96. Bharhani, M.S.; Borojevic, R.; Basak, S.; Ho, E.; Zhou, P.; Croitoru, K. IL-10 protects mouse intestinal epithelial cells from Fas-induced apoptosis via modulating Fas expression and altering caspase-8 and FLIP expression. *Am. J. Physiol. Gastrointest. Liver Physiol.* **2006**, *291*, G820–G829. [CrossRef]
97. Yang, S.; Wang, J.; Brand, D.D.; Zheng, S.G. Role of TNF-TNF Receptor 2 Signal in Regulatory T Cells and Its Therapeutic Implications. *Front. Immunol.* **2018**, *9*, 784. [CrossRef] [PubMed]
98. Kawakami, A.; Eguchi, K.; Matsuoka, N.; Tsuboi, M.; Koji, T.; Urayama, S.; Fujiyama, K.; Kiriyama, T.; Nakashima, T.; Nakane, P.K.; et al. Fas and Fas ligand interaction is necessary for human osteoblast apoptosis. *J. Bone Miner Res.* **1997**, *12*, 1637–1646. [CrossRef] [PubMed]
99. Hatakeyama, S.; Tomichi, N.; Ohara-Nemoto, Y.; Satoh, M. The immunohistochemical localization of Fas and Fas ligand in jaw bone and tooth germ of human fetuses. *Calcif. Tissue Int.* **2000**, *66*, 330–337. [CrossRef] [PubMed]
100. Katavić, V.; Lukić, I.K.; Kovacić, N.; Grcević, D.; Lorenzo, J.A.; Marusić, A. Increased bone mass is a part of the generalized lymphoproliferative disorder phenotype in the mouse. *J. Immunol.* **2003**, *170*, 1540–1547. [CrossRef]
101. Katavić, V.; Grcević, D.; Lukić, I.K.; Vucenik, V.; Kovacić, N.; Kalajzić, I.; Marusić, A. Non-functional Fas ligand increases the formation of cartilage early in the endochondral bone induction by rhBMP-2. *Life Sci.* **2003**, *74*, 13–28. [CrossRef]

antioxidants

Article

Protective Actions of α-Tocopherol on Cell Membrane Lipids of Paraquat-Stressed Human Astrocytes Using Microarray Technology, MALDI-MS and Lipidomic Analysis

Laura Sánchez-Sánchez [1,2,*], Roberto Fernández [1], Maria Dolores Ganfornina [2], Egoitz Astigarraga [1] and Gabriel Barreda-Gómez [1,*]

1 Research and Development Department, IMG Pharma Biotech S.L, 48160 Derio, Spain
2 Instituto de Biología y Genética Molecular (IBGM), Unidad de Excelencia, Universidad de Valladolid-CSIC, 47003 Valladolid, Spain
* Correspondence: laura.sanchez@imgpharma.com (L.S.-S.); gabriel.barreda@imgpharma.com (G.B.-G.); Tel.: +34-94-4316-577 (G.B.-G.); Fax: +34-94-6013-455 (G.B.-G.)

Abstract: Cellular senescence is one of the main contributors to some neurodegenerative disorders. The early detection of senescent cells or their related effects is a key aspect in treating disease progression. In this functional deterioration, oxidative stress and lipid peroxidation play an important role. Endogenous antioxidant compounds, such as α-tocopherol (vitamin E), can mitigate these undesirable effects, particularly lipid peroxidation, by blocking the reaction between free radicals and unsaturated fatty acid. While the antioxidant actions of α-tocopherol have been studied in various systems, monitoring the specific effects on cell membrane lipids at scales compatible with large screenings has not yet been accomplished. Understanding the changes responsible for this protection against one of the consequences of senescence is therefore necessary. Thus, the goal of this study was to determinate the changes in the lipid environment of a Paraquat-treated human astrocytic cell line, as a cellular oxidative stress model, and the specific actions of the antioxidant, α-tocopherol, using cell membrane microarray technology, MALDI-MS and lipidomic analysis. The stress induced by Paraquat exposure significantly decreased cell viability and triggered membrane lipid changes, such as an increase in certain species of ceramides that are lipid mediators of apoptotic pathways. The pre-treatment of cells with α-tocopherol mitigated these effects, enhancing cell viability and modulating the lipid profile in Paraquat-treated astrocytes. These results demonstrate the lipid modulation effects of α-tocopherol against Paraquat-promoted oxidative stress and validate a novel analytical high-throughput method combining cell cultures, microarray technology, MALDI-MS and multivariate analysis to study antioxidant compounds against cellular senescence.

Keywords: oxidative stress; lipid peroxidation; membrane microarrays; human cell line; astrocytes

Citation: Sánchez-Sánchez, L.; Fernández, R.; Ganfornina, M.D.; Astigarraga, E.; Barreda-Gómez, G. Protective Actions of α-Tocopherol on Cell Membrane Lipids of Paraquat-Stressed Human Astrocytes Using Microarray Technology, MALDI-MS and Lipidomic Analysis. *Antioxidants* **2022**, *11*, 2440. https://doi.org/10.3390/antiox11122440

Academic Editor: Tim Hofer

Received: 27 October 2022
Accepted: 8 December 2022
Published: 10 December 2022

1. Introduction

Lipid peroxidation is a physiological chain reaction process produced by free radicals [1,2] through enzymatic and non-enzymatic mechanisms [3]. In enzymatic mechanisms, lipid peroxidation can be produced by lipoxygenases (LOX), cyclooxygenases (COX), cytochrome c [4], or cytochrome P450 [5,6]. Whereas, in non-enzymatic mechanisms, the production is mediated by oxygen-free radicals [7], such as superoxide ($O_2^{\bullet -}$) or hydroxyl radicals ($OH^{\bullet}$), among others [8]. However, lipid peroxidation is regulated by a diverse array of antioxidant defense systems, such as the glutathione system or catalase and superoxide dismutase enzymes [6,9], as well as non-enzymatic compounds, such as vitamin E, beta-carotene or glutathione [6]. If an imbalance between the oxidant and antioxidant compounds is produced, it turns into a pathological process, which provokes the modification of physical properties of membranes, such as permeability, packing of lipids and proteins [10], and a loss of function [4] that leads to cell damage [7].

The non-enzymatic formation of lipid radicals is a three-phase reaction (initiation, propagation, and termination) which is primarily produced in polyunsaturated fatty acids (PUFAs) [11], due to its relative weak hydrogen-carbon bond [11,12], where a hydrogen radical ($H^\bullet$) is abstracted from the lipid chain [13]. The lipid radical can react with other molecules, giving rise to lipid hydroperoxides, which are extremely unstable and decompose themselves in secondary products [12,13]. The initiation phase requires an initiator molecule, such as iron or copper ions [8], reactive nitrogen species (RNS), or reactive oxygen species (ROS) [13].

The increased presence of reactive oxygen species entails the pathological process of oxidative stress, which has been related to several diseases [14] such as cancer, cardiovascular and neurodegenerative diseases [15]. As the brain is an organ with a huge energy consumption, it is more susceptible to oxidative stress. In addition, it is rich in phospholipids, and is especially enriched in PUFAs [16]. Oxidative stress generation can originate from different sources, such as NADPH oxidase enzymes [17,18], monoamine oxidase [19], peroxisomes, and lastly, the mitochondrial electron transport chain (mETC) [18,20,21] as the principal generator, primarily by complexes I and III [22].

Paraquat is a highly toxic herbicide that can lead to severe brain damage [23] due to its interference with NADP+ reduction [24,25]. It has also been reported as a complex I and III inhibitor [24]. In the first mechanism, Paraquat ions (PQ^{2+}) are reduced to their monocation radical form, which can react with oxygen and generate ROS [24,26,27], whereas in the last case, the inhibition of mitochondrial complexes provokes mETC dysfunction. Thus, as the principal toxic effect of paraquat is related to oxidative stress, its effects may be partially reversed using antioxidant compounds [25,28]. Concretely α-tocopherol, the vitamin E predominant form, as well as other tocopherols and tocotrienols, have lipoperoxyl radical scavenging activity [28–31]. Vitamin E is an effective antioxidant against oxidation mediated by free radicals [28,30]. The mechanism of action of this compound consists in donating a hydrogen atom to lipoperoxyl radicals generating its own radical [32].

As this antioxidant compound can have beneficial effects over oxidative stress conditions, levels of oxidation subproducts [33], such as monocyte superoxide anion concentrations, have been analyzed in humans [34]. However, the effects over lipids and lipoperoxyls have not yet been examined. In this work, we test the effects of stressing human astrocytic cells with Paraquat and the modification of these changes by the antioxidant α-tocopherol. Cell viability and lipid-profiling are performed. For the latter, we combine our Cell Membrane Microarray (CMMA) technology with MALDI mass spectrometry, as it allows for the simultaneous lipid-profiling of many samples [35]. Our analysis demonstrates the importance of the lipid composition and lipid homeostasis in cell viability upon oxidative stress and opens the door to high-throughput analysis of antioxidant compounds' effects on relevant membrane samples.

2. Materials and Methods

2.1. Drugs and Reagents

(±)-α-Tocopherol, Methyl Viologen dichloride hydrate, DMEM 1 g/L glucose, Penicillin–Streptomycin Solution Hybri-Max (P/S), Fetal Bovine Serum (FBS), L-glutamine (L-Glut), 2-mercaptobenzothiazole, and 1,5-diaminonaphthalene, trypan-blue were purchased from Sigma-Aldrich (Saint Louis, MI, USA).

2.2. Cell Culture and Treatments

The human astrocytic cell line 1321N1 was seeded at 10,000 cells/cm^2 and cultivated in 12 well plates in complete medium (DMEM medium with 1 g/L glucose, 5% Fetal Bovine Serum (FBS), 1% L-glutamine (L-Glut) and 1% Penicillin-Streptomycin Solution Hybri-Max (P/S)) for 24 h at 37 °C, 5% CO_2 and constant humidity. The cells were habituated to low serum conditions (DMEM 1 g/L glucose, 0.2% Charcoal-treated FBS, 1% L-Glut, and 1% P/S) for 12 h before treatments began. After adaptation, a pre-treatment of 3 h with or without α-Tocopherol (1 μM), prepared in a low serum medium, was performed before treatment, with or without Paraquat (500 μM), in low serum conditions for different times (24, 48, 72, 96, and 120 h). Cell viability assays were performed after each time interval as described below.

2.3. Viability Assay

To analyze the cellular viability in every condition and time point, Trypan-blue viability assays were performed. The cells were detached from the culture well by a mechanical method to avoid enzymatic treatments (changing media for PBS at 4 °C and tapping the plate laterally). The cell suspension was transferred to a microtube and diluted 1:1 with 0.4% Trypan blue solution. The cells were counted using a Neubauer chamber in an inverted microscope Olympus CKX41 (Olympus Corporation, Tokyo, Japan). Dead cells (with a compromised membrane) were stained in dark blue. The percentage of live cells, with respect to total cells, was calculated in order to determine the best time for MS analysis.

Viability data handling and analysis was carried out using Excel and GraphPad software (version 9.2). Briefly, cell viability data were presented as a percentage of cell growth. The identification of outliers was carried out by applying the following formulas:

$$Y_1 = \overline{X} - DF * SD \quad Y_2 = \overline{X} + DF * SD$$

* SD = standard deviation; DF = deviation factor;

Points were identified as outliers and excluded if Y_1 was higher than the point analyzed or Y_2 was lower than the point examined. We used a deviation factor of 1.25 in our analysis. The data were expressed as means of independent data points ± S.E.M. The results were analysed using one-way two-tailed two-way ANOVA with Tukey's post hoc. Statistical differences were indicated with p-values ≤ 0.05.

2.4. Cell Membrane Extraction and CMMA Fabrication

In order to fabricate Cell Membrane Microarrays (CMMA), the 1321N1 cell line was seeded at 15,000 cells/cm^2 and cultivated in 25 cm^2 culture flask following the same conditions explained previously. The cells were cultivated until 80% confluency was reached (obtaining 10^6 cells per flask), and then treated as described previously. The cells were detached from the flasks and homogenized using a Teflon-glass grinder (Heidolph RZR 2020, Schwabach, Germany) in 20 volumes of homogenization buffer (1 mM EGTA, 3 mM $MgCl_2$, and 50 mM Tris-HCl, pH 7.4). The crude homogenate was subjected to a 1500-rpm centrifugation (AllegraTM X 22R centrifuge, Beckman Coulter, CA, USA) for 5 min at 4 °C, and the resulting supernatant was collected and centrifuged at 18,000× g (Microfuge® 22R centrifuge, Beckman Coulter, CA, USA) for 15 min (4 °C). With this protocol, a fraction enriched in plasma membrane and membranes from internal organelles, including mitochondria, was obtained, as it has been demonstrated by detection of GLUT-4 transporter and Insulin Receptor β subunit [36,37], and in previous studies by cytochrome C detection [36], mETC enzymes and acetylcholinesterase activity [18,38,39]. The tubes were finally decanted, and the pellets were frozen at −80 °C, with the exception of one aliquot, which was used to determine the protein concentration. The protein concentration was determined by the Bradford method and adjusted to a final concentration of 5 mg/mL [40,41].

The membrane homogenates were resuspended in buffer and printed onto glass slides using a non-contact microarrayer (Nanoplotter NP 2.1, Gesim Bioinstruments, Radeberg, Germany) using a solenoid tip, placing 2 replicates of each sample (30 nL/spot), into pre-activated glass microscope slides. Printing was carried out under controlled humidity (relative humidity 60%) at a controlled temperature of 4 °C. The CMMAs were stored at −20 °C until usage. The CMMAs were validated before usage by different methods, including Bradford staining, for protein determination, enzyme activity assays (NADH oxidoreductase, succinate dehydrogenase, and cytochrome c oxidase), and radioligand binding assays [18,38–40].

2.5. MALDI-MS Lipidomic Analysis

The Cell Membrane Microarrays were covered with a suitable matrix with the aid of a standard glass sublimator (Ace Glass 8233,NJ, USA), producing a uniform film of approximately 0.2 mg/cm^2 [35]. For positive-ion and negative-ion modes, 2-mercaptobenzothiazole (MBT) and 1,5-diaminoaphtalene (DAN) were used, respectively [35,42]. The CMMAs were scanned, as in the MALDI imaging experiment. The area of the array was explored following a grid coordinate separated by 250 μm; as each spot has a diameter of 280 μm, six pixels were recorded at each spot. The mass spectrometer used was an LTQ-Orbitrap XL (Thermo Fisher Scientific, Waltham, Massachusetts, USA), equipped with a MALDI source with a N_2 laser (60 Hz, 100 μJ/pulse maximum power output). The laser spot is an ellipsoid of approximately 50–60 μm × 140–160 μm. Two microscans of 10 shots/pixel were used, with a laser power output of 20 μJ for MS+ and 30 μJ for MS- and a resolution of 250 μm. Data loading included spectra normalization by total ion current (TIC), spectra alignment, and peak picking, filtering all the *m/z* with intensity < 0.5% of the strongest peak in the spectrum.

The MALDI spectra were aligned using MATLAB (Matworks, Natick, Massachusetts, USA), lipids assignment was performed using the Lipid Maps LMSD database. For the MALDI data analysis, the MS+ and MS- data were normalized separately and then analyzed together. The matrix peaks and isotopic distribution were removed, and the remaining peaks were normalized against their total ion current (TIC). The MS+ and MS- data were standardized using the z-score method and using the following formula:

$$Z = \frac{X - \mu}{\sigma}$$

X: observed value; μ: mean of the sample; σ: standard deviation of the sample.

3. Results

3.1. Viability of Human Astrocytic Cells upon Different Treatments

The characterization of cell viability in the cultures was performed in order to decide the conditions in which to study the membrane lipid changes by CMMAs. The culture medium was maintained along the treatments, so that the cells would be entering senescence. The cultures were monitored until all of the cells were dead.

Paraquat treatment decreased cell viability at all time points, with respect to the control, changing from 83% to 68% at 24 h (Figure 1B, C). Moreover, with α-tocopherol pre-treatment, the viability improved significantly at 24 h (96%) (Figure 1C). At 72 h, the Paraquat-induced mortality reached 90% (10% viability), and the viability was rescued (up to 40%) if the cells were pre-treated with α-tocopherol. No viability differences between the control and the 3-h exposure to α tocopherol were observed when not combined with Paraquat. The cells in the control conditions reached death (98.6 %) at 96 h, while those treated with α-tocopherol only or with a pre-treatment of α-tocopherol, followed by Paraquat, survived longer (reaching death at 120 h).

Differences were observed at 24, 48, and 72 h between the Paraquat-treated and control situations and between the Paraquat-treated cells with or without α-tocopherol pre-treatment. The time chosen for mass spectrometry was 24 h, at which point viability is still not much compromised and the quantitative differences between treatments are larger.

3.2. Lipidomic Analysis in Cell Membrane Microarrys Is Able to Reveal Paraquat-Triggered Changes in Human Astrocytic Membranes

We used CMMAs composed of membranes from the human astrocytic cell line 1321N1, obtained in three different conditions: Control, Paraquat (500 μM, 24 h), and Paraquat preceded by α-tocopherol (1 μM, 3 h). Membrane arrays were developed for MALDI mass spectrometry analysis of their lipid fingerprint. The differences between the lipid fingerprint were analyzed between membranes from the control and Paraquat-treated cells, and between membranes in Paraquat conditions with or without α-tocopherol pre-treatment. Both MS+ and MS- modes were performed. Specific lipid adducts were detected in the Paraquat-treated samples, in contrast to the control situation; 66 lipid adducts were only present in the Paraquat-treated membranes with respect to the control situation. Nevertheless, the vast majority of the lipid adducts present in the Paraquat condition changed when the cells were pre-treated with α-tocopherol.

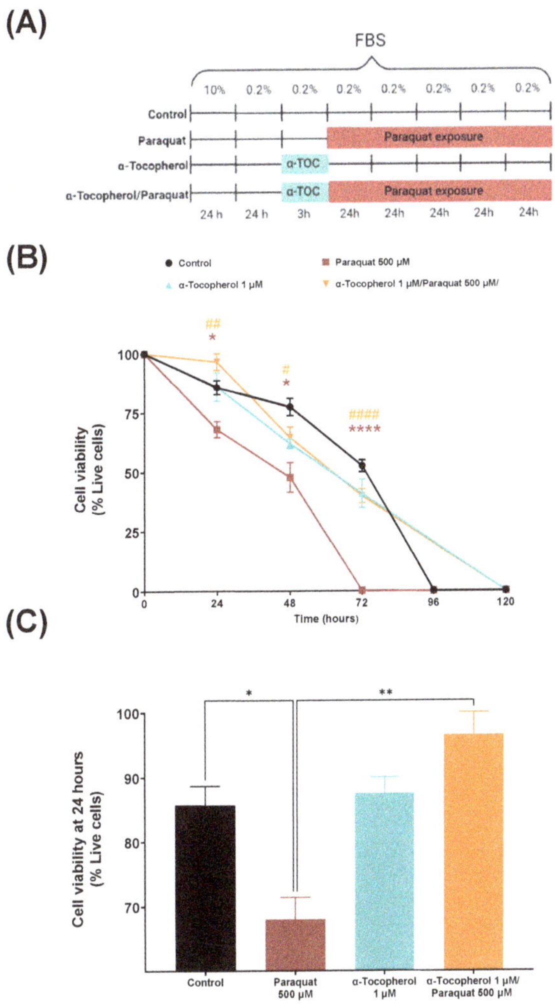

Figure 1. Effects of Paraquat on 1321N1 cells viability with or without pre-treatment with α-tocopherol. (**A**) Time course and design of treatments for the evaluation of cell culture viability. (**B**) Time course of cell viability in control conditions and after treatment with Paraquat (500 μM) with or without a pre-treatment with α-tocopherol (1 μM). Measurements were performed every 24 h using trypan blue cell viability assay. Asterisks symbols (*) refer to post-hoc comparisons with respect to control; hash symbols (#) refer to comparisons with respect to Paraquat. Colour of symbols allude to the condition been compared. (**C**) Cell viability results at 24 h expressed as percentage of live cells. Shapiro-Wilk test was performed to check the normality of distributions. One-way ANOVA analysis (two-tailed) and Tukey post-hoc were conducted, α set to 0.05. p-value < 0.05 (*), p-value < 0.01 (**), p-value < 0.0001 (****), moreover, p-value <0.05 (#), p-value < 0.01 (##), p-value < 0.0001 (####).

The membranes from the Paraquat-treated cells displayed very long chain ceramides (C > 26) Cer 40:0; O3, Cer 40:1; O3, and Cer 40:2; O3 (red arrows in Figure 2A), which were absent in the control membranes, and sphingomyelins generally increased their presence in the Paraquat treated samples (Figure 2A).

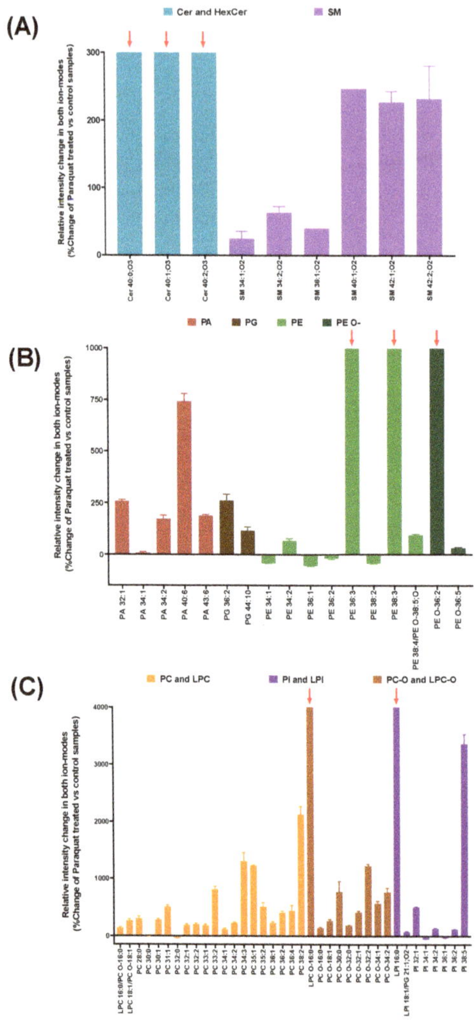

Figure 2. Relative intensity change in different lipid adducts when comparing Paraquat treated membranes with control samples. Data are shown as percentage of control values. (**A**) Changes in sphingolipids (Cer, HexCer and SM). (**B**,**C**)) Changes in glycerophospholipids. PA, PG, PE, and PE O- shown in (**B**). LPC, PC, and PI shown in (**C**). Abbreviations: Ceramides (Cer), hexosylceramides (HexCer), sphingomyelins (SM), glycerophosphates (PA), glycerophosphoethanolamines (PE and its ether form PE O-), glycerophosphoglycerols (PG), glycerophosphocholines and lisophosphocholines (PC and LPC), glycerophosphoinositols and lisofosfoinositols (PI and LPI), and ether forms of glycerophosphocholines and lisophosphocholines (PC O- and LPC O-). Oxygen number inside the lipid head and fatty acids in Ceramides and Sphingomyelins are indicated by O with a suffix. Bars with red arrows are only present in Paraquat-treated samples.

Long-chain glycerophosphates (PA with C > 26) (Figure 2B) also displayed a general increase, particularly evident in PA 40:6, with a very long chain with more than three unsaturations. Lipids of the phosphoethanolamines family are present as normal and as the ether version (PE and PE O-), where one of the fatty acids is connected to the glycerol molecule by an oxygen atom. In this regard, PE 36:3, PE 38:3, and PE O- 36:2 were only present in the Paraquat-treated membranes (red arrows in Figure 2B), while PE 36:2 appeared to decrease ($-17.62\% \pm 7.2$), as was the case for other PEs (Figure 2B). Moreover, two similar lipidic species with opposite behavior, upon Paraquat treatment, were observed in pairs differing in their degree of unsaturation: e.g., PE 38:2 showed a decrease of $-35.7\% \pm 1.4$, whereas PE 38:3 is one of the species only present in Paraquat treated samples. Finally, oxidated PE-O species of more than 30 carbons generally increased with Paraquat treatment. Very long chain phosphocholines (Figure 2C) are increased due to Paraquat treatment: the more unsaturation, the higher the change rates. In addition, the lysophosphocholines (lipid adducts with only one fatty acid) LPC 16:0 and LPC 18:1, presented an increase of 138% $\pm$ 25 and 261.8% $\pm$ 43.2, respectively, whereas the LPC 16:0 ether form appeared only upon Paraquat treatment (red arrows in Figure 2C). Phosphocholines with one or two unsaturations were increased in the membranes from the Paraquat treated cells, while their saturated forms were reduced or absent: e.g., PC32:1 (189.29% $\pm$ 28.1) and PC 32:2 (207.12% $\pm$ 27.0) compared with PC 32:0 ($-35.74\% \pm 1.4$). Finally, PI adducts were present as very long chain lipids, with a general increase observed in the forms with more unsaturations, while their saturated form or those with lower unsaturations were reduced or absent. (Figure 2C). Thus, the lipidomic analysis revealed a distinguishable fingerprint in cell membranes due to paraquat treatment, that we can identify using our CMMAs.

3.3. *Lipidomic Analysis in Cell Membrane Microarrys Is Able to Reveal the Effect of α-Tocopherol Pre-Treatment on the Paraquat-Triggered Signature*

The membranes from the Paraquat-treated cells with α-tocopherol pre-treatment, showed that unsaturated, very long ceramides Cer 40:1;O3 ($-8.77\% \pm 4.9$) and Cer 40:2;O3 ($-35.37\% \pm 23.94$) seemed to decrease, while their saturated version Cer40:0; O3 presented an increase of 271%. In sphingomyelins, SM 35:1;O2, SM 36:2;O2 and SM 41:2;O2 appeared only in the α-tocopherol pre-treated cells (red arrow in Figure 3A). In addition, a general increase was observed in the sphingomyelin adducts with one or two unsaturated ions (Figure 3A), while saturated ones remained absent.

In contrast, saturated glycerophosphates generally increased; the lower the number of double bonds, the higher the change rate (Figure 3B), while phosphoglycerols showed a general decrease due to the antioxidant pre-treatment. Examples of two similar lipidic species with opposite behavior upon α-tocopherol pre-treatment also occurred. Such is the case of PG 36:1 (15.49% $\pm$ 20.32), which appeared to increase, whereas PG 36:2 did not change ($-3.81\% \pm 3.84$). This pattern can also be observed between the different phosphoethanolamine adducts; ether PE generally decreased, while normal adducts displayed a mild increase. The exception is PE O-40:5, an ether PE only present when Paraquat treatment was preceded by α-tocopherol.

With the antioxidant treatment preceding Paraquat, some very long chain phosphocholines (PC 35:4, PC 35:5) disappeared from the membranes (black arrows in Figure 3C). Ether phosphocholines generally showed a slight increase, with the exception of PC O-40:5, which was only present upon α-tocopherol pre-treatment. Lysophosphocholines between 16 and 18 carbons slightly increased (Figure 3C). Finally, the PI adducts showed a general, mild decrease, whereas LPI 20:4 seemed to increase (81.3% $\pm$ 41.41) (Figure 3C).

To summarize, clear differences triggered by Paraquat in the lipid fingerprint were detected in our system in ceramides, sphingomyelins, glycolipids, and unsaturated phospholipids. In contrast, unsaturated ceramides and phospholipids were reduced when the cells were treated with α-tocopherol before the Paraquat stress (Figure 4).

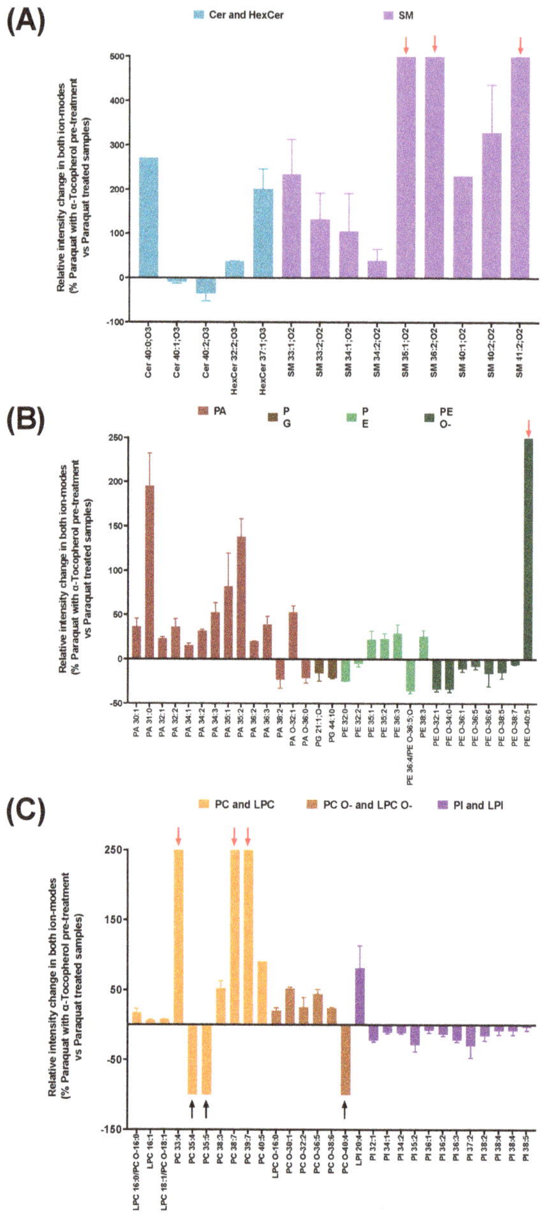

Figure 3. Relative intensity change in lipid adducts when comparing Paraquat-treated membranes preceded by an α-tocopherol pre-treatment with samples treated with Paraquat only. Data are shown as percentage of control values. (**A**) Change in sphingolipids (Cer, HexCer and SM). (**B**,**C**)) Changes in glycerophospholipids. PA, PG, PE, and PE O shown in (**B**). PC, PC O-, LPC, LPC O-, PI, and LPI shown in (**C**). Abbreviations: Ceramides (Cer), hexosylceramides (HexCer), spingomyelins (SM), glycerophosphatidic acid (PA), glycerophosphoethanolamine (PE), glycerophosphoglycerols (PG), glycerophosphocholines and lysophosphocholines (PC and LPC), glycerophosphoinositols and lysophosphoinositols (PI and LPI). Ether forms indicated by the O-suffix. Oxygen number inside lipid head and fatty acids in Ceramides and Sphingomyelins is indicated by O with a suffix.

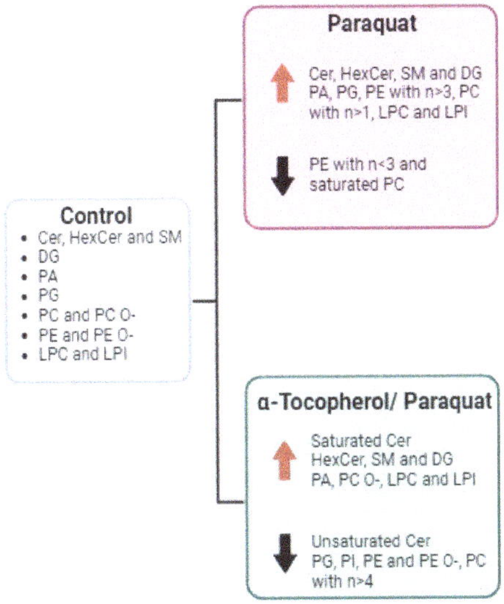

Figure 4. Graphic summary of lipid changes observed in CMMAs upon oxidative stress preceded or not by antioxidant treatment.

4. Discussion

In the present study, we describe a method based on CMMA to observe lipidome changes in membranes from brain cells that are triggered by prooxidant and antioxidant treatments. The starting point is that α-tocopherol, with antioxidant properties against membrane lipid peroxidation, could protect from Paraquat damage [43]. We found that cell viability is enhanced when an α-tocopherol treatment precedes exposure to Paraquat. It is known that tocopherol and tocotrienol compounds prevent lipid peroxidation due to their ROS scavenger properties [44], protecting unsaturated fatty acids and lipid mediators [31]. In addition, α-tocopherol regulates the expression of genes implicated in apoptosis or antioxidant defenses, such as Bcl2-L1 or γ-glutamylcystein synthetase [44,45], or in lipid homeostasis, such as phospholipase A2 [46], which has 1-O-acyl ceramide synthase activity, using ceramides as an acceptor [47]. Thus, we predicted that lipid changes due to Paraquat exposure can be mitigated by this antioxidant pre-treatment.

We show that changes in the membrane lipid fingerprint between experimental conditions can be analyzed with our CMMA technology, using a significantly small amount of material per sample. This allows for the production of thousands of microarrays with negligible total sample amounts. Moreover, using this high-throughput technology allows for lipidomic analysis to be performed in many samples at the same time, therefore opening the path to future large-scale screenings. CMMAs have previously been used in different lipidomic studies to analyze the lipid fingerprint of nerve and peripheral tissue in animal models [35,40]. An additional advantage of this technology is that protein expression or activity assays can be performed, as the protein-lipid relation and protein functionality is maintained [35,39–42]. Thus, the combination of membrane microarray technology with mass spectrometry results in a powerful technique to determine the effects of different compounds over lipid composition.

It is known that the ceramide content remains low in non-dividing cells, but can increase due to exposure to stress conditions, such as serum starvation, chemical compound exposure, or oxidative stress conditions [48]. These sphingolipids, composed of sphingosine group and fatty acid, are primarily produced by de novo synthesis [49], the salvage pathway, or sphingomyelinase pathways [50]. We are able to detect an increase in ceramides in Paraquat-treated membranes, as all adducts increase their presence, whereas, with α-tocopherol pre-treatment, the unsaturated ceramides decrease with respect to their values in Paraquat treatment, or remain absent. These bioactive lipid adducts can participate in diverse signaling pathways themselves or by their hydrolyzation to sphingosine through ceramidase enzymes [51]. The increased levels observed in the membranes from the Paraquat-treated cells, with respect to the control, can act as second messengers in apoptosis pathways [52,53], increasing cell death, as it is observed in Figure 1. Moreover, the increase in ceramide production might be due to an activation of the de novo pathway, and this activation may lead to an increase in other sphingolipids by ceramides transformation, such as sphingomyelins [49] by the action of the SMS enzyme. However, the detected ceramides had ultra-long chain (ULC) fatty acids, the presence of which is related to inflammation and different diseases [54], including neurodegenerative disorders [54,55]. With α-tocopherol exposure, sphingomyelin adducts, such as SM 35:1; O2, either increase or appear as new species, with respect to the membranes from cells treated only with Paraquat. In contrast, some ceramide species decrease with antioxidant pretreatment (Figure 3), which can lead to higher cell survival (Figure 1).

Phospholipids are particularly sensitive to oxidative stress exposure. Phosphoethanolamines and their ethers present changes upon Paraquat treatment. The observed increase in phosphoethanolamine ethers (PE O-) might be an adaptive response of the cell as they have antioxidant activity [56]. They can be precursors of plasmalogen, a lipid species whose oxidative products do not propagate lipid peroxidation [57], protecting other phospholipids, lipids, and lipoproteins from oxidative stress [58]. The observed decrease in most ether PE species when α-tocopherol precedes Paraquat treatment can reflect a return to basal conditions, where this antioxidant mechanism is not necessary [59]. The antioxidant properties of the PE O- species are mainly attributed to the preferential oxidation of the vinyl ether bond, which results in the protection of the polyunsaturated fatty acids [60]. In addition, in agreement with these observations, Paraquat treatment results in greater increases in ether forms of phosphatidylcholines, such as LPC O-16:0 and PC O-32:2, than their non-ether analogs (Figure 2C). Furthermore, as shown in other studies, [59], an increased presence of PE 38:3 is observed in cell membranes upon Paraquat treatment. These species only appear in Paraquat-treated samples with or without pre-treatment with α-tocopherol. Nevertheless, these lipid adducts containing long-chain polyunsaturated fatty acids (LCPUFAs) intervene in brain inflammatory reactions. In Paraquat-treated membranes, PCs containing LCPUFAs are increased, while with α tocopherol pre-treatment, some of them disappear [61].

We have detected an increase in LPI 20:4 when Paraquat treatment is preceded by an α-tocopherol pre-treatment. This lipidic adduct, which contains arachidonic acid (AA), might increase in the membranes due to lower AA release from the membrane phospholipids. Such an effect can be mediated by a reduction in phospholipase A2 activity [62], as α-tocopherol is known to be a modulator of phospholipase A2 [46] or by an increase in phospholipase A1 or 2-acyl LPI transferase activity [63]. In addition, excessive free AA has been related to neurotoxic effects [64]; thus, the higher relative content of lipid adducts with AA may be related to a lower free AA content. This lipid adduct can be an agonist of GPR55, a G-protein coupled receptor, whose biological activities include the modulation of immune cells and insulin secretion, and also have a potential mitogen activity in cancer cells [63]. In contrast, lysophosphocholines stimulate intracellular ROS production and ATM/Chk2, ATR/Chk1, as well as Akt activation in endothelial cells [65]. LPC 16:0, which contains palmitoyl molecule, is present in the samples treated with Paraquat and has a slight increase when samples are pre-treated with α-tocopherol. As palmitoylation is particularly important for cell membrane stabilization [66,67], its increase in our samples may support cellular viability (Figure 1).

In conclusion, the data obtained from the MALDI mass spectrometry performed in our CMMA system on astrocytic human cell membranes after prooxidant and antioxidant treatments provide distinguishable and meaningful lipid fingerprints. These differences emerge from the relative increase in ultra-long chain glycerophospholipids, unsaturated ceramides, and lysophospholipids, caused by Paraquat treatment. Tocopherol pre-treatment changes these effects by reducing the presence of unsaturated ceramides, PIs, and PE adducts, and increasing LPC and LPI, which contain palmitoyl and arachidonic acid, respectively. A correct membrane lipidome is essential for proper membrane fluidity and functionality; therefore, the effects of antioxidant compounds over them might exert an important influence on cell viability. Moreover, CMMA technology allows for the performance of this MS analysis over membranes with a negligible sample amount, in contrast to usual MS. Future high-throughput studies will allow the simultaneous identification of compounds influencing lipid profiles and their adverse or beneficiary effects on cell membranes.

Author Contributions: All authors had full access to all the data in the study and take responsibility for the integrity of the data and accuracy of the data analysis. Conceptualization, G.B.-G. and M.D.G.; methodology, L.S.-S., M.D.G. and G.B.-G.; software, R.F.; formal analysis, L.S.-S.; writing—original draft preparation, L.S.-S.; writing—review and editing, G.B.-G., M.D.G. and E.A.; supervision, G.B.-G., M.D.G. and E.A.; funding acquisition, G.B.-G., M.D.G. and E.A. All authors have read and agreed to the published version of the manuscript.

Funding: This research was funded by the Ministry of Economy and Competitiveness (grants DIN2019-010902 and PTQ 17-09409 to L.S.-S. and R.F.) and by Ministerio de Ciencia e Innovación (grant PID2019-110911RB-I00/AEI/10.13039/501100011033 to MD.G). The APC was funded by MD.G.

Institutional Review Board Statement: Not applicable.

Informed Consent Statement: Not applicable.

Data Availability Statement: The data are contained within the article.

Acknowledgments: We thank M. Corraliza-Gómez and Teresa Bermejo for technical help with cell culture experiments, and D. Sanchez for personal support and ideas.

Conflicts of Interest: All coauthors of IMG Pharma Biotech S.L. have no conflicts of interest in the present study for publication. E.A. and G.B.-G. are listed as inventors on patent (EP2048534A4). The remaining authors declare no competing interests.

References

1. Yin, H.; Xu, L.; Porter, N.A. Free Radical Lipid Peroxidation: Mechanisms and Analysis. *Chem. Rev.* **2011**, *111*, 5944–5972. [CrossRef] [PubMed]
2. Girotti, A.W. Nitric Oxide-Elicited Resistance to Antitumor Photodynamic Therapy via Inhibition of Membrane Free Radical-Mediated Lipid Peroxidation. *Photochem. Photobiol.* **2021**, *97*, 653–663. [CrossRef] [PubMed]
3. Niki, E.; Yoshida, Y.; Saito, Y.; Noguchi, N. Lipid Peroxidation: Mechanisms, Inhibition, and Biological Effects. *Biochem. Biophys. Res. Commun.* **2005**, *338*, 668–676. [CrossRef] [PubMed]
4. Niki, E. Lipid Peroxidation Products as Oxidative Stress Biomarkers. *Biofactors* **2008**, *34*, 171–180. [CrossRef] [PubMed]
5. Gaschler, M.M.; Stockwell, B.R. Lipid Peroxidation in Cell Death. *Biochem. Biophys. Res. Commun.* **2017**, *482*, 419–425. [CrossRef]
6. Senoner, T.; Dichtl, W. Oxidative Stress in Cardiovascular Diseases: Still a Therapeutic Target? *Nutrients* **2019**, *11*, 2090. [CrossRef] [PubMed]
7. Ramana, K.V.; Srivastava, S.; Singhal, S.S. Lipid Peroxidation Products in Human Health and Disease 2014. *Oxid. Med. Cell Longev.* **2014**, *2014*, 162414. [CrossRef] [PubMed]
8. Halliwell, B.; Chirico, S. Lipid Peroxidation: Its Mechanism, Measurement, and Significance. *Am. J. Clin. Nutr.* **1993**, *57*, 715S–724S. [CrossRef]
9. Jaganjac, M.; Cindrić, M.; Jakovčević, A.; Žarković, K.; Žarković, N. Lipid Peroxidation in Brain Tumors. *Neurochem. Int.* **2021**, *149*, 105118. [CrossRef]
10. Catalá, A. Lipid Peroxidation of Membrane Phospholipids Generates Hydroxy-Alkenals and Oxidized Phospholipids Active in Physiological and/or Pathological Conditions. *Chem. Phys. Lipids* **2009**, *157*, 1–11. [CrossRef]
11. Venkataraman, S.; Schafer, F.Q.; Buettner, G.R. Detection of Lipid Radicals Using EPR. *Antioxid. Redox Signal.* **2004**, *6*, 631–638. [CrossRef] [PubMed]
12. Greene, L.E.; Lincoln, R.; Cosa, G. Spatio-Temporal Monitoring of Lipid Peroxyl Radicals in Live Cell Studies Combining Fluorogenic Antioxidants and Fluorescence Microscopy Methods. *Free Radic. Biol. Med.* **2018**, *128*, 124–136. [CrossRef] [PubMed]
13. Guéraud, F.; Atalay, M.; Bresgen, N.; Cipak, A.; Eckl, P.M.; Huc, L.; Jouanin, I.; Siems, W.; Uchida, K. Chemistry and Biochemistry of Lipid Peroxidation Products. *Free Radic. Res.* **2010**, *44*, 1098–1124. [CrossRef] [PubMed]
14. del Rio, D.; Stewart, A.J.; Pellegrini, N. A Review of Recent Studies on Malondialdehyde as Toxic Molecule and Biological Marker of Oxidative Stress. *Nutr. Metab. Cardiovasc. Dis.* **2005**, *15*, 316–328. [CrossRef] [PubMed]
15. Luo, J.; Mills, K.; le Cessie, S.; Noordam, R.; van Heemst, D. Ageing, Age-Related Diseases and Oxidative Stress: What to Do Next? *Ageing Res. Rev.* **2020**, *57*, 100982. [CrossRef]

16. Chen, Z.; Zhong, C. Oxidative Stress in Alzheimer's Disease. *Neurosci. Bull.* **2014**, *30*, 271–281. [CrossRef]
17. Yang, S.; Lian, G. ROS and Diseases: Role in Metabolism and Energy Supply. *Mol. Cell Biochem.* **2020**, *467*, 1–12. [CrossRef]
18. Elexpe, A.; Sánchez-Sánchez, L.; Tolentino-Cortez, T.; Astigarraga, E.; Torrecilla, M.; Barreda-Gómez, G. Analysis of Mitochondrial Function in Cell Membranes as Indicator of Tissue Vulnerability to Drugs in Humans. *Biomedicines* **2022**, *10*, 980. [CrossRef]
19. Youdim, M.B.H.; Edmondson, D.; Tipton, K.F. The Therapeutic Potential of Monoamine Oxidase Inhibitors. *Nat. Rev. Neurosci.* **2006**, *7*, 295–309. [CrossRef]
20. Bedard, K.; Krause, K.H. The NOX Family of ROS-Generating NADPH Oxidases: Physiology and Pathophysiology. *Physiol. Rev.* **2007**, *87*, 245–313. [CrossRef]
21. Brieger, K.; Schiavone, S.; Miller, F.J.; Krause, K.H. Reactive Oxygen Species: From Health to Disease. *Swiss Med. Wkly.* **2012**, *142*, w13659. [CrossRef] [PubMed]
22. Holley, A.K.; Dhar, S.K.; Xu, Y.; Clair, D.K.S. Manganese Superoxide Dismutase: Beyond Life and Death. *Amino Acids* **2012**, *42*, 139. [CrossRef] [PubMed]
23. Yang, W.; Tiffany-Castiglioni, E. Paraquat-Induced Apoptosis in Human Neuroblastoma SH-SY5Y Cells: Involvement of P53 and Mitochondria. *J. Toxicol. Environ. Health A* **2008**, *71*, 289–299. [CrossRef] [PubMed]
24. Blanco-Ayala, T.; Andérica-Romero, A.C.; Pedraza-Chaverri, J. New Insights into Antioxidant Strategies against Paraquat Toxicity. *Free Radic. Res.* **2014**, *48*, 623–640. [CrossRef]
25. Han, J.; Zhang, Z.; Yang, S.; Wang, J.; Yang, X.; Tan, D. Betanin Attenuates Paraquat-Induced Liver Toxicity through a Mitochondrial Pathway. *Food Chem Toxicol.* **2014**, *70*, 100–106. [CrossRef]
26. Vicente, J.A.; Peixoto, F.; Lopes, M.L.; Madeira, V.M. Differential Sensitivities of Plant and Animal Mitochondria to the Herbicide Paraquat. *J. Biochem. Mol. Toxicol.* **2001**, *15*, 322–330. [CrossRef]
27. Reczek, C.R.; Birsoy, K.; Kong, H.; Martínez-Reyes, I.; Wang, T.; Gao, P.; Sabatini, D.M.; Chandel, N.S. A CRISPR Screen Identifies a Pathway Required for Paraquat-Induced Cell Death. *Nat. Chem. Biol.* **2017**, *13*, 1274. [CrossRef]
28. Zhang, P.-Y.; Xu, X.; Li, X.-C.; Zhang, P. Cardiovascular Diseases: Oxidative Damage and Antioxidant Protection. *Eur. Rev. Med. Pharmacol. Sci.* **2014**, *18*, 3091–3096.
29. Jiang, Q. Natural Forms of Vitamin E: Metabolism, Antioxidant, and Anti-Inflammatory Activities and Their Role in Disease Prevention and Therapy. *Free Radic. Biol. Med.* **2014**, *72*, 76–90. [CrossRef]
30. Niki, E.; Noguchi, N. Antioxidant Action of Vitamin E in Vivo as Assessed from Its Reaction Products with Multiple Biological Oxidants. *Free Radic. Res.* **2021**, *55*, 352–363. [CrossRef]
31. Zingg, J.M. Vitamin E: Regulatory Role on Signal Transduction. *IUBMB Life* **2019**, *71*, 456–478. [CrossRef]
32. Miyazawa, T.; Burdeos, G.C.; Itaya, M.; Nakagawa, K.; Miyazawa, T. Vitamin E: Regulatory Redox Interactions. *IUBMB Life* **2019**, *71*, 430–441. [CrossRef] [PubMed]
33. Peña-Bautista, C.; Baquero, M.; Vento, M.; Cháfer-Pericás, C. Free Radicals in Alzheimer's Disease: Lipid Peroxidation Biomarkers. *Clin. Chim. Acta* **2019**, *491*, 85–90. [CrossRef] [PubMed]
34. Devaraj, S.; Tang, R.; Adams-Huet, B.; Harris, A.; Seenivasan, T.; de Lemos, J.A.; Jialal, I. Effect of High-Dose α-Tocopherol Supplementation on Biomarkers of Oxidative Stress and Inflammation and Carotid Atherosclerosis in Patients with Coronary Artery Disease. *Am. J. Clin. Nutr.* **2007**, *86*, 1392. [CrossRef] [PubMed]
35. Fernández, R.; Garate, J.; Tolentino-Cortez, T.; Herraiz, A.; Lombardero, L.; Ducrocq, F.; Rodríguez-Puertas, R.; Trifilieff, P.; Astigarraga, E.; Barreda-Gómez, G.; et al. Microarray and Mass Spectrometry-Based Methodology for Lipid Profiling of Tissues and Cell Cultures. *Anal. Chem.* **2019**, *91*, 15967–15973. [CrossRef]
36. Yamamoto, N.; Yamashita, Y.; Yoshioka, Y.; Nishiumi, S.; Ashida, H. Rapid Preparation of a Plasma Membrane Fraction: Western Blot Detection of Translocated Glucose Transporter 4 from Plasma Membrane of Muscle and Adipose Cells and Tissues. *Curr. Protoc. Protein Sci.* **2016**, *85*, 29.18.1–29.18.12. [CrossRef]
37. Nishiumi, S.; Ashida, H. Rapid Preparation of a Plasma Membrane Fraction from Adipocytes and Muscle Cells: Application to Detection of Translocated Glucose Transporter 4 on the Plasma Membrane. *Biosci. Biotechnol. Biochem.* **2007**, *71*, 2343–2346. [CrossRef]
38. Rienda, B.; Elexpe, A.; Tolentino-Cortez, T.; Gulak, M.; Bruzos-Cidón, C.; Torrecilla, M.; Astigarraga, E.; Barreda-Gómez, G. Analysis of Acetylcholinesterase Activity in Cell Membrane Microarrays of Brain Areas as a Screening Tool to Identify Tissue Specific Inhibitors. *Analytica* **2021**, *2*, 25–36. [CrossRef]
39. Elexpe, A.; Nieto, N.; Fernández-Cuétara, C.; Domínguez-Fernández, C.; Morera-Herreras, T.; Torrecilla, M.; Miguélez, C.; Laso, A.; Ochoa, E.; Bailen, M.; et al. Study of Tissue-Specific Reactive Oxygen Species Formation by Cell Membrane Microarrays for the Characterization of Bioactive Compounds. *Membranes* **2021**, *11*, 943. [CrossRef]
40. Manuel, I.; Barreda-Gomez, G.; González De San Román, E.; Veloso, A.; Fernández, J.A.; Giralt, M.T.; Rodriguez-Puertas, R. Neurotransmitter Receptor Localization: From Autoradiography to Imaging Mass Spectrometry. *ACS Chem. Neurosci.* **2015**, *6*, 362–373. [CrossRef]
41. Hebert-Chatelain, E.; Desprez, T.; Serrat, R.; Bellocchio, L.; Soria-Gomez, E.; Busquets-Garcia, A.; Pagano Zottola, A.C.; Delamarre, A.; Cannich, A.; Vincent, P.; et al. A Cannabinoid Link between Mitochondria and Memory. *Nature* **2016**, *539*, 555–559. [CrossRef] [PubMed]

42. Astigarraga, E.; Barreda-Gómez, G.; Lombardero, L.; Fresnedo, O.; Castaño, F.; Giralt, M.T.; Ochoa, B.; Rodríguez-Puertas, R.; Fernández, J.A. Profiling and Imaging of Lipids on Brain and Liver Tissue by Matrix-Assisted Laser Desorption/ Ionization Mass Spectrometry Using 2-Mercaptobenzothiazole as a Matrix. *Anal. Chem.* **2008**, *80*, 9105–9114. [CrossRef] [PubMed]
43. Suntres, Z.E.; Hepworth, S.R.; Shek, P.N. Protective Effect of Liposome-Associated α-Tocopherol against Paraquat-Induced Acute Lung Toxicity. *Biochem. Pharmacol.* **1992**, *44*, 1811–1818. [CrossRef] [PubMed]
44. Azzi, A.; Gysin, R.; Kempná, P.; Ricciarelli, R.; Villacorta, L.; Visarius, T.; Zingg, J.M. The Role of α-Tocopherol in Preventing Disease: From Epidemiology to Molecular Events. *Mol. Asp. Med.* **2003**, *24*, 325–336. [CrossRef] [PubMed]
45. Zakharova, I.O.; Bayunova, L.V.; Zorina, I.I.; Sokolova, T.V.; Shpakov, A.O.; Avrova, N.F. Insulin and α-Tocopherol Enhance the Protective Effect of Each Other on Brain Cortical Neurons under Oxidative Stress Conditions and in Rat Two-Vessel Forebrain Ischemia/Reperfusion Injury. *Int. J. Mol. Sci.* **2021**, *22*, 11768. [CrossRef] [PubMed]
46. Duncan, K.R.; Suzuki, Y.J. Vitamin E Nicotinate. *Antioxidants* **2017**, *6*, 20. [CrossRef]
47. Shayman, J.A.; Tesmer, J.J.G. Lysosomal Phospholipase A2. *Biochim Biophys. Acta Mol. Cell Biol. Lipids* **2019**, *1864*, 932. [CrossRef]
48. Nikolova-Karakashian, M.N.; Rozenova, K.A. Ceramide in Stress Response. *Adv. Exp. Med. Biol.* **2010**, *688*, 86–108. [CrossRef]
49. Hannun, Y.A.; Obeid, L.M. Principles of Bioactive Lipid Signalling: Lessons from Sphingolipids. *Nat. Rev. Mol. Cell Biol.* **2008**, *9*, 139–150. [CrossRef]
50. Ho, Q.W.C.; Zheng, X.; Ali, Y. Ceramide Acyl Chain Length and Its Relevance to Intracellular Lipid Regulation. *Int. J. Mol. Sci.* **2022**, *23*, 9697. [CrossRef]
51. Thomas, J.M.; Sudhadevi, T.; Basa, P.; Ha, A.W.; Natarajan, V.; Harijith, A. The Role of Sphingolipid Signaling in Oxidative Lung Injury and Pathogenesis of Bronchopulmonary Dysplasia. *Int. J. Mol. Sci.* **2022**, *23*, 1254. [CrossRef] [PubMed]
52. Jazvinšćak Jembrek, M.; Hof, P.R.; Šimić, G. Ceramides in Alzheimer's Disease: Key Mediators of Neuronal Apoptosis Induced by Oxidative Stress and Aβ Accumulation. *Oxid. Med. Cell Longev.* **2015**, *2015*, 346783. [CrossRef] [PubMed]
53. James, B.N.; Oyeniran, C.; Sturgill, J.L.; Newton, J.; Martin, R.K.; Bieberich, E.; Weigel, C.; Maczis, M.A.; Palladino, E.N.D.; Lownik, J.C.; et al. Ceramide in apoptosis and oxidative stress in allergic inflammation and asthma. *J. Allergy Clin. Immunol* **2021**, *147*, 1936. [CrossRef] [PubMed]
54. Zwara, A.; Wertheim-Tysarowska, K.; Mika, A. Alterations of Ultra Long-Chain Fatty Acids in Hereditary Skin Diseases—Review Article. *Front. Med.* **2021**, *8*, 730855. [CrossRef] [PubMed]
55. Panchal, M.; Gaudin, M.; Lazar, A.N.; Salvati, E.; Rivals, I.; Ayciriex, S.; Dauphinot, L.; Dargère, D.; Auzeil, N.; Masserini, M.; et al. Ceramides and Sphingomyelinases in Senile Plaques. *Neurobiol. Dis.* **2014**, *65*, 193–201. [CrossRef] [PubMed]
56. Poleschuk, T.S.; Sultanov, R.M.; Ermolenko, E.V.; Shulgina, L.V.; Kasyanov, S.P. Protective Action of Alkylglycerols under Stress. *Stress* **2020**, *23*, 213–220. [CrossRef]
57. Sindelar, P.J.; Guan, Z.; Dallner, G.; Ernster, L. The Protective Role of Plasmalogens in Iron-Induced Lipid Peroxidation. *Free Radic. Biol. Med.* **1999**, *26*, 318–324. [CrossRef]
58. Brites, P.; Waterham, H.R.; Wanders, R.J.A. Functions and Biosynthesis of Plasmalogens in Health and Disease. *Biochim. Biophys. Acta Mol. Cell Biol. Lipids* **2004**, *1636*, 219–231. [CrossRef]
59. Drechsler, R.; Chen, S.-W.; Dancy, B.C.R.; Mehrabkhani, L.; Olsen, C.P. HPLC-Based Mass Spectrometry Characterizes the Phospholipid Alterations in Ether-Linked Lipid Deficiency Models Following Oxidative Stress. *PLoS ONE* **2016**, *11*, e0167229. [CrossRef]
60. Messias, M.C.F.; Mecatti, G.C.; Priolli, D.G.; de Oliveira Carvalho, P. Plasmalogen Lipids: Functional Mechanism and Their Involvement in Gastrointestinal Cancer. *Lipids Health Dis.* **2018**, *17*, 1–12. [CrossRef]
61. Janssen, C.I.F.; Kiliaan, A.J. Long-Chain Polyunsaturated Fatty Acids (LCPUFA) from Genesis to Senescence: The Influence of LCPUFA on Neural Development, Aging, and Neurodegeneration. *Prog. Lipid Res.* **2014**, *53*, 1–17. [CrossRef] [PubMed]
62. Mîinea, C.; Kuruvilla, R.; Merrikh, H.; Eichberg, J. Altered Arachidonic Acid Biosynthesis and Antioxidant Protection Mechanisms in Schwann Cells Grown in Elevated Glucose. *J. Neurochem.* **2002**, *81*, 1253–1262. [CrossRef] [PubMed]
63. Yamashita, A.; Oka, S.; Tanikawa, T.; Hayashi, Y.; Nemoto-Sasaki, Y.; Sugiura, T. The Actions and Metabolism of Lysophosphatidylinositol, an Endogenous Agonist for GPR55. *Prostaglandins Lipid Mediat.* **2013**, *107*, 103–116. [CrossRef] [PubMed]
64. Okuda, S.; Saito, H.; Katsuki, H. Arachidonic Acid: Toxic and Trophic Effects on Cultured Hippocampal Neurons. *Neuroscience* **1994**, *63*, 691–699. [CrossRef] [PubMed]
65. Chang, M.C.; Lee, J.J.; Chen, Y.J.; Lin, S.I.; Lin, L.D.; Liou, E.J.W.; Huang, W.L.; Chan, C.P.; Huang, C.C.; Jeng, J.H. Lysophosphatidylcholine Induces Cytotoxicity/Apoptosis and IL-8 Production of Human Endothelial Cells: Related Mechanisms. *Oncotarget* **2017**, *8*, 106177. [CrossRef]
66. Charollais, J.; van der Goot, F.G. Palmitoylation of Membrane Proteins (Review). *Mol. Membr. Biol.* **2009**, *26*, 55–66. [CrossRef]
67. Koca, T.; Canatan, D.; Örmeci, A.R.; Koca, Y.S.; Duman, H.; Baykal, A.; Akçam, M. Amino Acids and Fatty Acids in Patients with Beta Thalassemia Major. *Acta Bio. Med. Atenei Parm.* **2020**, *91*, 79. [CrossRef]

antioxidants

MDPI

Article

Impact of Glyphosate on the Development of Insulin Resistance in Experimental Diabetic Rats: Role of NFκB Signalling Pathways

Monisha Prasad [1], Mansour K. Gatasheh [2], Mohammad A. Alshuniaber [3], Rajapandiyan Krishnamoorthy [3], Ponnulakhmi Rajagopal [4], Kalaiselvi Krishnamoorthy [1], Vijayalakshmi Periyasamy [5], Vishnu Priya Veeraraghavan [1,*] and Selvaraj Jayaraman [1,*]

1 Centre of Molecular Medicine and Diagnostics (COMManD), Department of Biochemistry, Saveetha Institute of Medical & Technical Sciences, Saveetha Dental College & Hospitals, Saveetha University, Chennai 600077, India
2 Department of Biochemistry, College of Science, King Saud University, Riyadh 11451, Saudi Arabia
3 Department of Food Science and Nutrition, College of Food and Agriculture Sciences, King Saud University, Riyadh 11451, Saudi Arabia
4 Department of Central Research Laboratory, Meenakshi Ammal Dental College and Hospitals, Chennai 600095, India
5 Department of Biotechnology and Bioinformatics, Holy Cross College, Trichy 620002, India
* Correspondence: vishnupriya@saveetha.com (V.P.V.); selvarajj.sdc@saveetha.com (S.J.)

Citation: Prasad, M.; Gatasheh, M.K.; Alshuniaber, M.A.; Krishnamoorthy, R.; Rajagopal, P.; Krishnamoorthy, K.; Periyasamy, V.; Veeraraghavan, V.P.; Jayaraman, S. Impact of Glyphosate on the Development of Insulin Resistance in Experimental Diabetic Rats: Role of NFκB Signalling Pathways. *Antioxidants* **2022**, *11*, 2436. https://doi.org/10.3390/antiox11122436

Academic Editor: Tim Hofer

Received: 1 October 2022
Accepted: 24 November 2022
Published: 9 December 2022

Abstract: Glyphosate, an endocrine disruptor, has an adverse impact on human health through food and also has the potential to produce reactive oxygen species (ROS), which can lead to metabolic diseases. Glyphosate consumption from food has been shown to have a substantial part in insulin resistance, making it a severe concern to those with type 2 diabetes (T2DM). However, minimal evidence exists on how glyphosate impacts insulin-mediated glucose oxidation in the liver. Hence the current study was performed to explore the potential of glyphosate toxicity on insulin signaling in the liver of experimental animals. For 16 weeks, male albino Wistar rats were given 50 mg, 100 mg and 250 mg/kg b. wt. of glyphosate orally. In the current study, glyphosate exposure group was linked to a rise in fasting sugar and insulin as well as a drop in serum testosterone. At the same time, in a dose dependent fashion, glyphosate exposure showed alternations in glucose metabolic enzymes. Glyphosate exposure resulted in a raise in H_2O_2 formation, LPO and a reduction in antioxidant levels those results in impact on membrane integrity and insulin receptor efficacy in the liver. It also registered a reduced levels of mRNA and protein expression of insulin receptor (IR), glucose transporter-2 (GLUT2) with concomitant increase in the production of proinflammatory factors such as JNK, IKKβ, NFkB, IL-6, IL-1β, and TNF-α as well as transcriptional factors like SREBP1c and PPAR-γ leading to pro-inflammation and cirrhosis in the liver which results in the development of insulin resistance and type 2 diabetes. Our present findings for the first time providing an evidence that exposure of glyphosate develops insulin resistance and type 2 diabetes by aggravating NFkB signaling pathway in liver.

Keywords: glyphosate; type 2 diabetes; GLUT2 expression; NFκB signalling pathways; rats

1. Introduction

Diabetes mellitus (DM) has been estimated to affect 578 million people worldwide in next few years, rising to 700 million by the 2045. It is a serious, chronic condition that adversely affects people, families, and societies worldwide. Its primary features include chronic hyperglycemia, a total or partial impairment in insulin action, and abnormalities in the metabolism of proteins, carbohydrates, and lipids [1]. Diabetes is a metabolic condition that is greatly influenced by nutrition, genetics, and environment. The presence of certain environmental pollutants may further increase the risk of developing diabetes.

Herbicides are weed-control substances (insecticides) that have the potential to cause substantial environmental damage [2,3]. Because there is a strong relationship between these chemical molecules and diseases, particularly metabolic disorders, they have both good and detrimental impact on humans [4]. People are typically exposed to a variety of herbicides through the use of fresh fruits and vegetables as weed switch plant goods [5]. Herbicides are absorbed by the gastrointestinal/respiratory system and also via skin. These herbicides can metabolise in human organs, mainly in the liver tissue, due to their strong affinity and stability. They have a long persistence in the environment and because of their high toxicity and stability, they pollute; therefore, they require special attention these days with regards to human toxicity that results in the development of diabetes [6].

The liver is a vital organ for the metabolism of energy as well as the detoxification of xenobiotics, environmental hormones, and hazardous chemicals. The inhibition of gluconeogenesis and enhanced storage of glucose as glycogen are brought on by insulin signalling, particularly in hepatocytes. Additionally, NF-κB, a DNA-binding protein factor, is involved in the transcription of many pro-inflammatory and inflammatory molecules, including cytokines, chemokines, cell adhesion molecules (CAM), and various enzymes [7]. Specifically, in metabolic organs such as liver muscle and adipose tissue, the expression of cytokines and inflammatory molecules have a substantial impact on the aetiology of diabetes. Examples include compromised insulin signalling in hepatocytes, elevated free fatty acids, systemic insulin resistance, and glucose intolerance due to constitutively active (CA) IKK expression in the liver brought on by these hazardous substances or the metabolic imbalance [8]. Inflammation and insulin resistance are caused by environmental factors such as toxins or other modulators that can impact metabolic signalling gene expression and enhance NFκB signalling via proinflammatory cytokines [9].

Glyphosate is one of the phosphonate and systemic herbicide used globally to control weed in farming and agriculture [10]. It is marketed in the brand names "Roundup" and "Ranger Pro", and it's often assorted with supposedly innocuous ingredients like surfactants. This herbicide, which comes in a variety of salt forms impedes the conversion of shikimate-3-phosphate to 5-endopyruvylshimikate 3-phosphate by 5-enolpyruvylshikimate-3-phosphate synthase (shikimic acid pathway), thereby hampering the aromatic amino acids production such as tyrosine, phenylalanine and tryptophan [11]. This shikimic acid pathway, which is exclusive to plants, bacteria, and fungi, is consequently eliminated by glyphosate. Glyphosate, being a chelating agent interferes with nutrients required for a variety of plant processes, has the potential to alter plant metabolism, which also has implications for humans. Ingestion of glyphosate orally through food, drinking water and other ecological channels has been associated to endocrine disruption, cancer, mutagenicity and reproductive damage [12–14].

Previous studies have shown that glyphosate is hazardous to the reproductive system particularly, the exploration of glyphosate has numerous adverse consequences on the male reproductive system in both humans and experimental animal models [15]. Furthermore, it was found to be a strong human endocrine disruptor that affects serum testosterone and is subsequently linked to higher glucose and insulin levels, obesity, the distribution of fat in the upper body, and is strongly linked to insulin resistance and the other metabolic syndrome [16]. Additionally, insulin resistance is linked to the castration of male rats, and the condition is corrected with subsequent testosterone replacement, demonstrating the connection between testosterone insufficiency and diabetes [17]. As instance, Muthusamy et al. [18] established that insulin and its receptor gene expression were disrupted in testosterone deficiency-induced experimental animals that resulted in impaired glucose oxidation in skeletal muscles, liver, and adipose tissue. In addition to its effects on endocrine function, glyphosate exposure in experimental animals results in hyperglycemia, reactive oxygen species (ROS), and related metabolic problems [19–22]. Despite these findings, it is unclear what mechanisms underlie glyphosate's particular effects on insulin signalling and glucose homeostasis in hepatocytes. This work aims to investigate into the mechanism of glyphosate-induced toxic effects on insulin signalling in rat liver.

2. Materials and Methods

2.1. Chemicals

All chemicals and reagents used in the present study were of molecular and analytical grade; and they were purchased from Sigma Chemical Company, Krishgen Biosystem, DBC Diagnostics Biochem, Invitrogen, New England Biolabs (NEB), Eurofins Genomics and Santa Cruz Biotechnology. Glyphosate (Product no. 45521) was purchased from Sigma Chemical Company, St. Louis, MI, USA. ELISA kits for insulin, IL-6 and IL-1 were bought from Krishgen Biosystem, Mumbai, India. INS GENLISA ELISA kit was bought from Krishgen Biosystem, Mumbai, India and testosterone ELISA kit from DBC Diagnostics Biochem, London, ON, Canada. The total RNA isolation kit was supplied by Invitrogen, USA. The RT enzyme was provided by New England Biolabs (NEB), Ipswich, MA, USA and the Go Taq Green master mix was provided by Promega, Madison, WI, USA. The primers for sterol regulatory element-binding transcription factor 1 (SREBP1c), interleukins (IL-6, IL-1β), insulin receptor (IR), Inhibitory kappaB kinase beta (IKKβ), c-Jun N-terminal kinases (JNK), peroxisome proliferator-activated receptor gamma (PPAR-γ), glucose transporter-2 (GLUT2), tumour necrosis factor (TNF-α), Nuclear factor kappa B (NFκB) and β-actin were supplied by Eurofins Genomics India, Pvt. Ltd., Bangalore, India. TNF-α and IL-6 antibodies were provided by Santa Cruz Biotechnology, Santa Cruz, CA, USA.

2.2. Animals and Induction of Diabetes with Glyphosate

Eight week old adult male Wistar albino rats (*Rattus norvegicus*) weighing 180–200 g were used in the present study. Rats were maintained under the standard environmental conductions as per the guidelines of institutional animal ethics committee (IAEC) with temperature (24 ± 2 °C), 50% humidity and continual 12 h dark and 12 h light cycle. They were fed with standard pellet and water ad libitum. The study was approved (IAEC No. BRULAC/SDCH/SIMATS/IAEC/02-2019/015) by Biomedical Research Unit and Laboratory Animal Centre (BRULAC), Saveetha Dental College and Hospitals, SIMATS, Chennai-77 and the study conducted in accordance with the "Guide for the care and use of laboratory animals".

Glyphosate dose selection in the present study was based on the previous reports [23,24]. Glyphosate at the dose of 50, 100 and 250 mg/kg b.wt was dissolved in water and treated orally. The experiment lasted for 16 weeks during which the fasting blood glucose levels of the experimental group were greater than 200 mg/dL were served as diabetic rats.

2.3. Experimental Time Line

Rats were randomly divided into four groups of six rats each.

Group 1. Normal control rats treated with vehicle (water only)
Group 2. Normal rats received oral administration of glyphosate dissolved in water (50 mg/kg b. wt./day) for 16 weeks.
Group 3. Normal rats received oral administration of glyphosate dissolved in water (100 mg/kg b. wt./day) for 16 weeks.
Group 4. Normal rats received oral administration of glyphosate dissolved in water (250 mg/kg b. wt./day) for 16 weeks.

On the last day of experiment the animals in various groups were sacrificed by cervical decapitation under sodium thiopentone (40 mg/kg b. wt.) anesthesia. The blood was drawn via venipuncture and the sera was separated by centrifugation and then stored at −80 °C. 20 mL of isotonic NaCl solution was injected into the left ventricle to remove blood from various parts of the body [25]. The liver was promptly dissected to be used in the experiment immediately.

2.4. Determination of Fasting Sugar, Serum Insulin and Testosterone

The animals were starved overnight the day before sacrifice after receiving glyphosate treatment for 16 weeks. The blood was drawn the next day from the rat tail tip to estimate the fasting blood glucose (FBG) levels using On-Call Plus blood glucose test strips. The data were displayed in milligrammes per decilitre on the metre display window.

ELISA technique was used to measure the levels of insulin and testosterone in rat serum. As per instruction provided in the manual, the detection range and coefficient of variation were set. The testosterone concentration was measured in ng/mL and the serum insulin concentration was measured in pg/mL.

2.5. Determination of Oral Glucose Tolerance (OGT)

For OGT, the control and glyphosate-exposed rats remained fasted overnight. After an oral glucose load (10 mL/kg; 50% *w/v*), we monitored fasting blood glucose level using glucose test strips at 60-min, 120-min and 180-min time intervals. The glucose level was measured as the 0-min value prior to administration of glucose. The values were measured in mg/dL.

2.6. Determination of Insulin Tolerance (IT)

Insulin (0.75 U/kg b.wt. i.p.) was given to each group two days before the animals were slaughtered. At 0 h, the level of blood glucose were tested at different time intervals. The results were given in mg/dL of blood.

2.7. HOMA-IR and QUICKI

The insulin sensitivity index was investigated using the HOMA-IR and QUICKI calculation methods developed by Matthews et al. [26] and Katz et al. [27]. The equations fasting blood glucose+fasting serum insulin/405 and 1/log fasting serum insulin+log fasting blood glucose, respectively, were used to calculate HOMA-IR and QUICKI.

2.8. Measurement of Oxidative Stress Marker

Using the methodologies of Fraga et al. [28] and Jiang et al. [29], thiobarbituric acid reactive substances (TBARS) and lipid peroxidation (LPO) in the tissue were evaluated calorimetrically. The results were represented in mM/mg tissue.

2.9. Measurement of Antioxidant Enzymes

Superoxide dismutase (SOD) was quantified using the procedure mentioned in Kakkar et al. [30] and the findings were expressed as units/mg protein. Sinha's method [31] was used to quantify the catalase activity (CAT), and expressed as units/mg protein.

2.10. Determination of Specific Enzymes Involved in Carbohydrate Metabolism

Hexokinase (HK) activity was measured using the method of Brandstrup et al. [32]. The enzyme's activity was calculated in mol glucose phosphorylated/h/mg protein. The enzymatic activity of pyruvate kinase (PK) in tissues was measured using Valentine and Tanaka's [33] technique. The results were expressed as µmol pyruvate formed/min/mg protein.

The activity of hepatic glucose-6-phosphatase (G6P) was examined using the Koida and Oda's technique [34]. The G6P activity was measured in moles of inorganic phosphorus liberated/min/mg protein. Gancedo and Gancedo's [35] method was used to measure the activity of hepatic fructose-1,6-bisphosphatase (F1,6BP). The activity of the enzyme was measured in moles of inorganic phosphorus liberated/hh/mg protein.

2.11. Total RNA, cDNA Synthesis and Real-Time PCR

The total RNA was isolated from the liver of the rats according to the procedure described in Fourney et al. [36]. About 2 µg of RNA was taken and reverse transcribed into cDNA using the RT kit (Seraing, Belgium). The primers utilized in this study are listed in Table 1. The SYBR green mastermix (Takara, Japan) and housekeeping gene (β-actin) were used in a Real-Time PCR system (Bio-Rad C1000 Touch, thermal cycler, Bio-Rad Laboratories Ltd. (Bio-Rad House, Herts, UK) to amplify the interest genes under the following reaction conditions: initial denaturation at 95 °C for 5 min, followed by 40 cycles of denaturation: 95 °C for 30 s, annealing: 60 °C for 1 min and extension: 72 °C for 30 s each. The analyses of melt and amplification curves were carried out to calculate the relative quantification.

Table 1. Primer details.

S. No.	Gene Details	Primer Details	Reference
1.	GLUT2	Forward: 5′-CTC GGG CCT TAC GTG TTC TTC CTT-3′ Reverse: 5′-TGG TTC CCT TCT GGT CTG TTC CTG-3′	[37]
2.	IR	Forward: 5′-GCC ATC CCG AAA GCG AAG ATC-3′ Reverse: 5′-TCT GGG GAG TCC TGA TTG CAT-3′	[37]
3.	SREBP1c	Forward: 5′-GGA GCC ATG GAT TGC ACA TT-3′ Reverse: 5′-GCT TCC AGA GAG GAG CCC AG-3′	[38]
4.	PPAR-γ	Forward: 5′-CCT GAA GCT CCA AGA ATA CC-3′ Reverse: 5′-GAT GCT TTA TCC CCA CAG AC-3′	[39]
5.	NFκB	Forward: 5′-CAT GAA GAG AAG ACA CTG ACC ATG GAA A-3′ Reverse: 5′-TGG ATA GAG GCT AAG TGT AGA CAC G-3′	[40]
6.	JNK	Forward: 5′-TCA GAA TCC GAA CGA GAC AAA AT-3′ Reverse: 5′-AAG CCA GAG TCC TTC ACA GAC AA-3′	[41]
7.	IL-6	Forward: 5′-GTG AGA AGT ATG AGA AGT GTG A-3′ Reverse: 5′-GCA GGA TGA GAA TGA TCT TTG-3′	[42]
8.	IKKβ	Forward: 5′-TGG CAT GGA AAC GGA TAA CTG A-3′ Reverse: 5′-CTG GAA CTC TGT GCC TGT GGA A-3′	[43]
9.	TNF-α	Forward: 5′-GTC GTA GCA AAC CAC CAA GC-3′ Reverse: 5′-TGT GGG TGA GGA GCA CAT AG-3′	[44]
10.	IL-1β	Forward: 5′- GCA ATG GTC GGG ACA TAG TT-3′ Reverse: 5′-AGA CCT GAC TTG GCA GAG A-3′	[44]
11.	β-actin	Forward: 5′-AAG TCC CTC ACC CTC CCA AAA G-3′ Reverse: 5′-AAG CAA TGC TGT CAC CTT CCC-3′	[35]

2.12. Protein Analysis of Pro-Inflammatory Cytokines and Transcription Factors

Commercial rat ELISA kits purchased from Krishgen Biosystem, Mumbai 400018, India (for serum IL-6 and IL-1β), Ray Biotech, Illinois, GA, USA (for TNF-α) and My BioSource, Inc., San Diego, CA, USA (transcription factors such as PPAR-γ and SREBP1c) were used to measure the protein concentration in experimental rats as directed by the manufacturer's manual. SREBP-1c, IL-1β, IL-6 and TNF-α levels were measured in pg/mL whereas PPAR-γ levels were measured in ng/mL.

2.13. Histopathological Staining

By embedding the liver tissue in 10% neutral buffered formalin, followed by sectioning and staining it with hematoxylin and eosin dye, the morphology of the liver was studied [45,46]. Then, using a LKB ultra-microtome, semi-thin sections (0.5–1 μ) were cut, stained with toluidine blue and shot using an Olympus light microscope and a Nikon digital camera at a magnification of X200.

2.14. Immunohistochemical Staining

Immunohistochemistry was performed on paraffin embedded liver slices of control and glyphosate-treated rat tissues using a microwave-based antigen retrieval technique. The segments were incubated overnight at 4 °C in a dark, humid chamber with primary polyclonal anti-TNF-α and anti-IL-6 (1:100 dilution) antibodies, followed by 45 min incubation with secondary antibodies under the same conditions. After that, the slides were washed in 1 M PBS to eliminate any remaining secondary antibodies and incubated with horse-radish peroxidase (HRP) for 45 min in a humidified environment before being rinsed for 5 min in 1 M PBS. Then, the slides were incubated for 6 min at 37 °C with the 3,3′-diaminobenzidine (DAB) substrate chromogen 3.3, followed by a 5-min wash in water. The specimens were dried, cleaned and mounted after being counterstained with hematoxylin for 40 s. To ensure the quality of the staining, positive and negative controls were used for each immunohistochemical staining procedure. To establish the overall distribution of the main protein, the cells were initially examined at a modest magnification 100×).

2.15. Statistical Analysis

Using computer-based software, the significance difference between the control group and treatment sets was assessed by one-way analysis of variance (ANOVA) and Duncan's multiple range tests using computational software Graph Pad Prism version 5. The significance of Duncan's test was determined at the level of $p < 0.05$.

3. Results

3.1. Impact of Glyphosate on FBG, Insulin and Testosterone

Fasting blood glucose (Figure 1a) and serum insulin (Figure 1b) were significantly raised in the glyphosate exposed groups compared to control. Conversely, serum testosterone concentration (Figure 1c) was found to be reduced in a dose dependent manner. These results clearly indicate that glyphosate has the toxic effect at 50–250 mg/kg b.wt. dose to induce hyperglycemia and insulin resistance in 16 weeks duration.

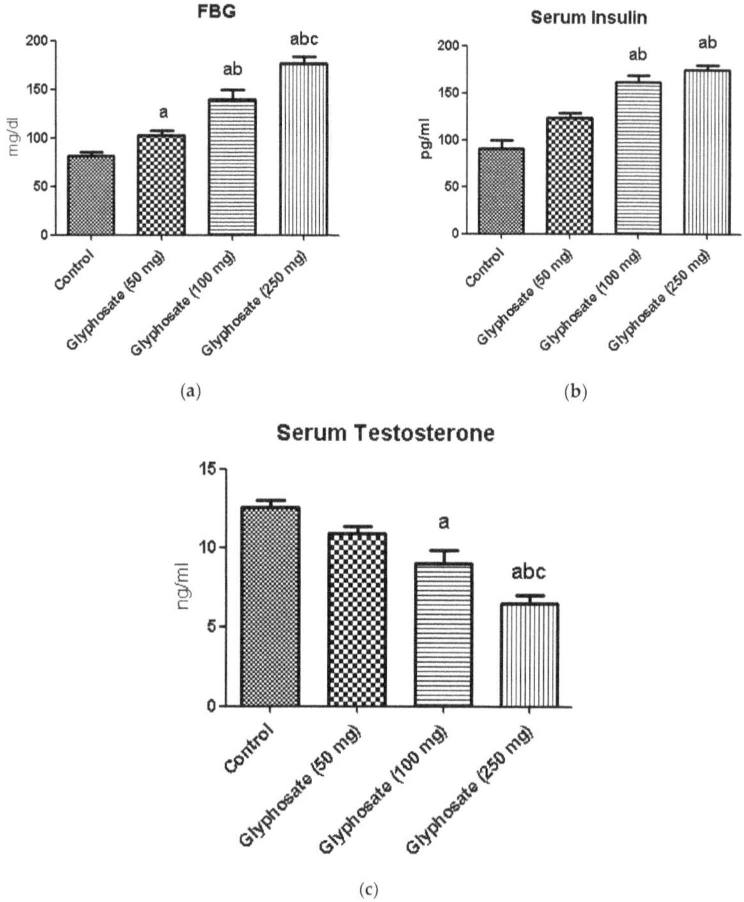

Figure 1. (**a–c**): The effect of glyphosate on FBG (**a**), serum insulin (**b**) and the serum testosterone (**c**) in control and glyphosate treated rats. Data were expressed as mean ± SEM where $n = 6$ rats for each group. Values significant at $p < 0.05$ were analyzed with a—control, b—glyphosate (50 mg/kg b.wt.), and c—glyphosate (100 mg/kg b.wt.).

3.2. Impact of Glyphosate on OGT

Table 2 depicts the OGT levels in experimental and control rats with time. Blood glucose levels increased after taking glucose orally; this rise is dose-dependent and peaks at one hour. At 180 min, the glucose level in control rats had returned to normal (76 mg/dL), whereas it was significantly higher in the glyphosate-induced groups, demonstrating that the glyphosate had caused glucose intolerance.

Table 2. OGT in glyphosate exposed diabetic animals.

Groups	0 h	60 min	120 min	180 min
Control	69 ± 1.2	72 ±1.4	75 ±3.2	76 ± 4.2
Glyphosate (50 mg)	110 ± 1.5 ^{a}	115 ± 2.2 ^{a}	120 ± 3.5 ^{a}	121 ± 3.4 ^{a}
Glyphosate (100 mg)	132 ± 2.5 ^{ab}	138 ± 5.9 ^{ab}	149 ± 6.2 ^{ab}	150 ± 6.2 ^{ab}
Glyphosate (250 mg)	138 ± 0.5 ^{ab}	140 ± 6.7 ^{ab}	152 ± 7.8 ^{ab}	159 ± 7.7 ^{ab}

The effect of glyphosate on OGT in control and glyphosate treated rats. Data were expressed as mean ± SEM where n = 6 rats for each group. Values significant at $p < 0.05$ were analyzed with ^{a}—control, ^{b}—glyphosate (50 mg/kg b.wt.).

3.3. Impact of Glyphosate on IT

Table 3 shows the ITT levels of normal and diabetic (Glyphosate treated) rats after oral administration of insulin (0.75 U/kg b.wt. i.p.). In diabetic control rats, the highest increase in serum insulin level was seen after 30 min and it was persisted high for over an hour has caused changes in insulin function.

Table 3. Level insulin tolerance in glyphosate exposed animals.

Group	0 h	15 min	30 min	45 min	60 min
Control	68 ± 2.9	70 ±4.2	62 ±1.2	69 ±3.2	70 ±2.9
Glyphosate (50 mg)	109 ± 7.29 ^{a}	110 ± 6.29 ^{a}	109 ±4.2 ^{a}	110 ±4.9 ^{a}	101 ±5.2 ^{a}
Glyphosate (100 mg)	129 ± 1.9 ^{a}	117 ± 1.5 ^{a}	116 ±5.2 ^{ab}	110 ±6.7 ^{a}	112 ± 6.5 ^{a}
Glyphosate (250 mg)	127 ± 6.2 ^{b}	122 ±5.9 ^{ab}	129 ±6.2 ^{ab}	127 ±7.9 ^{ab}	112 ±6.4 ^{a}

The effect of glyphosate on insulin tolerance in control and glyphosate treated rats. Data were expressed as mean ± SEM where n = 6 rats for each group. Values significant at $p < 0.05$ were analyzed with ^{a}—control, ^{b}—glyphosate (50 mg/kg b.wt.).

3.4. Impact of Glyphosate on HOMA-IR and QUICKI

Table 4 illustrates the HOMA-IR and QUICKI calculations for rats exposed to glyphosate and rats in the control group. HOMA-IR was significantly increased in glyphosate-induced diabetic rats compared to control rats, whereas QUICKI values decreased. These significant differences in the insulin-sensitivity check index between normal and diabetic rats show that glyphosate impacts insulin signaling, which in turn affects glucose metabolism and leads to glucose and insulin intolerances.

Table 4. Levels of HOMA-IR and QUICKI in glyphosate explore diabetic rats.

Groups	HOMA-IR	QUICKI
Control	2.76 ± 0.11	0.92 ± 0.05
Glyphosate (50 mg)	9 ± 0.52 ^{a}	0.8 ± 0.03 ^{a}
Glyphosate (100 mg)	10± 0.42 ^{a}	0.62 ± 0.02 ^{ab}
Glyphosate (250 mg)	12± 0.25 ^{ab}	0.6 ± 0.03 ^{ab}

The effect of glyphosate on HOMA-IR and QUICKI in control and glyphosate treated rats. Data were expressed as mean ± SEM where n = 6 rats for each group. Values significant at $p < 0.05$ were analyzed with ^{a}—control, ^{b}—glyphosate (50 mg/kg b.wt.).

3.5. Impact of Glyphosate on Oxidative Stress Markers

The changes in the levels of hepatic thiobarbituric acid (TBARS) and lipid peroxidation (LPO) in control and experimental rats are shown in Figure 2a,b. The levels of thiobarbituric acid reactive substances, and lipid peroxidation were significantly increased ($p < 0.05$) in glyphosate-treated rats when compared with normal control rats.

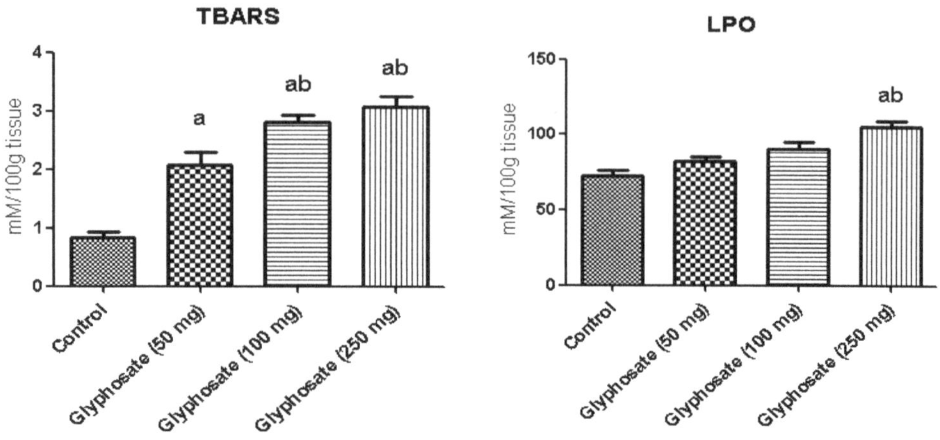

(a) (b)

Figure 2. (**a**,**b**): The effect of glyphosate on oxidative stress markers (TBARS and LPO) in control and glyphosate treated rats. Data were expressed as mean ± SEM where $n = 6$ rats for each group. Values significant at $p < 0.05$ were analyzed with a—control, b—glyphosate (50 mg/kg b.wt.).

3.6. Impact of Glyphosate on SOD and CAT Enzymes Activity

Figure 3a,b illustrates the activities of enzymatic antioxidants namely superoxide dismutase (SOD) and catalase (CAT) in the liver of control and experimental rats. A significant ($p < 0.05$) depletion in the activities of enzymatic antioxidants in glyphosate treated rats was observed. Treatment of glyphosate decreased the levels of these enzymatic antioxidants in the liver indicate its toxicity.

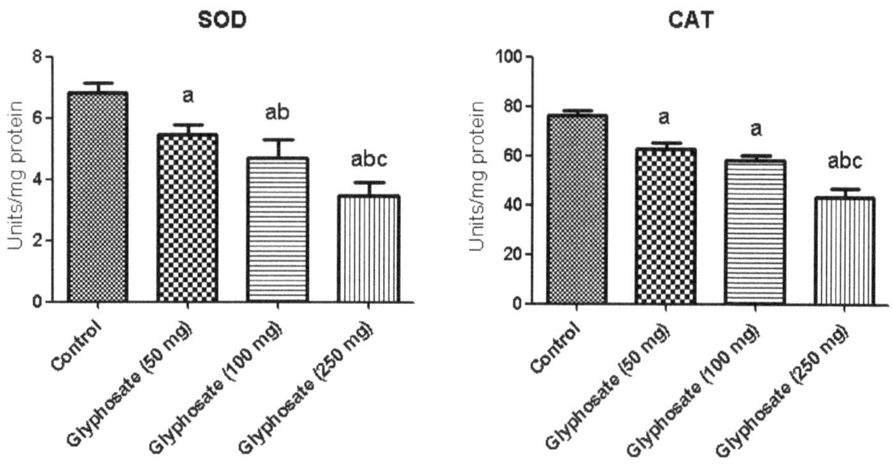

(a) (b)

Figure 3. (**a**,**b**): The effect of glyphosate on antioxidant enzymes (SOD and CAT) in control and glyphosate treated rats. Data were expressed as mean ± SEM where $n = 6$ rats for each group. Values significant at $p < 0.05$ were analyzed with a—control, b—glyphosate (50 mg/kg b.wt.), and c—glyphosate (100 mg/kg b.wt.).

3.7. Impact of Glyphosate on Glycolytic and Gluconeogenic Enzymes

Figure 4a–d depicts the activities of carbohydrate metabolizing enzymes in liver of normal and glyphosate-treated diabetic rats. The activities of glycolysis enzyme hexokinase and pyruvate kinase was found to be decreased whereas the activities of gluconeogenic enzymes like glucose-6-phosphatase and fructose-1,6-bisphosphatase were significantly increased in diabetic rats compared to normal rats.

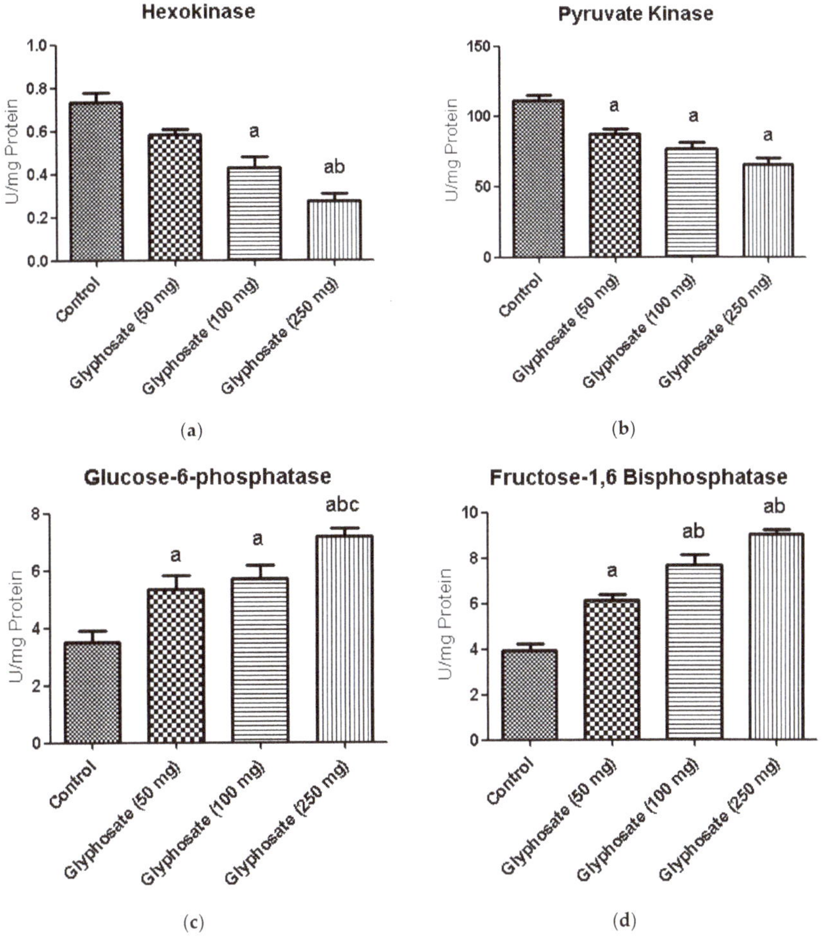

Figure 4. (**a**–**d**): The effect of glyphosate on the activity of glycolytic (hexokinase and pyruvate kinase) and gluconeogenic (glucose-6-phosphatase and fructose-1,6-bisphosphatase) enzymes in control and glyphosate treated rats. Data were expressed as mean ± SEM where $n = 6$ rats for each group. Values significant at $p < 0.05$ were analyzed with a—control, b—glyphosate (50 mg/kg b.wt.), and c—glyphosate (100 mg/kg b.wt.).

3.8. Impact of Glyphosate on the mRNA Expression of Insulin Signalling Molecules

The glyphosate effect on IR and GLUT2 mRNA levels in the liver is depicted in Figure 5a,b. Insulin-stimulated signal transduction mechanisms for hepatic glucose generation, including insulin receptors and downstream mediators, have been reduced in hepatic insulin resistance. We found a significant decrease in IR and GLUT2 mRNA expression in glyphosate-induced rats in this study.

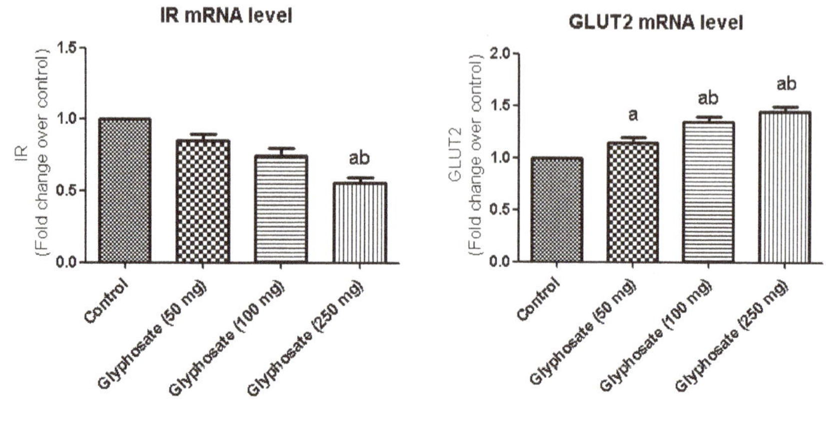

(a) (b)

Figure 5. (**a**,**b**): The effect of glyphosate on mRNA levels of IR and GLUT2 in control and glyphosate treated rats. IR and GLUT 2 mRNA expression were analyzed by Real Time-PCR using gene specific primers and results were expression in fold-change over control. Data were expressed as mean ± SEM where $n = 6$ rats for each group. Values significant at $p < 0.05$ were analyzed with a—control, b—glyphosate (50 mg/kg b.wt.).

3.9. Impact of Glyphosate on the mRNA Expression of Transcriptional Factors Like PPAR-γ and SREBP1c

Major transcriptional factors of hepatic insulin signalling like SREBP1c and PPAR-γ which commonly seen in the regulation of insulin signalling, were investigated using qRT-PCR (Figure 6a,b) in the present study. SREBP1c mRNA expression was significantly higher in glyphosate-treated diabetic rats (Figure 6a) when compare to control, although PPAR-γ mRNA expression was significantly reduced in experimental rats as compared to normal rats liver (Figure 6b).

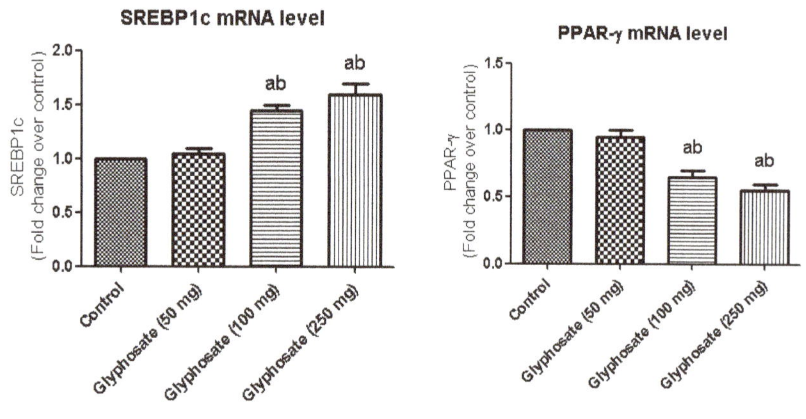

(a) (b)

Figure 6. (**a**,**b**): Impact of glyphosate on mRNA levels of transcriptional factors like SREBP1c and PPAR-γ in control and glyphosate treated rats. SREBP1c and PPAR-γ mRNA expression were analyzed by Real Time-PCR using gene specific primers and results were expression in fold-change over control. Data were expressed as mean ± SEM where $n = 6$ rats for each group. Values significant at $p < 0.05$ were analyzed with a—control, b—glyphosate (50 mg/kg b.wt.).

3.10. *Impact of Glyphosate on the mRNA Expression of JNK Pathway Related Molecules*

The JNK pathway is known to be active in diabetes and may have a role in the evolution of diabetes. JNK, NFκB and IKKβ mRNA levels were evaluated in the liver of glyphosate-treated rats. In glyphosate exposed rats, the expression of JNK, NFκB and IKKβ signalling molecules was shown to be significantly higher (Figure 7a–c) indicating the glyphosate's ability to induce inflammation.

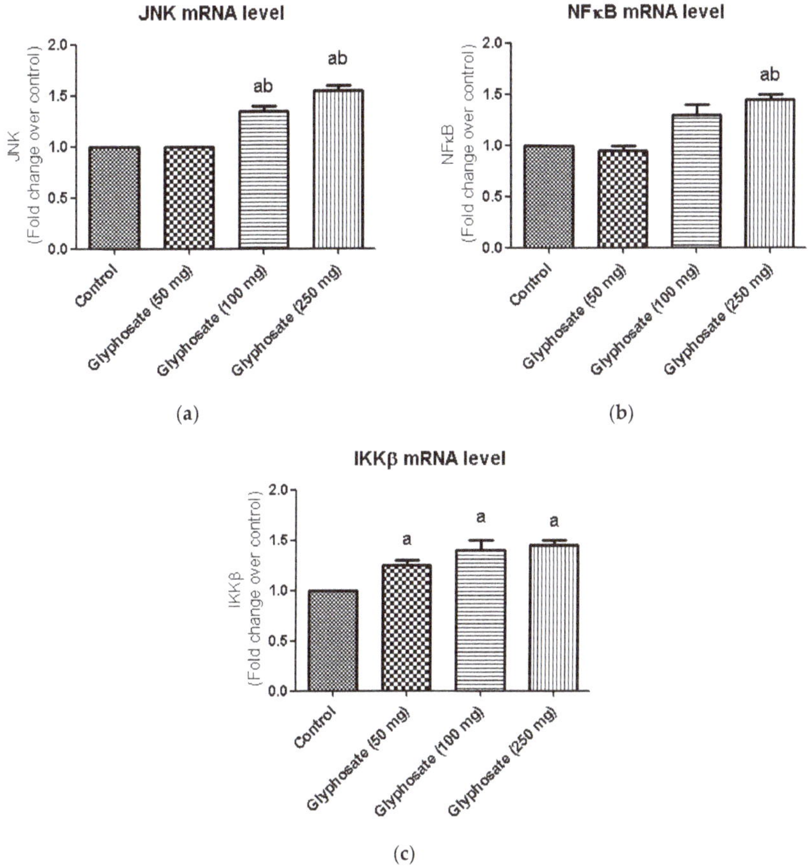

Figure 7. (**a**–**c**): Impact of glyphosate on the mRNA expression levels of JNK, NFκB and IKKβ in control and glyphosate treated rats. JNK, NFκB and IKKβ mRNA expression were analyzed by Real Time-PCR using gene specific primers and results were expression in fold-change over control. Data were expressed as mean ± SEM where $n = 6$ rats for each group. Values significant at $p < 0.05$ were analyzed with a—control, b—glyphosate (50 mg/kg b.wt.).

3.11. *Impact of Glyphosate on the mRNA Expression of Pro-Inflammatory Cytokines*

Inflammatory reactions are triggered by the production of pro-inflammatory cytokines such as TNF-α IL-1β, and IL-6 which are linked in the development of a number of metabolic diseases. Hence, we examine the mRNA expression levels of TNF-α, IL-1β, and IL-6 in glyphosate-treated group (Figure 8a–c). In this study, the hepatic mRNA expression levels of TNF-α and IL-6 were increased significantly ($p < 0.05$) in glyphosate-induced rats when compared to normal rats (Figure 8a,c); whereas, the IL-1β mRNA levels were unaltered in all the groups (Figure 8b).

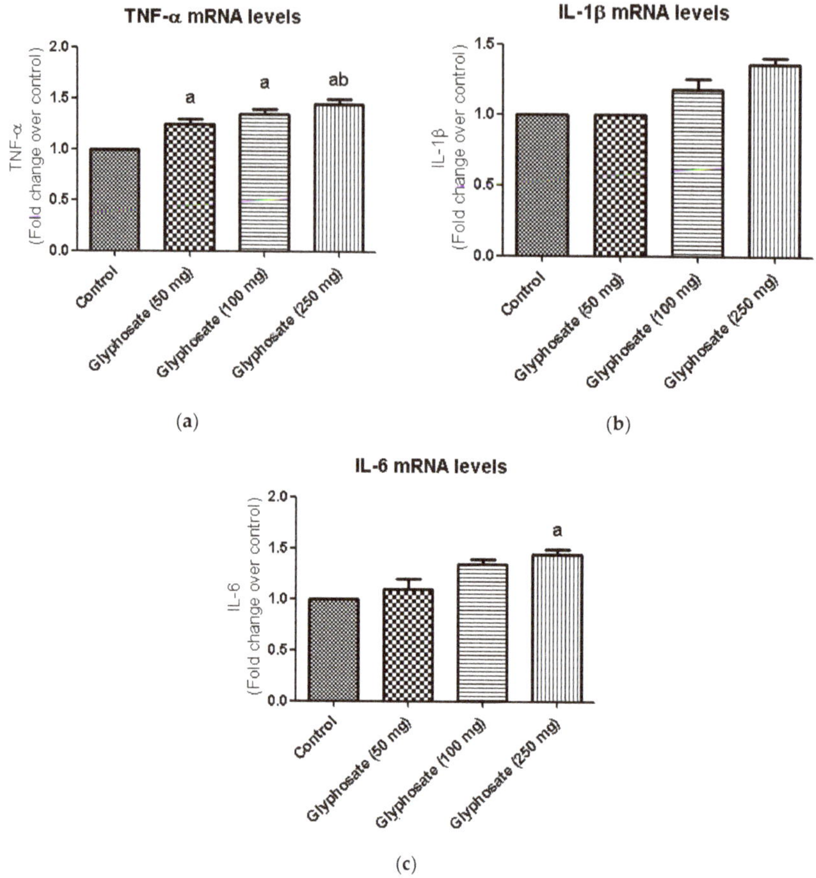

Figure 8. (**a**–**c**): The effect of glyphosate on mRNA levels of pro-inflammatory cytokines (TNF-α, IL-1β, and IL-6) in control and glyphosate treated rats. TNF-α, IL-1β, and IL-6 mRNA expression were analyzed by Real Time-PCR using gene specific primers and results were expression in fold-change over control. Data were expressed as mean ± SEM where $n = 6$ rats for each group. Values significant at $p < 0.05$ were analyzed with a—control, b—glyphosate (50 mg/kg b.wt.).

3.12. Impact of Glyphosate on the Protein Analysis of Pro-Inflammatory Cytokines and Transcription Factors

The development of diabetes is linked to chronic inflammation and regulation of many transcriptional factors. We used the ELISA technique to quantify the protein of proinflammatory cytokines like IL-6, IL-1β and TNF-α and the transcriptional factors SREBP1c and PPAR-γ (Figure 9a–c). When compared to the control rats, the glyphosate-treated group had considerably higher protein concentration of proinflammatory and transcriptional factors in glyphosate treated groups whereas the PPAR-γ level was significantly ($p < 0.05$) decreased showing that glyphosate caused low-grade liver inflammation (Figure 9d,e).

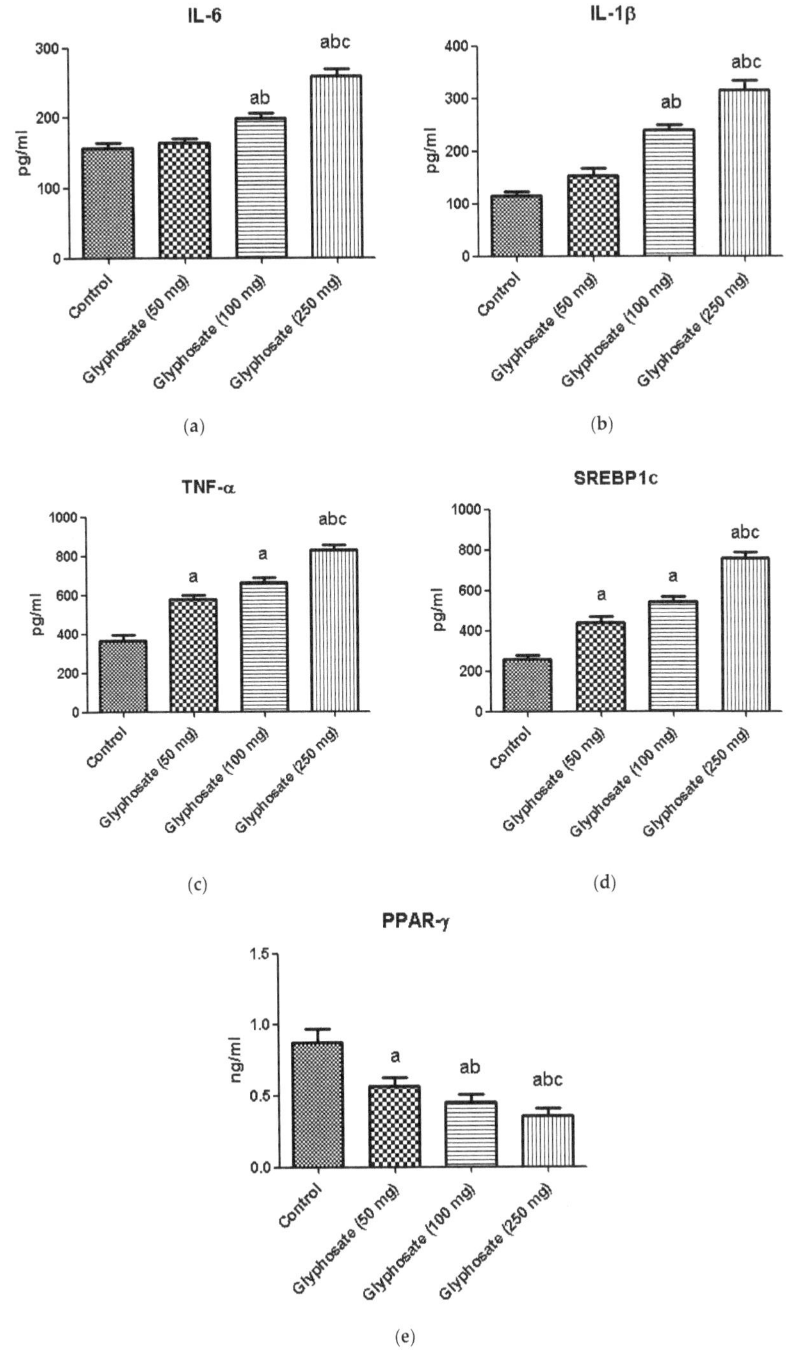

Figure 9. (**a**–**e**): The effect of glyphosate on protein levels of pro-inflammatory cytokines (IL-6, IL-1β, and TNF-α) and transcriptional factors (SREBP1c and PPAR-γ) in control and glyphosate treated rats. Data were expressed as mean ± SEM where $n = 6$ rats for each group. Values significant at $p < 0.05$ were analyzed with a—control, b—glyphosate (50 mg/kg b.wt.) and c—glyphosate (100 mg/kg b.wt.).

3.13. Impact of Glyphosate on the Expression of Pro-Inflammatory Proteins in Sections of the Liver Using Immunohistochemistry

Figures 10a–d and 11a–d depict the results of an immunohistochemical study of TNF-α and IL-6 in the liver tissues of rats exposed to glyphosate and control rats. When compared to control rats, the glyphosate-exposed group showed greater staining for inflammatory mediators such TNF-α and IL-6. Microscopic analysis of liver sections revealed light staining, indicating a weak expression of TNF-α and IL-6 in the samples from the normal group. The glyphosate-induced group of rat livers, on the other hand, showed a progressive increase in the mild to moderate expression of TNF-α and IL-6 along with an increase in high fluorescence intensity. This clearly indicates that glyphosate can cause mild to severe hydropic degeneration and focal necrosis as well as inflammation via TNF-α and IL-6.

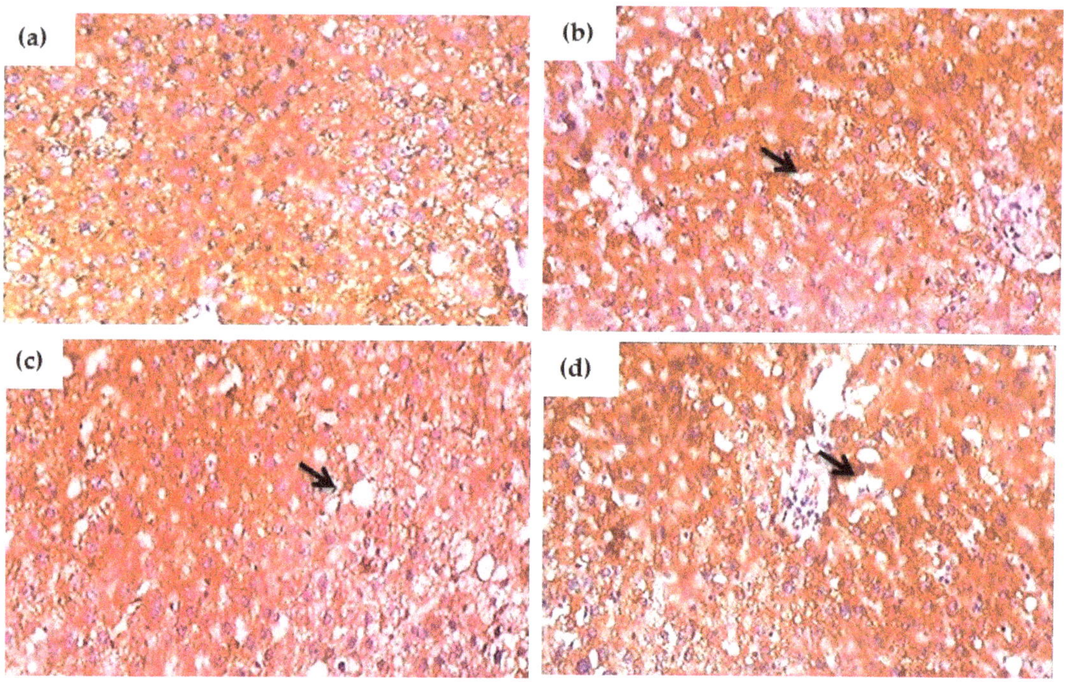

Figure 10. (**a**–**d**). Immunohistochemical evaluation of the effect of glyphosate on TNF-α protein expression in the liver sections of experimental rats. This image depicts exemplary IHC photomicrographs (100× magnification) of TNF-α staining intensity (black arrows) in the liver tissues. (**a**) Normal control rats showing small and large diameter hepatocytes; (**b**) glyphosate-exposed rats (50 mg/kg b.wt.) showed an mild expression of TNF-α with change in intensity that reflexed mild hydropic degeneration; (**c**) glyphosate-treated rats (100 mg/kg b.wt.) showed an elevated expression of TNF-α with increased hydropic degeneration indicated in the black arrow; and (**d**) glyphosate -treated rats (250 mg/kg b.wt.) showed an elevated expression of TNF-α with moderate expression and sever hydropic degeneration when compared to control rats.

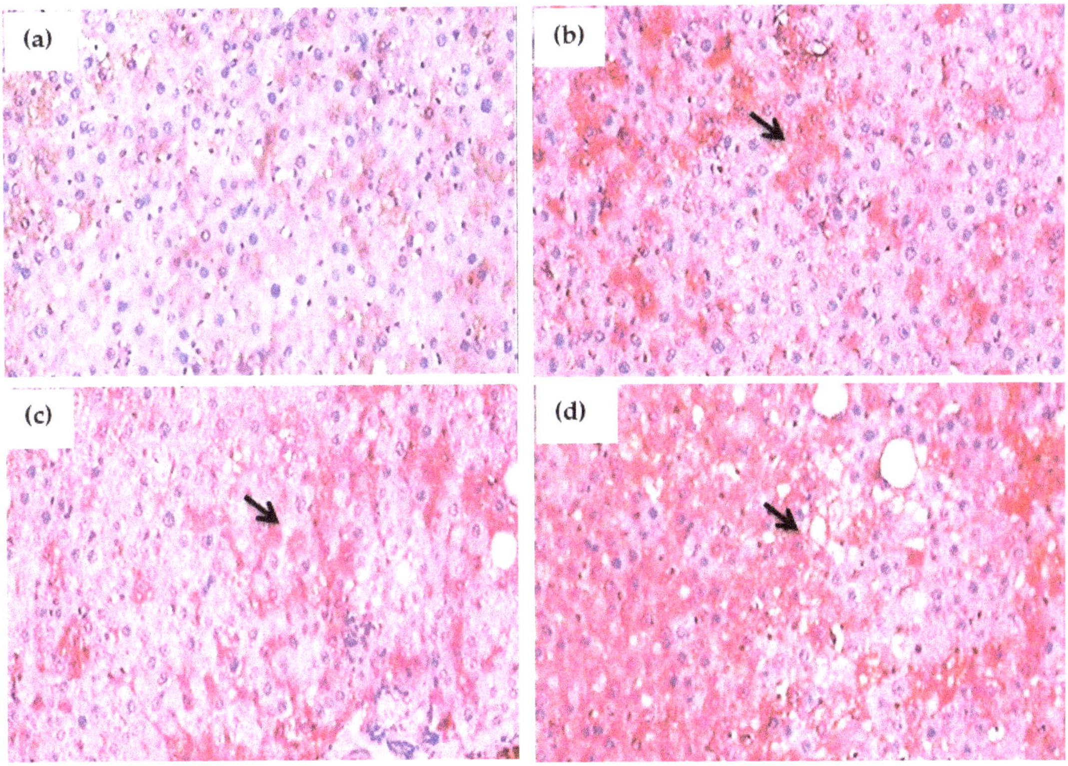

Figure 11. (**a–d**): Immunohistochemical evaluation of the effect of glyphosate on IL-6 protein expression in the liver sections of experimental rats. This image depicts typical IHC photomicrographs (100× magnification) of IL-6 staining intensity (black arrows) in the liver tissues of experimental group. (**a**) Normal control rats showing small and large diameter hepatocytes; (**b**) glyphosate-treated rats (50 mg/kg b.wt.) showed IL-6 expression with mild hydropic degeneration of hepatocytes; (**c**) glyphosate-treated rats (100 mg/kg b.wt.) showed elevated expression of IL-6 with moderate hydropic degeneration and focal necrosis; and (**d**) glyphosate-treated rats (250 mg/kg b.wt.) showed increased expression of IL-6 with high level of hydropic degeneration and focal necrosis when compared to control rats.

3.14. Histopathological Observation

The histological segments of the liver were examined with H&E staining is shown in Figure 12a–d. When compared to control rats, glyphosate-induced animals had an enlarged liver with more inflammation and fibrosis. Glyphosate exposed rats showed the change on hepatocyte morphology in the dose depended manner, as the 50 mg/kg b.wt. glyphosate explore showed the mild white patches around the nuclei (Figure 12b) compare to control with normal hepatocyte nuclei located inside the cells (Figure 12a). Glyphosate explore rats with 100 mg/kg b.wt. showed the microvesicular steatosis characterized by the presence of numerous small vesicles of fat that do not displace the nucleus (Figure 12c). There was the high number of large fat globule that displaces the nucleus as macrovesicular steatosis appears in Figure 12d.

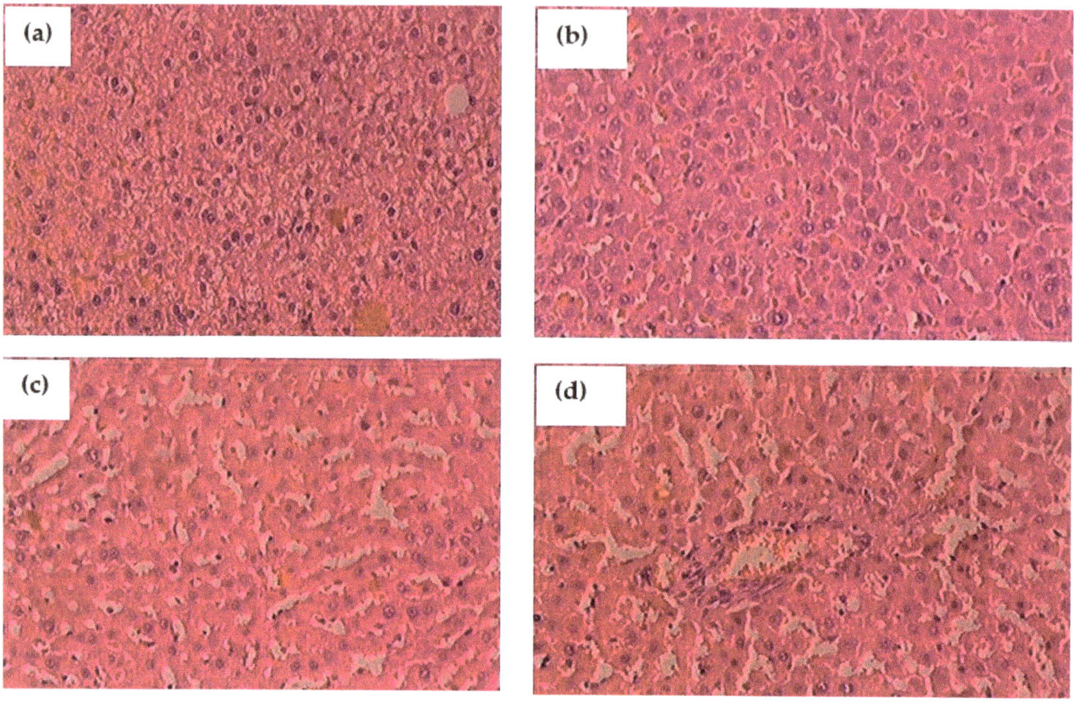

Figure 12. (**a**–**d**): The impact of glyphosate exposure on liver histopathology of experimental animals using hematoxylin and eosin (H&E) stain. (**a**) In normal rats, the hepatocyte nuclei are round and located inside the cells; (**b**) glyphosate-exposed rats (50 mg/kg b.wt.) showed a slight change in hepatocyte morphology when compared to control rats as the white patches around the nuclei depict lipid vesicles present inside the hepatocytes; (**c**) glyphosate-treated rats (100 mg/kg b.wt.) showed a significant fibrosis architecture of the liver with moderate lipid vesicles when compared to control rats; and (**d**) When compared to control rats, glyphosate (250 mg/kg b.wt.) rats showed elevated levels of liver fibrosis and inflammatory cells with large scale of white patches, demonstrating glyphosate toxicity in the liver.

4. Discussion

The most widely used herbicide in the world, glyphosate, has been linked to numerous environmental problems. Environmental exposure to glyphosate-based herbicides has recently been associated to hormone disturbance, kidney damage, liver toxicity, and other tissue damage at concentrations below regulatory boundaries. Additionally, these environmental pollutants activate NF-κB signalling, which plays a central role in the development of many metabolic diseases [46–48]. Therefore, the aim of the current study was to ascertain how glyphosate influence insulin signalling and glucose metabolism through NF-κB signalling in the liver of adult male albino rats.

Any substance that creates a toxic environment causes ectopic lipid build-up in the skeletal muscle and liver, obstructing insulin action. Following adipocyte failure, macrophage infiltration and increased lipolysis significantly affects hepatic lipid and glucose metabolism in a variety of ways. While this increases β cell function and hyperinsulinemia compensates for these dysregulated routes at first, glucolipotoxicity and genetic factors eventually contribute to β cell dysfunction and type 2 diabetes mellitus (T2DM) development [49]. Similarly, when comparing glyphosate-exposed rats to control rats, we discovered a significant increase in blood glucose and insulin levels. Furthermore, as an endocrine disruptor, glyphosate exposure affects testosterone production in rats, which was also seen in the current study. In previous studies, adult male rats treated with glyphosate had lower levels of blood testosterone and total sperm content, as well as more defective sperm [50,51]. This study demonstrated that exposure to glyphosate, affected testosterone levels through raising blood sugar and insulin levels which may be related to the generation of reactive oxygen species (ROS), which in turn affect these blood parameters.

In the current study, glyphosate-treated rats revealed an upsurge in blood glucose and serum insulin levels and these effects could be related to insulin resistance caused by the activation of hypothalamic-pituitary-adrenal (HPA) axis along with oxidative stress allied with hepatic insulin resistance [52,53]. Subsequently, rats exposed to glyphosate showed significant changes in the critical parameters such as OGT, ITT, HOMA-IR and QUICKI indexes in the current study, showing that glyphosate affects liver metabolism by creating ROS. In living systems, ROS-induced oxidative stress has both beneficial and harmful effects, including involvement in diabetogenesis and the development of diabetes associated complications [25,54].

ROS are produced by organophosphate herbicides like glyphosate, which leads to oxidative stress in the liver and other tissues, contributing to the toxic consequences of these xenobiotics. Oxidative stress-induced ROS can cause lipid peroxidation in the liver and brain, as well as damages DNA which in turn brings about tissue damage. Similarly, recent data on glyphosate-induced oxidative stress in several species are available. We investigated oxidative stress indicators as a marker for glyphosate toxicity in the liver [55–57]. In the present study, high amounts of LPO and TBARS were seen in glyphosate-treated animals, which might be due to hyperglycaemia. This in turn could have further accelerated the production of ROS and dysregulate glucose and fat metabolism. These oxidation processes can create free radicals, which can cause cell damage in a chain reaction. Antioxidants inhibit chain reactions by eliminating the free radical intermediates and by being oxidized themselves, they inhibit additional oxidation processes. Hence, we analyzed the levels of antioxidants in the glyphosate-treated rats in the present study.

SOD is an essential antioxidant enzyme that serves as the protective agent against ROS by scavenging superoxide radicals. It helps to prevent the development of oxygen-free radicals in tissues. The reduction in SOD activity in glyphosate-treated rat liver might have resulted in oxidative DNA or mitochondrial damage in cells. CAT is a heme protein that catalyzes the reduction of H_2O_2 to oxygen and water thereby protects the cell from ROS-induced toxicity caused by H_2O_2. It also includes Fe in its active core which shields the cell from oxidative damage [58,59]. Fe deficiency may be the cause of reduced CAT enzyme activity in the liver tissue of glyphosate-exposed rats [60]. Furthermore, through affecting insulin production and blood glucose balance in rats, glyphosate can elevate ROS production and metabolic imbalance.

Glyphosate can generate oxidative stress, which causes hyperglycaemia by stimulating the sympathetic nervous system and the hypothalamic-pituitary-adrenal (HPA) axis. As a result, glucagon, catecholamines and growth hormone are released, which boosts gluconeogenesis, glycogenolysis and insulin resistance. In this study also the mechanism by which glyphosate might have altered the glucose metabolic pathways in rats [53,61]. Hence, the activities of critical glycolytic and gluconeogenic enzymes were assessed in the present study. In glyphosate-exposed rats, glucose metabolic enzymes such as HK, PK, G6P and F1,6BP were significantly altered. The decline in the activities of HK and PK in glyphosate-treated rat liver might be attributed to reduced glycolysis and glucose uptake for oxidation [62,63]. Due to insulin and glucose intolerance, which raises lipogenesis and blood glucose levels, the glyphosate-treated groups showed significant increase in G6P and F1,6BP activity [64]. Under physiological conditions, insulin typically limits gluconeogenesis; hence, the rise in these gluconeogenic enzymes may be triggered by an issue with insulin action.

The initiation of the hepatic insulin signalling mechanism is mediated by autophosphorylation, stimulation, and activation of scaffold signalling molecules such IRS1 and IRS2 [65]. The role of both isoforms in stable glucose regulation is the same. The glyphosate-treated group in the present study had significantly lower levels of IR mRNA, is suggestive of a shift in insulin transduction. Hyperinsulinemia could have contributed to the reduced IR mRNA expression levels in the glyphosate-treated groups. As the transit of insulin from the endoplasmic reticulum (ER) to the plasma membrane requires glycines retained in the insulin receptor, a lack of insulin receptors induces hyperglycaemia and diabetes. Another factor is that cytochrome *c* oxidase (COX) is the enzyme responsible for the last step of ATP production in mitochondria. Substitutions for conserved glycines severely impede COX's oxidative phosphorylation. This could explain that the fact that glyphosate is toxic to mitochondria [66,67]. As shown in this study, glyphosate activates cellular stress response pathways and creates ROS, which may lead to enhanced oxidation of membrane molecules. This shift in the antioxidant and oxidative stress balance damages biomolecules including DNA, proteins and lipids, leading to non-alcoholic fatty liver disease (NAFLD), which is a major contributor to hepatic insulin resistance [68,69].

An organophosphate like glyphosate, which has a phosphate group, binds with anticholinesterase (AChE) through a covalent bond between the glyphosate phosphate and the oxygen of serine at the active site of AChE. This causes an irreversible phosphorylation of inactive AChE and increases acetylcholine (Ach) activity. Extreme addition of Ach at cholinergic sites causes nicotinic effects, such as metabolic shifts, resulting in significantly increased oxygen and glucose requirements, as well as ATP requirements, throughout the muscles. As a result of the increased energy demand, NADH, which is a glycolysis pathway product, is oxidized with ambient oxygen, increasing free radical formation (ROS) and increased production of GLUT2. These ROS cause the glucose transporter to be altered in the lysosome by inducing the insulin stimulation at PI3-kinase to phosphorylate Rac GTPase. As a result, intravascular glucose and GLUT2 levels stay high, resulting in a defective GLUT2 and insulin action [70]. GLUT2 regulates the majority of glucose uptake in hepatocytes, which is influenced by the amount of circulating glucose in the bloodstream. In the current study, the GLUT2 mRNA expression was significantly amplified in the liver of glyphosate-treated group. These findings show that glyphosate disrupts glucose homeostasis through the production of ROS which leads to insulin resistance.

The primary function of hepatic insulin is to influence lipid metabolism. SREBP-1c, a transcription factor controlled by insulin, is essential for de novo lipogenesis, the process by which sugar is converted into fat. As a result, glyphosate-exposed rats exhibited elevated liver SREBP1c mRNA levels. The present work has provided a strong evidence that glyphosate alters lipid metabolism in response to oxidative stress that results in the detrimental changes on lipid markers in glyphosate treated rats [71].

Glucose homeostasis is also influenced by ligand-activated transcription factors from the peroxisome proliferator-activated receptors superfamily of nuclear hormone receptors (PPARs). In type 2 diabetes patients, PPAR-γ activation lead to a significant improvement in insulin and glucose indices, owing to an increase in whole body insulin sensitivity [72]. It plays an imperative function in the regulation of lipid metabolism in adult adipocytes by enhancing fatty acid entrapment. PPAR-γ activation has been linked to positive impact on the expression and release of a wide spectrum of cytokines and disruptions in their production may lead to metabolic disorders. The mechanisms of PPAR-mediated insulin sensitivity are complicated and particular effects on skeletal muscle, fat and liver are considered to be involved [73]. In the present study, mRNA expression of PPAR-γ in glyphosate exposed rats was found to be reduced which could be due to glyphosate induced increase in the LPO and H_2O_2 in the liver that leads to decreased expression of PPAR-γ compared to control rats. In accordance with present study, it has been reported that glyphosate impede the activation of PPAR-γ via enhanced lipid peroxidation, decrease preadipocyte proliferation and differentiation and thereby resulting in oxidative stress, which is suggestive of its potential to disrupt cellular physiology [74].

SREBP1c and PPAR-γ are known to control lipid metabolism altered in hepatotoxicity and might lead to NAFLD in response to glyphosate exposure. As they affect the stress pathway, these components may be involved in a crosstalk network with JNK and participates in metabolic a pathway that links inflammation to metabolic disorders like insulin resistance through PPAR-γ and NF-κB. JNK activation via IKKβ and NF-κB in response to pro-inflammatory indicators may also lead to disease progression in insulin resistance and atherosclerosis. In the present study, it was clearly observed that glyphosate exposure led to a significant increase in the mRNA levels of JNK, IKKβ and NFκB, as well as pro-inflammatory markers (TNF-α, IL-6 and IL-1β) when compared to that of control rats. These data indicates a shift in insulin signalling that might have led to inflammation and systemic insulin resistance [75,76]. Exposure of rats to Roundup (a herbicide containing glyphosate) caused chronic inflammation in their liver and adipose tissue as reported by Pandey et al. [77]. When adult male rats were treated with varied dosages (0, 5, 10, 25, 50, 100 and 250 mg/kg bodyweight [bw]) of Roundup, it resulted in increased levels of C-reactive protein, cytokines IL-1, TNF-, IL-6 and inflammatory response marker, as well as prostaglandin-endoperoxide synthase in the liver and adipose tissue. Along with our study, it has been shown that short-term Roundup exposure promotes liver scarring, multi-organ inflammation and liver dysfunction in adult male rats. Taken together, our present findings strongly bring experimental evidence that glyphosate exposure causes liver inflammation-induced insulin resistance in rats (Figure 13). Future research need to concentrate on understanding the genetic pathways as well as the therapeutic aspects of glyphosate-regulated insulin signalling aiding T2DM research.

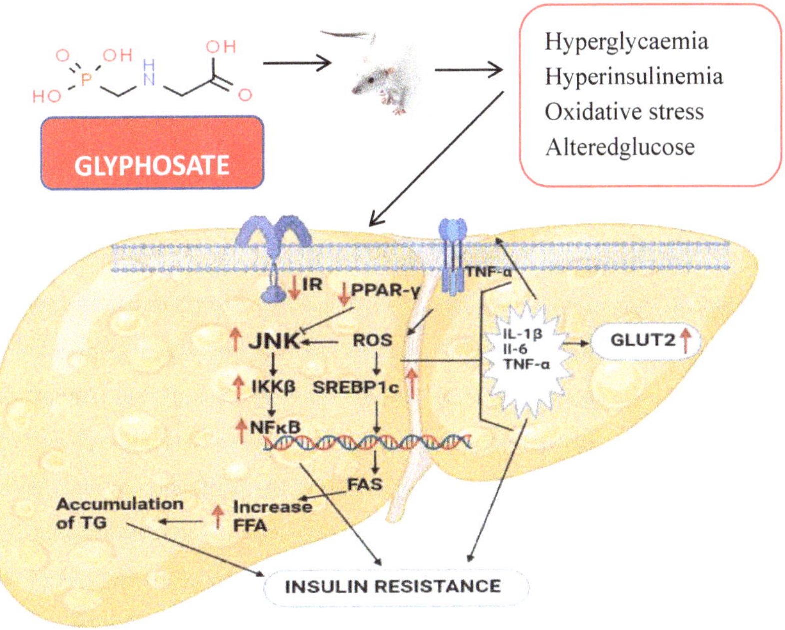

Figure 13. The diagram shows the effect of glyphosate on hepatic insulin signalling and inflammation in male albino rats. Glyphosate exposure results in hyperglycaemia, hyperinsulinemia and oxidative stress in rats that alter the insulin signalling in the liver. Hepatic glucose metabolism was also disturbed as a result, and pro-inflammatory cytokines were increased, leading to NAFLD and insulin resistance.

5. Conclusions

Glyphosate aggravated ROS, altered antioxidant activity, disrupted the glucose homeostasis, altered insulin action in liver via NFκB signaling in liver. As the primary organ for metabolism and detoxification, the liver is affected by oxidative stress caused by glyphosate, which could change crucial regulatory pathways including insulin signalling and JNK, resulting in insulin resistance and inflammation. Further research is needed to understand how glyphosate affects inflammation-related glucose homeostasis, which is important for T2DM/insulin signalling.

Author Contributions: Conceptualization, S.J., M.P. and V.P.V.; methodology, M.P., S.J. and K.K.; software, M.K.G. and M.A.A.; validation, S.J., and P.R., R.K. and K.K.; formal analysis, V.P., V.P.V. and S.J.; investigation, M.P. and S.J.; visualization, S.J. and V.P.V.; supervision, S.J. Writing—original draft, M.P. and K.K.; Writing—review and editing, all coauthors; project administration, S.J. All authors have read and agreed to the published version of the manuscript.

Funding: The authors are grateful to the Researchers Supporting Project number (RSP-2021/393), King Saud University, Riyadh, Saudi Arabia.

Institutional Review Board Statement: The present in vivo study was after proper approval from the Institutional animal ethical committee at Saveetha Dental College and Hospitals, Saveetha Institute of Medical and Technical Sciences (IAEC No: BRULAC/SDCH/SIMATS/IAEC/02-2019/015). Male Wistar rats obtained from Biogen animal facility, Bangalore. Animals were acclimatized for 10 days before the commencement of experiments and fed with a standard pellet diet and water ad libitum. The experiment was done following the CPCSEA guidelines.

Informed Consent Statement: Not applicable.

Data Availability Statement: The data presented in this study are available in this article.

Acknowledgments: The authors are grateful to the Researchers Supporting Project number (RSP-2021/393), King Saud University, Riyadh, Saudi Arabia.

Conflicts of Interest: The authors declare no conflict of interest.

References

1. Saeedi, P.; Petersohn, I.; Salpea, P.; Malanda, B.; Karuranga, S.; Unwin, N.; Colagiuri, S.; Guariguata, L.; Motala, A.A.; Ogurtsova, K.; et al. Global and regional diabetes prevalence estimates for 2019 and projections for 2030 and 2045: Results from the International Diabetes Federation Diabetes Atlas. IDF Diabetes Atlas Committee (2019), 9th ed. *Diabetes Res. Clin. Pract.* **2019**, *157*, 107843.
2. Bonciu, E. Cytological effects induced by Agil herbicide to onion. *J. Hortic. For. Biotechnol.* **2012**, *16*, 68–72.
3. Mesi, A.; Kopliku, D.; Golemi, S. The use of higher plants as bio-indicators of environmental pollution—A new approach for toxicity screening in Albania. *Mediterr. J. Soc. Sci.* **2012**, *3*, 237–248.
4. Silveira, M.A.; Ribeiro, D.L.; Dos Santos, T.A.; Vieira, G.M.; Cechinato, C.N.; Kazanovski, M.; Grégio d'Arce, L.P. Mutagenicity of two herbicides widely used on soybean crops by the Allium cepa test. *Cytotechnology* **2016**, *68*, 1215–1222. [CrossRef]
5. Kraehmer, H.; van Almsick, A.; Beffa, R.; Dietrich, H.; Eckes, P.; Hacker, E.; Hain, R.; Strek, H.J.; Stuebler, H.; Willms, L. Herbicides as weed control agents: State of the art: II. Recent achievements. *Plant Physiol.* **2014**, *166*, 1132–1148. [CrossRef]
6. Nicolopoulou-Stamati, P.; Maipas, S.; Kotampasi, C.; Stamatis, P.; Hens, L. Chemical Pesticides and Human Health: The Urgent Need for a New Concept in Agriculture. *Front. Public Health* **2016**, *4*, 148. [CrossRef]
7. Cai, D.; Yuan, M.; Frantz, D.F.; Melendez, P.A.; Hansen, L.; Lee, J.; Shoelson, S.E. Local and systemic insulin resistance resulting from hepatic activation of IKK-beta and NF-kappaB. *Nat. Med.* **2005**, *11*, 183–190.
8. He, Z.; King, G.L. Microvascular complications of diabetes. *Endocrinol. Metab. Clin. N. Am.* **2004**, *33*, 215–238. [CrossRef]
9. Zambrano, S.; De Toma, I.; Piffer, A.; Bianchi, M.E.; Agresti, A. NF-κB oscillations translate into functionally related patterns of gene expression. *eLife* **2016**, *5*, e09100.
10. Rothschild, D.; Weissbrod, O.; Barkan, E.; Kurilshikov, A.; Korem, T.; Zeevi, D.; Costea, P.I.; Godneva, A.; Kalka, I.N.; Bar, N.; et al. Environment dominates over host genetics in shaping human gut microbiota. *Nature* **2018**, *555*, 210–215. [CrossRef]
11. Komives, T.; Schroder, P. On glyphosate. *Ecocycles* **2016**, *2*, 1–8.
12. Thongprakaisang, S.; Thiantanawat, A.; Rangkadilok, N.; Suriyo, T.; Satayavivad, J. Glyphosate induces human breast cancer cells growth via estrogen receptors. *Food Chem. Toxicol. Int. J. Publ. Br. Ind. Biol. Res. Assoc.* **2013**, *59*, 129–136. [CrossRef]
13. Silva, V.; Montanarella, L.; Jones, A.; Fernández-Ugalde, O.; Mol, H.; Ritsema, C.J.; Geissen, V. Distribution of glyphosate and aminomethylphosphonic acid (AMPA) in agricultural topsoils of the European Union. *Sci. Total Environ.* **2018**, *621*, 1352–1359. [CrossRef]
14. Mertens, M.; Höss, S.; Neumann, G.; Afzal, J.; Reichenbecher, W. Glyphosate, a chelating agent-relevant for ecological risk assessment? *Environ. Sci. Pollut. Res. Int.* **2018**, *25*, 5298–5317. [CrossRef]
15. Williams, G.M.; Berry, C.; Burns, M.; de Camargo, J.L.; Greim, H. Glyphosate rodent carcinogenicity bioassay expert panel review. *Crit. Rev. Toxicol.* **2016**, *46*, 44–55.

16. Kapoor, D.; Malkin, C.J.; Channer, K.S.; Jones, T.H. Androgens, insulin resistance and vascular disease in men. *Clin. Endocrinol.* **2005**, *63*, 239–250.
17. Andersson, B.; Mårin, P.; Lissner, L.; Vermeulen, A.; Björntorp, P. Testosterone concentrations in women and men with NIDDM. *Diabetes Care* **1994**, *17*, 405–411. [CrossRef]
18. Muthusamy, T.; Dhevika, S.; Murugesan, P.; Balasubramanian, K. Testosterone deficiency impairs glucose oxidation through defective insulin and its receptor gene expression in target tissues of adult male rats. *Life Sci.* **2007**, *81*, 534–542. [CrossRef]
19. Vandenberg, L.N.; Blumberg, B.; Antoniou, M.N.; Benbrook, C.M.; Carroll, L.; Colborn, T.; Everett, L.G.; Hansen, M.; Landrigan, P.J.; Lanphear, B.P.; et al. Is it time to reassess current safety standards for glyphosate-based herbicides? *J. Epidemiol. Community Health* **2017**, *71*, 613–618. [CrossRef]
20. Nardi, J.; Moras, P.B.; Koeppe, C.; Dallegrave, E.; Leal, M.B.; Rossato-Grando, L.G. Prepubertal subchronic exposure to soy milk and glyphosate leads to endocrine disruption. *Food Chem. Toxicol. Int. J. Publ. Br. Ind. Biol. Res. Assoc.* **2017**, *100*, 247–252. [CrossRef]
21. Martinez, A.; Al-Ahmad, A.J. Effects of glyphosate and aminomethylphosphonic acid on an isogeneic model of the human blood-brain barrier. *Toxicol. Lett.* **2019**, *304*, 39–49. [CrossRef]
22. Trasande, L.; Aldana, S.I.; Trachtman, H.; Kannan, K.; Morrison, D.; Christakis, D.A.; Whitlock, K.; Messito, M.J.; Gross, R.S.; Karthikraj, R.; et al. Glyphosate exposures and kidney injury biomarkers in infants and young children. *Environ. Pollut.* **2020**, *256*, 113334. [CrossRef]
23. Pandey, A.; Rudraiah, M. Analysis of endocrine disruption effect of Roundup®in adrenal gland of male rats. *Toxicol. Rep.* **2015**, *2*, 1075–1085. [CrossRef]
24. Liu, J.B.; Li, Z.F.; Lu, L.; Wang, Z.Y.; Wang, L. Glyphosate damages blood-testis barrier via NOX1-triggered oxidative stress in rats: Long-term exposure as a potential risk for male reproductive health. *Environ. Int.* **2022**, *159*, 107038. [CrossRef]
25. Matthews, D.R.; Hosker, J.P.; Rudenski, A.S.; Naylor, B.A.; Treacher, D.F.; Turner, R.C. Homeostasis model assessment: Insulin resistance and beta-cell function from fasting plasma glucose and insulin concentrations in man. *Diabetologia* **1985**, *28*, 412–419. [CrossRef]
26. Katz, A.; Nambi, S.S.; Mather, K.; Baron, A.D.; Follmann, D.A.; Sullivan, G.; Quon, M.J. Quantitative insulin sensitivity check index: A simple, accurate method for assessing insulin sensitivity in humans. *J. Clin. Endocrinol. Metab.* **2000**, *85*, 2402–2410. [CrossRef]
27. Fraga, C.G.; Leibovitz, B.E.; Tappel, A.L. Lipid peroxidation measured as thiobarbituric acid-reactive substances in tissue slices: Characterization and comparison with homogenates and microsomes. *Free. Radic. Biol. Med.* **1988**, *4*, 155–161. [CrossRef]
28. Jiang, Z.Y.; Hunt, J.V.; Wolff, S.P. Ferrous ion oxidation in the presence of xylenol orange for detection of lipid hydroperoxide in low density lipoprotein. *Anal. Biochem.* **1992**, *202*, 384–389. [CrossRef]
29. Kakkar, P.; Das, B.; Viswanathan, P.N. A modified spectrophotometric assay of superoxide dismutase. *Indian J. Biochem. Biophys.* **1984**, *21*, 130–132.
30. Sinha, A.K. Colorimetric assay of catalase. *Anal. Biochem.* **1972**, *47*, 389–394. [CrossRef]
31. Brandstrup, N.; Kirk, J.E.; Bruni, C. The hexokinase and phosphoglucoisomerase activities of aortic and pulmonary artery tissue in individuals of various ages. *J. Gerontol.* **1957**, *12*, 166–171. [CrossRef]
32. Valentine, W.N.; Tanaka, K.R. Pyruvate kinase: Clinical aspects. *Methods Enzym.* **1966**, *9*, 468–473.
33. Koide, H.; Oda, T. Pathological occurrence of glucose-6-phosphatase in serum in liver diseases. *Clin. Chim. Acta* **1959**, *4*, 554–561.
34. Gancedo, J.M.; Gancedo, C. Fructose-1,6-diphosphatase, phosphofructokinase and glucose-6-phosphate dehydrogenase from fermenting and non fermenting yeasts. *Arch. Mikrobiol.* **1971**, *76*, 132–138. [CrossRef]
35. Fourney, R.M.; Day, M.J. Randall RJRS. Northern blotting: Efficient RNA staining and transfer. *Focus* **1988**, *10*, 5–7.
36. Jayashree, S.; Indumathi, D.; Akilavalli, N.; Sathish, S.; Selvaraj, J.; Balasubramanian, K. Effect of Bisphenol-A on insulin signal transduction and glucose oxidation in liver of adult male albino rat. *Environ. Toxicol. Pharmacol.* **2013**, *35*, 300–310. [CrossRef]
37. Bizeau, M.E.; MacLean, P.S.; Johnson, G.C.; Wei, Y. Skeletal Muscle Sterol Regulatory Element Binding Protein-1c Decreases with Food Deprivation and Increases with Feeding in Rats. *J. Nutr.* **2003**, *133*, 1787–1792.
38. Mahmoud, A.M.; Abdel-Rahman, M.M.; Bastawy, N.A.; Eissa, H.M. Modulatory effect of berberine on adipose tissue PPAR, adipocytokines and oxidative stress in high fat diet/streptozotocin-induced diabetic rats. *J. Appl. Pharm. Sci.* **2017**, *7*, 1–10.
39. Al-Rasheed, N.M.; Fadda, L.M.; Al-Rasheed, N.M.; Ali, H.M.; Yacoub, H.I. Down-Regulation of NFκB, Bax, TGF-α, Smad-2mRNA expression in the Livers of Carbon Tetrachloride Treated Rats using Different Natural Antioxidants. *Braz. Arch. Biol. Technol.* **2016**, *59*, e16150553.
40. Zhou, H.; Li, Y.J.; Wang, M.; Zhang, L.H.; Guo, B.Y.; Zhao, Z.S.; Meng, F.L.; Deng, Y.G.; Wang, R.Y. Involvement of RhoA/ROCK in myocardial fibrosis in a rat model of type 2 diabetes. *Acta Pharmacol. Sin.* **2011**, *32*, 999–1008. [CrossRef]
41. Lu, L.; Zhang, Q.; Pu, L.J.; Xu, X.W.; Zhang, R.Y.; Zhang, J.S.; Hu, J.; Yang, Z.K.; Lü, A.K.; Ding, F.H.; et al. Elevation of tumor necrosis factor-alpha, interleukin-1beta and interleukin-6 levels in aortic intima of Chinese Guizhou minipigs with streptozotocin-induced diabetes. *Chin. Med. J.* **2007**, *120*, 479–484.
42. Qiu, L.L.; Wang, C.; Yao, S.; Li, N.; Hu, Y.; Yu, Y.; Xia, R.; Zhu, J.; Ji, M.; Zhang, Z.; et al. Fenvalerate induces oxidative hepatic lesions through an overload of intracellular calcium triggered by the ERK/IKK/NF-κB pathway. *FASEB J.* **2019**, *33*, 2782–2795. [CrossRef]

43. Dange, R.B.; Agarwal, D.; Teruyama, R.; Francis, J. Toll-like receptor 4 inhibition within the paraventricular nucleus attenuates blood pressure and inflammatory response in a genetic model of hypertension. *J. Neuroinflamm.* **2015**, *12*, 31. [CrossRef]
44. Peinnequin, A.; Mouret, C.; Birot, O.; Alonso, A.; Mathieu, J.; Clarençon, D.; Agay, D.; Chancerelle, Y.; Multon, E. Rat proinflammatory cytokine and cytokine related mRNA quantification by real-time polymerase chain reaction using SYBR green. *BMC Immunol.* **2004**, *3*, 3.
45. Gabe, M. *Techniques Histologiques*; Massie e Cie: Paris, France, 1968; p. 1113.
46. Gill, J.P.K.; Sethi, N.; Mohan, A.; Datta, S.; Girdhar, M. Glyphosate toxicity for animals. *Environ. Chem. Lett.* **2018**, *16*, 401–426. [CrossRef]
47. Mesnage, R.; Renney, G.; Séralini, G.E.; Ward, M.; Antoniou, M.N. Multiomics reveal non-alcoholic fatty liver disease in rats following chronic exposure to an ultra-low dose of Roundup herbicide. *Sci. Rep.* **2017**, *7*, 39328. [CrossRef]
48. Samuel, V.T.; Shulman, G.I. The pathogenesis of insulin resistance: Integrating signaling pathways and substrate flux. *J. Clin. Investig.* **2016**, *126*, 12–22.
49. Dallegrave, E.; Mantese, F.D.; Oliveira, R.T.; Andrade, A.J.; Dalsenter, P.R.; Langeloh, A. Pre-and postnatal toxicity of the commercial glyphosate formulation in Wistar rats. *Arch. Toxicol.* **2007**, *81*, 665–673.
50. Romano, R.M.; Romano, M.A.; Bernardi, M.M.; Furtado, P.V.; Oliveira, C.A.D. Prepubertal exposure to commercial formulation of the herbicide glyphosate alters testosterone levels and testicular morphology. *Arch. Toxicol.* **2010**, *84*, 309–317.
51. Mechanick, J.I. Metabolic mechanisms of stress hyperglycemia. *JPEN J. Parenter. Enteral. Nutr.* **2006**, *30*, 157–163.
52. Tizhe, E.; Ibrahim, N.; Fatihu, M.; Ambali, S.; Igbokwe, I.; Tizhe, U. Pancreatic function and histoarchitecture in Wistar rats following chronic exposure to Bushfire®: The mitigating role of zinc. *J. Int. Med. Res.* **2018**, *46*, 3296–3305. [CrossRef]
53. Rajesh, P.; Sathish, S.; Srinivasan, C.; Selvaraj, J.; Balasubramanian, K. Phthalate is associated with insulin resistance in adipose tissue of male rat: Role of antioxidant vitamins. *J. Cell. Biochem.* **2013**, *114*, 558–569. [CrossRef]
54. Ponnulakshmi, R.; Shyamaladevi, B.; Vijayalakshmi, P.; Selvaraj, J. In silico and in vivo analysis to identify the antidiabetic activity of beta sitosterol in adipose tissue of high fat diet and sucrose induced type-2 diabetic experimental rats. *Toxicol. Mech. Methods* **2019**, *29*, 276–290. [CrossRef]
55. Bagchi, D.; Bagchi, M.; Hassoun, E.A.; Stohs, S.J. In vitro and in vivo generation of reactive oxygen species, DNA damage and lactate dehydrogenase leakage by selected pesticides. *Toxicology* **1995**, *104*, 129–140. [CrossRef]
56. Palmeira, C.M.; Moreno, A.J.; Madeira, V.M. Thiols metabolism is altered by the herbicides paraquat, dinoseb and 2,4-D: A study in isolated hepatocytes. *Toxicol. Lett.* **1995**, *81*, 115–123. [CrossRef]
57. Milić, M.; Žunec, S.; Micek, V.; Kašuba, V.; Mikolić, A.; Lovaković, B.T.; Semren, T.Ž.; Pavičić, I.; Čermak, A.; Pizent, A.; et al. Oxidative stress, cholinesterase activity, and DNA damage in the liver, whole blood, and plasma of Wistar rats following a 28-day exposure to glyphosate. *Arh. Hig. Rada Toksikol.* **2018**, *69*, 154–168. [CrossRef]
58. Rikans, L.E.; Yamano, T. Mechanisms of cadmium mediated acute hepatotoxicity. *J. Biochem. Mol. Toxicol.* **2000**, *14*, 110–117.
59. Chance, B.; Greenstein, D.S.; Roughton, R.J.W. The mechanism of catalase action 1-steady state analysis. *Arch. Biochem. Biophys.* **1952**, *37*, 301–339.
60. Jurczuk, M.; Brzoska, M.M.; Moniuszko-Jakoniuk, J.; Galazyn-Sidorczuk, M.; Kulikowska-Karpinska, E. Antioxidant enzymes activity and lipid peroxidation in liver and kidney of rats exposed to cadmium and ethanol. *Food Chem. Toxicol.* **2004**, *42*, 429–438.
61. Djeffal, A.; Messarah, M.; Boumendjel, A.; Kadeche, L.; Feki, A.E. Protective effects of vitamin C and selenium supplementation on methomyl-induced tissue oxidative stress in adult rats. *Toxicol. Ind. Health* **2015**, *31*, 31–43.
62. Baquer, N.Z.; Gupta, D.; Raju, J. Regulation of metabolic pathways in liver and kidney during experimental diabetes: Effects of antidiabetic compounds. *Indian J. Clin. Biochem.* **1998**, *13*, 63–80. [CrossRef]
63. Vats, V.; Yadav, S.P.; Grover, J.K. Effect of *T. foenumgraecum* on glycogen content of tissues and the key enzymes of carbohydrate metabolism. *J. Ethnopharmacol.* **2003**, *85*, 237–242. [CrossRef]
64. Chakrabarti, S.; Biswas, T.K.; Rokeya, B.; Ali, L.; Mosihuzzaman, M.; Nahar, N.; Khan, A.K.; Mukherjee, B. Advanced studies on the hypoglycemic effect of Caesalpinia bonducella F. in type 1 and 2 diabetes in Long Evans rats. *J. Ethnopharmacol.* **2003**, *84*, 41–46. [CrossRef]
65. Mugabo, Y.; Lim, G.E. Scaffold proteins: From coordinating signaling pathways to metabolic regulation. *Endocrinology* **2018**, *159*, 3615–3630.
66. Uren Webster, T.M.; Santos, E.M. Global transcriptomic profiling demonstrates induction of oxidative stress and of compensatory cellular stress responses in brown trout exposed to glyphosate and Roundup. *BMC Genom.* **2015**, *16*, 32. [CrossRef]
67. da Silva Rosa, S.C.; Nayak, N.; Caymo, A.M.; Gordon, J.W. Mechanisms of muscle insulin resistance and the cross-talk with liver and adipose tissue. *Physiol. Rep.* **2020**, *8*, e14607.
68. Hamdaoui, L.; Naifar, M.; Mzid, M.; Ben Salem, M.; Chtourou, A.; Ayedi, F.; Sahnoun, Z.; Rebai, T. Nephrotoxicity of Kalach 360 SL: Biochemical and histopathological findings. *Toxicol. Mech. Methods* **2016**, *26*, 685–691.
69. Yazdinezhad, A.; Abbasian, M.; Hojjat Hosseini, S.; Naserzadeh, P.; Agh-Atabay, A.H.; Hosseini, M.J. Protective effects of Ziziphora tenuior extract against chlorpyrifos induced liver and lung toxicity in rat: Mechanistic approaches in subchronic study. *Environ. Toxicol.* **2017**, *32*, 2191–2202. [CrossRef]
70. Namba, T.; Nolte, C.T.; Jackrel, J.; Grob, D. Poisoning due to organophosphate insecticides: Acute and chronic manifestations. *Am. J. Med.* **1971**, *50*, 475–492.

71. Ren, X.; Dai, P.; Perveen, A.; Tang, Q.; Zhao, L.; Jia, X.; Li, Y.; Li, C. Effects of chronic glyphosate exposure to pregnant mice on hepatic lipid metabolism in offspring. *Environ. Pollut.* **2019**, *254 Pt A*, 112906. [CrossRef]
72. Blaschke, F.; Takata, Y.; Caglayan, E.; Law, R.E.; Hsueh, W.A. Obesity, peroxisome proliferator-activated receptor, and atherosclerosis in type 2 diabetes. *Arter. Thromb Vasc. Biol* **2006**, *26*, 28–40.
73. Leonardini, A.; Laviola, L.; Perrini, S.; Natalicchio, A.; Giorgino, F. Cross-Talk between PPARgamma and Insulin Signaling and Modulation of Insulin Sensitivity. *PPAR Res.* **2009**, *2009*, 818945. [CrossRef]
74. Martini, C.N.; Gabrielli, M.; Brandani, J.N.; Vila Mdel, C. Glyphosate Inhibits PPAR Gamma Induction and Differentiation of Preadipocytes and is able to Induce Oxidative Stress. *J. Biochem. Mol. Toxicol.* **2016**, *30*, 404–413.
75. Jayaraman, S.; Devarajan, N.; Rajagopal, P.; Babu, S.; Ganesan, S.K.; Veeraraghavan, V.P.; Palanisamy, C.P.; Cui, B.; Periyasamy, V.; Chandrasekar, K. β-Sitosterol Circumvents Obesity Induced Inflammation and Insulin Resistance by down-Regulating IKKβ/NF-κB and JNK Signaling Pathway in Adipocytes of Type 2 Diabetic Rats. *Molecules* **2021**, *26*, 2101. [CrossRef]
76. Prasad, M.; Jayaraman, S.; Rajagopal, P.; Veeraraghavan, V.P.; Kumar, P.K.; Piramanayagam, S.; Pari, L. Diosgenin inhibits ER stress-induced inflammation in aorta via iRhom2/TACE mediated signaling in experimental diabetic rats: An in vivo and in silico approach. *Chem.-Biol. Interact.* **2022**, *358*, 109885. [CrossRef]
77. Pandey, A.; Dhabade, P.; Kumarasamy, A. Inflammatory Effects of Subacute Exposure of Roundup in Rat Liver and Adipose Tissue. *Dose-Response A Publ. Int. Hormesis Soc.* **2019**, *17*, 1559325819843380. [CrossRef]

antioxidants

MDPI

Article

Antiproliferation Effects of Marine-Sponge-Derived Methanol Extract of *Theonella swinhoei* in Oral Cancer Cells In Vitro

Jun-Ping Shiau [1,†], Ya-Ting Chuang [2,†], Jen-Yang Tang [3,4], Shu-Rong Chen [5], Ming-Feng Hou [1,6], Jiiang-Huei Jeng [7,8,9], Yuan-Bin Cheng [5,*] and Hsueh-Wei Chang [2,10,*]

1 Division of Breast Oncology and Surgery, Department of Surgery, Kaohsiung Medical University Hospital, Kaohsiung Medical University, Kaohsiung 80708, Taiwan
2 Graduate Institute of Medicine, College of Medicine, Kaohsiung Medical University, Kaohsiung 80708, Taiwan
3 School of Post-Baccalaureate Medicine, Kaohsiung Medical University, Kaohsiung 80708, Taiwan
4 Department of Radiation Oncology, Kaohsiung Medical University Hospital, Kaoshiung Medical University, Kaohsiung 80708, Taiwan
5 Department of Marine Biotechnology and Resources, National Sun Yat-sen University, Kaohsiung 80424, Taiwan
6 Department of Biomedical Science and Environmental Biology, College of Life Science, Kaohsiung Medical University, Kaohsiung 80708, Taiwan
7 School of Dentistry, College of Dental Medicine, Kaohsiung Medical University, Kaohsiung 80708, Taiwan
8 Department of Dentistry, Kaohsiung Medical University Hospital, Kaohsiung 80708, Taiwan
9 Department of Dentistry, National Taiwan University Hospital, Taipei 100225, Taiwan
10 Center for Cancer Research, Kaohsiung Medical University, Kaohsiung 80708, Taiwan
* Correspondence: jmb@mail.nsysu.edu.tw (Y.-B.C.); changhw@kmu.edu.tw (H.-W.C.); Tel.: +886-07-525-2000 (ext. 5212) (Y.-B.C.); +886-7-312-1101 (ext. 2691) (H.-W.C.)
† These authors contributed equally to this work.

Abstract: The purpose of this study aimed to assess the antiproliferation effects of methanol extract of *T. swinhoei* (METS) and explore the detailed responses of oral cancer cells compared to normal cells. METS effectively inhibits the cell proliferation of oral cancer cells but does not affect normal cell viability, exhibiting preferential antiproliferation function. METS exerted more subG1 accumulation, apoptosis induction, cellular and mitochondrial oxidative stress, and DNA damage than normal cells, reverted by oxidative stress inhibitor *N*-acetylcysteine. This METS-caused oxidative stress was validated to attribute to the downregulation of glutathione. METS activated both extrinsic and intrinsic caspases. DNA double-strand breaks (γH2AX) and oxidative DNA damage (8-hydroxy-2-deoxyguanosine) were stimulated by METS. Therefore, for the first time, this investigation shed light on exploring the functions and responses of preferential antiproliferation of METS in oral cancer cells.

Keywords: marine sponge; oral cancer; oxidative stress; natural product

Citation: Shiau, J.-P.; Chuang, Y.-T.; Tang, J.-Y.; Chen, S.-R.; Hou, M.-F.; Jeng, J.-H.; Cheng, Y.-B.; Chang, H.-W. Antiproliferation Effects of Marine-Sponge-Derived Methanol Extract of *Theonella swinhoei* in Oral Cancer Cells In Vitro. *Antioxidants* **2022**, *11*, 1982. https://doi.org/10.3390/antiox11101982

Academic Editor: Tim Hofer

Received: 19 September 2022
Accepted: 30 September 2022
Published: 4 October 2022

1. Introduction

Oral cancer patients show high morbidity and mortality [1] and show high incidence worldwide [2]. Surgery, radiation, and chemotherapy are common clinical therapies for oral cancer, but chemoradiation occasionally generates side effects [3]. It is necessary to identify new anticancer drugs for oral cancer treatment.

Marine natural products are rich resources for anti-cancer agents [4–7]. Marine sponges are the largest biomass in the ocean and contain diverse bioactive compounds [8–11] for cancer, inflammation, viral, and antibiotic treatments [10,12]. Sponge *Niphates* sp.-derived proteinaceous factions show antibacterial properties [13]. *Spongosorites halichondriodes* extracts show antiinflammation effects [14]. Accordingly, the anticancer effects on several kinds of marine sponges warrant detailed re-examination.

Several extraction studies recently isolated several pure compounds from the marine sponge *Theonella swinhoei* (*T. swinhoei*) [15–23]. However, most studies [17,23] focused on bioactive compound isolations of the marine sponge *T. swinhoei*. They provided IC_{50} values for certain cancer cell lines such as liver (HepG2) [18], colon (HCT-16) [19], lymphoma (P388) [20], and breast (MCF7 and MDA-MB-231) [22,23] without investigating its antiproliferation mechanism. However, the antiproliferation studies of *Theonella* natural products were not reported in oral cancer cells and evaluated for their safety in normal cells.

Crude extracts of natural products contain several bioactive compounds. Different compounds may modulate different targets, providing the benefits of the potential for multiple targeting of cancer [24]. Such multiple-target strategy of natural products has been applied to inflammation [25] and cancer [26] treatments. Notably, sponges contain natural products such as terpenoids, glycosides, phenols, fatty acids, peptides, sterols, and others [27]. Accordingly, the anticancer effects of *Theonella* crude extract warrants a detailed investigation of oral cancer cells.

This investigation evaluates the antiproliferation effects of methanol extracts of *T. swinhoei* (METS) and explores the antiproliferation mechanism of oral cancer cells. Several oral cancer and normal cells were included to examine their proliferation, oxidative stress, apoptosis, and DNA damage responses.

2. Materials and Methods

2.1. METS Preparation

The sponge *Theonella swinhoei* was collected by scuba diving in Orchid Island, Taitung County, Taiwan, in April 2011. A voucher specimen (OISP-1) was deposited at the Department of Marine Biotechnology and Resources, National Sun Yat-sen University, Kaohsiung, Taiwan. The animal material was extracted by ethanol thrice to provide a crude extract (469.0 g). This extract was aliquoted into ethyl-acetate-soluble and water-soluble fractions, and the former portion was further partitioned between 75% methanol (aq) and hexanes. The 75% methanol soluble layer (43.9 g) of *T. swinhoei* was termed METS.

2.2. HPLC Analysis of METS

The Shimadzu HPLC system contains an LC-40D solvent delivery module, a DGU-405 degassing unit, an SPD-M40A photo diode array detector, a CTO-40S column oven, and a CBM-40 controller. The Phenomenex (Torrance, CA, USA) Luna 5 μ C18(2) 100 A analytical column was chosen for separation. The chromatography conditions were shown as follows. Solution A: 0.1% trifluoroacetic acid (aq); solution B: MeOH; flow rate: 1.0 mL/min; 0 min: 85% solution B, 0–10 min: 85% to 95% solution B, 10–20 min: 95% solution B, 20–30 min: 95% to 100% solution B. Injection weight: 10 μg.

2.3. Cell Cultures and Reagents

Oral cancer cell lines such as tongue-derived CAL 27 (ATCC, Manassas, VA, USA) and Gingiva-derived Ca9-22 (HSRRB, Ibaraki, Osaka, Japan) were used. To assess the drug safety of METS, a non-malignant cell line derived from gingival epithelial Smulow-Glickman (S-G) was chosen as normal control cells [28–30]. The culture medium was prepared by mixing DMEM with F12 (3:2) (Gibco, Grand Island, NY, USA), 10% fetal bovine serum, and P/S antibiotics [31]. Cells were seeded at 4×10^4/well for a 12-well plate for growth overnight and subjected to drug treatment for flow cytometry experiments.

N-acetylcysteine (NAC; 10 mM, 1 h pretreatment) (Sigma-Aldrich, St. Louis, MO, USA) [32–34] was used to check the influence of oxidative stress in METS experiments.

2.4. Cell Viability Assay

MTS reagent (Promega Corporation, Madison, WI, USA), a tetrazolium dye, was used to detect cell viability. Ca9-22, CAL 27, and S-G cells were plated at 4, 4, and 6×10^3/well for a 96-well plate. Subsequently, cells were incubated overnight. Finally, cells were then treated with drugs for 24 h. Finally, MTS reagents were added and incubated for 1 h before ELISA reader detection (490 nm) [35].

2.5. Cytometric Cell Cycle Assay

Cell cycle phases are proportional to the DNA content, which was stained using 1 μg/mL of 7-aminoactinmycin D (7AAD) (Biotium, Hayward, CA, USA) for 30 min for 75% ethanol fixed cells [36]. DNA content was assessed using a flow cytometer (Guava easyCyte, Luminex, TX, USA), and cell cycle phases were analyzed using Flow Jo 10 software (Becton-Dickinson; Franklin Lakes, NJ, USA).

2.6. Cytometric Apoptosis Assay (Annexin V/7AAD)

Phosphatidylserine located at the outer plasma membrane of apoptotic cells is proportional to the annexin V detection, which is stained by annexin V-FITC/7AAD (1:1000/1 μg/mL) [37] (Strong Biotech, Taipei, Taiwan) for 1 h. Guava easyCyte flow cytometer assessed these intensities. The counts for annexin V (+)/7AAD (+ or −) regions were classified as apoptotic (+) cells.

2.7. Peptide-Based Apoptosis Assay (Caspase 3/7)

The caspases-Glo® 3/7 luminescent kit (Promega; Madison, WI, USA) utilizes proluminescent peptide substrates to detect caspase 3/7 activity [38]. When the caspase 3/7 was activated in apoptotic cells, the proluminescent peptide substrates were converted to luminescent signals monitored by a luminometer (Berthold Technologies GmbH & Co., Bad Wildbad, Germany). The caspase 3/7 intensities were calibrated by adjusting the respective cell viabilities.

2.8. Cytometric Apoptosis Assay (Caspases 3, 8, and 9)

Activities of caspases 3, 8, and 9 are proportional to the staining intensities by incubating their specific peptide substrates (PhiPhiLux-G1D2, CaspaLux8-L1D2, and CaspaLux9-M1D2) (OncoImmunin; Gaithersburg, MD, USA) according to user's instruction [39,40]. The caspases 3, 8, and 9 activated reactions can generate green fluorescence after peptide cleavage, i.e., caspase activation. The Guava easyCyte flow cytometer assessed these intensities. Their positive (+) region counts were classified as caspases 3, 8, and 9 (+) cells.

2.9. Cytometric Reactive Oxygen Species (ROS), Mitochondrial Superoxide (MitoSOX), and Glutathione (GSH) Assays

Levels of ROS, MitoSOX, and GSH are proportional to the staining intensities by incubating their specific dyes such as 2′,7′-dichlorodihydrofluorescein diacetate (H2DCFDA) (Sigma-Aldrich, St. Louis, MO, USA) [31], MitoSOX™ Red, and 5-chloromethylfluorescein diacetate (CMF-DA) (Thermo Fisher Scientific, Carlsbad, CA, USA) [41] in the darkness conditions (10 μM/30 min, 50 nM/30 min, and 5 μM/20 min at 37 °C). The Guava easyCyte flow cytometer assessed these intensities. Their counts for (+) regions were classified as ROS, MitoSOX, and GSH (+) cells.

2.10. Cytometric γH2AX/7AAD and 8-Hydroxy-2-deoxyguanosine (8-OHdG) Assays

Levels of γH2AX and 8-OHdG [42] are proportional to the fluorescent intensities generated by antibody detection to the 75% ethanol fixed cells. Primary antibodies for γH2AX [42] (Santa Cruz Biotechnology, Santa Cruz, CA, USA) and FITC-8-OHdG antibody (Santa Cruz Biotechnology, Santa Cruz, CA, USA) were used to incubate with these fixed cells at 4 °C, 1 h. Notably, secondary antibody (Cell Signaling Technology, Danvers, MA, USA) and 7AAD (5 μg/mL, 30 min) were further applied for γH2AX detection. The Guava easyCyte flow cytometer assessed these intensities. Their counts for (+) regions were classified as γH2AX and 8-OHdG (+) cells.

2.11. Statistical Analysis

The one-way ANOVA analysis and post hoc test (JMP software, SAS Institute Inc., Cary, NC, USA) provides the connecting letters for determining the significant difference for multiple comparison. Connecting letters without an overlap indicate a significant difference, while it differs non-significantly when the connecting letters are overlapping. Examples were mentioned in the figure legend to explain significance.

3. Results

3.1. HPLC Analysis of METS

HPLC-PDA fingerprint profiles of METS and the isolated major compound, theonellapeptolide 1d, were provided (Figure 1A). The retention time of theonellapeptolide 1d was found at 14.205 min, which overlapped the major peak of METS. The linear equations ($y = 10^7x - 79228$, $R^2 = 0.9995$) of theonellapeptolide 1d was deduced by the HPLC peak area in four different concentrations (Figure 1B). In METS, theonellapeptolide 1d accounts for 21.0% of the whole amount.

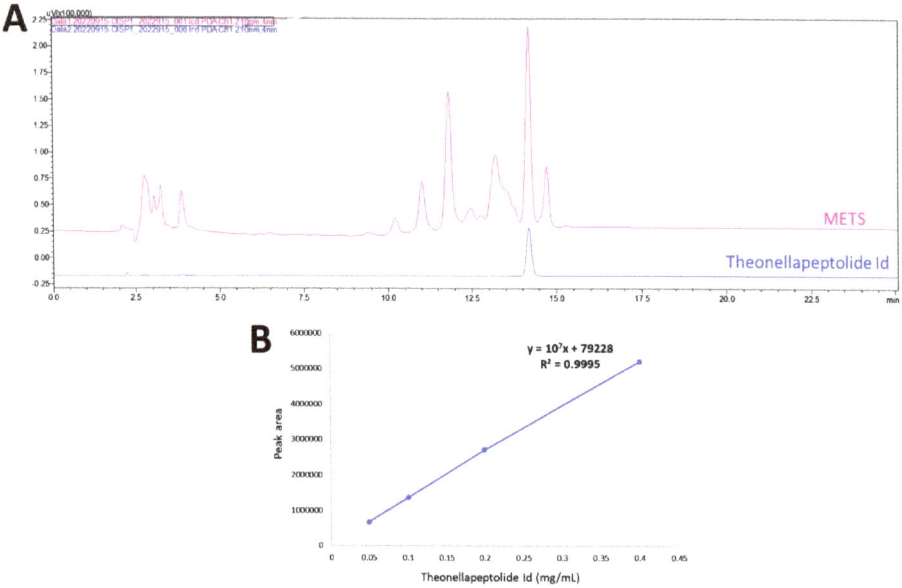

Figure 1. Fingerprint profile. (**A**) Fingerprint profiles of METS (pink, upper line) and the theonellapeptolide 1d (blue, lower line). (**B**) The calibration curve of theonellapeptolide 1d ($n = 3$).

3.2. METS Causes More Antiproliferation to Oral Cancer Cells than Normal Cells

METS decreased the cell viability for oral cancer cells (Ca9-22 and CAL 27) (Figure 2A). The drug safety of METS was examined using the viability of normal cells (S-G). These METS-treated normal cells exhibited healthy viability like the control. Additionally, NAC, a ROS removal agent, was chosen to test the contribution of ROS in antiproliferation caused by METS. Under different concentrations of METS, NAC alleviated the METS-promoted antiproliferation against oral cancer cells (Figure 2B), revealing the oxidative-stress-dependent action of METS.

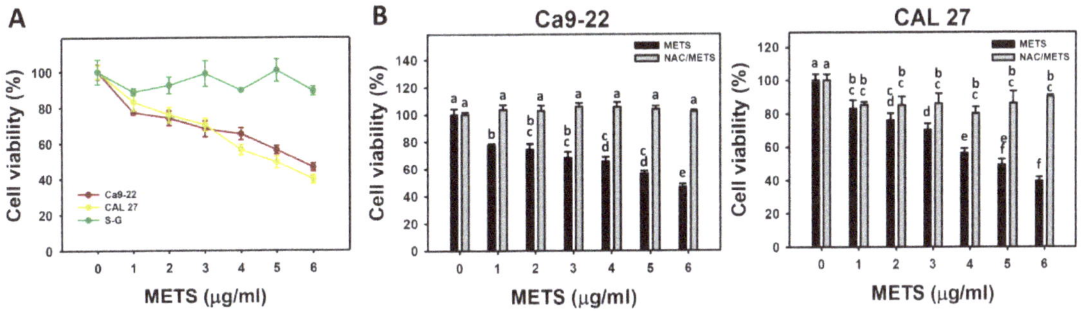

Figure 2. METS causes antiproliferation of oral cancer cells. (**A**) Cell viability. Cells were treated with METS for 24 h. Oral cancer (Ca9-22 and CAL 27) and normal cells (S-G) were included. (**B**) NAC response on cell viability of METS treatment. Untreated control and NAC pretreatment were followed by posttreatment of vehicle control (0 μg/mL) and METS (1, 2 ,3, 4, 5, and 6 μg/mL) for 24 h, i.e., control, NAC, METS, and NAC/METS, respectively. Data = mean ± SD (n = 3). Data showing non-overlapping lower-case letters have significant results ($p < 0.05$) as analyzed using the one-way ANOVA analysis and post hoc test. In the example of Figure 1B (Ca9-22 cells), the lower-case letters for METS at 0, 1, 4, and 6 μg/mL are "a, b, cd, and e", revealing significant differences between each other. By contrast, the lower-case letters for METS at 1, 2, and 3 μg/mL are "b, bc, and bc", indicating non-significant differences since they overlap with the same letter "b".

3.3. METS Causes More subG1 Accumulation to Oral Cancer Cells than Normal Cells

The subG1 events of histograms were used to measure primarily apoptosis. METS caused the accumulation of subG1 events in oral cancer cells (Ca9-22 and CAL 27) under dose and time course experiments (Figure 3A,B). CAL 27 cells showed a higher extent of subG1 events than Ca9-22 cells. In contrast, METS exhibited lower subG1 events in normal S-G cells than oral cancer cells. Accordingly, METS exerts a greater extent of subG1 in oral cancer than in normal cells.

The G1 events were increased, and G2/M events were decreased at 6 μg/mL of METS for Ca9-22 cells. By contrast, the G1 events were reduced, and G2/M events were increased at 6 μg/mL of METS for CAL 27 and S-G cells. Accordingly, METS exerts differential cell cycle disturbance for various cell lines.

Additionally, NAC was used to test the contribution of ROS in cell cycle disturbance caused by METS. NAC inhibited subG1 increment of METS-treated oral cancer cells (Figure 3B), revealing that oxidative stress modulated METS-triggered subG1 increment. NAC inhibited G1 increment and G2/M decrement of METS-treated Ca9-22 cells. In contrast, NAC inhibited G1 decrement and G2/M increment of METS-treated CAL 27 cells. For normal S-G cells, NAC enhanced G1 decrement and G2/M increment at 6 μg/mL of 24 h METS treatment.

3.4. METS Causes More Annexin V-Based Apoptosis to Oral Cancer Cells than Normal Cells

SubG1 increment is an apoptosis-like change, and it needs further validation for apoptosis. The annexin V (+) events of histograms were used to measure apoptosis. Annexin V (+) events were dose- and time-dependently increased by METS in oral cancer cells (Ca9-22 and CAL 27) (Figure 4). However, it indicated lower annexin (+) events in normal S-G cells than in oral cancer cells by METS treatment. Accordingly, METS exerts a greater extent of apoptosis induction in oral cancer cells than normal cells.

Additionally, NAC was used to test the contribution of ROS in annexin V increment caused by METS. NAC inhibited annexin V increment of METS-treated oral cancer cells (Figure 4B), particularly for 24 h. It reveals that oxidative stress modulated METS-triggered apoptosis.

in normal cells (Figure 5C). Accordingly, METS exert more caspase 3 and 3/7 activations in oral cancer cells than in normal cells.

Additionally, NAC was used to test the contribution of ROS in caspase 3 and 3/7 activation caused by METS. NAC moderately inhibited caspase 3 activations of Ca9-22 cells at 24 h METS treatment (Figure 5B). NAC dramatically inhibited caspase 3 activations of CAL 27 cells at 12 and 24 h METS treatment. NAC also moderately inhibited 3/7 activations of Ca9-22 and CAL 27 cells at 6 μg/mL of 24 h METS treatment (Figure 5B,C). In contrast, normal S-G cells showed low changes in caspase 3 and 3/7 activities. These results reveal that oxidative stress modulated METS-triggered caspase 3 and 3/7 activations.

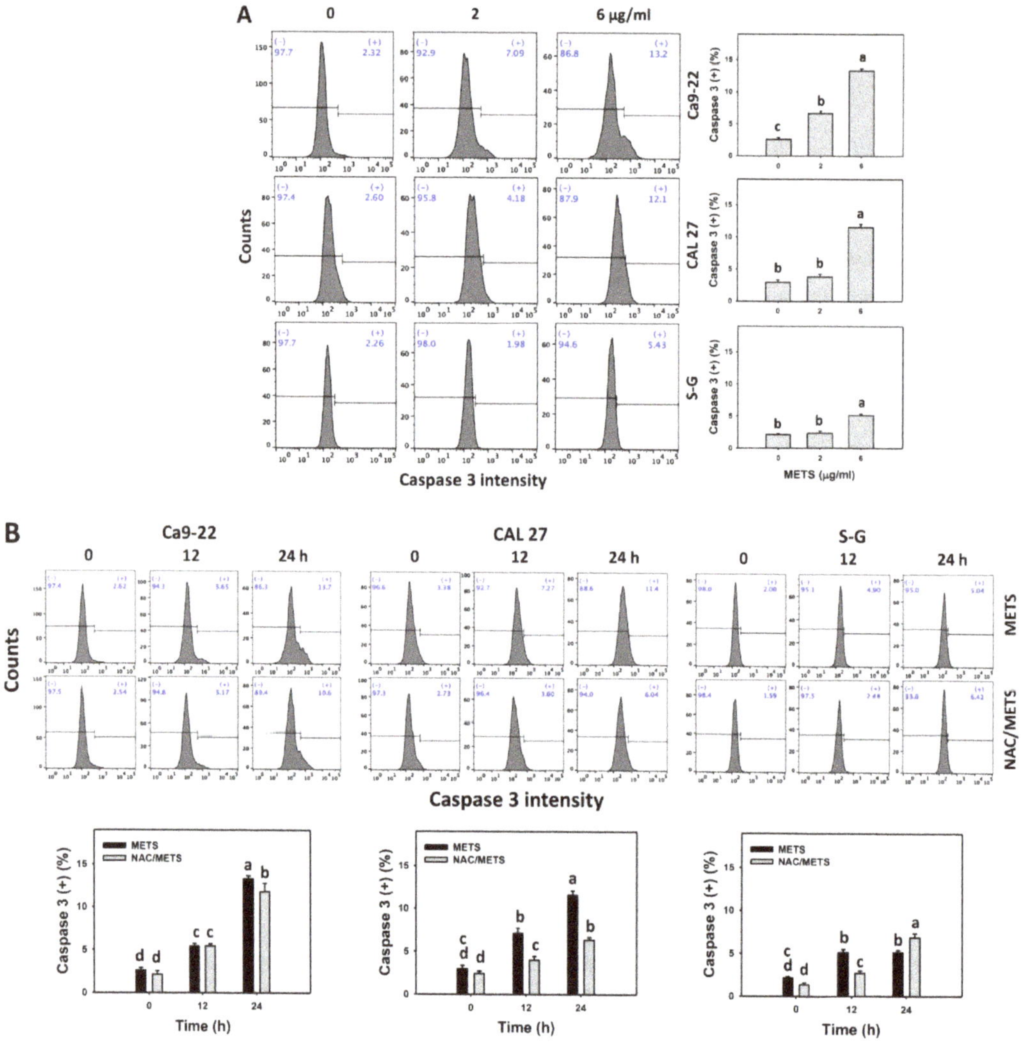

Figure 5. *Cont.*

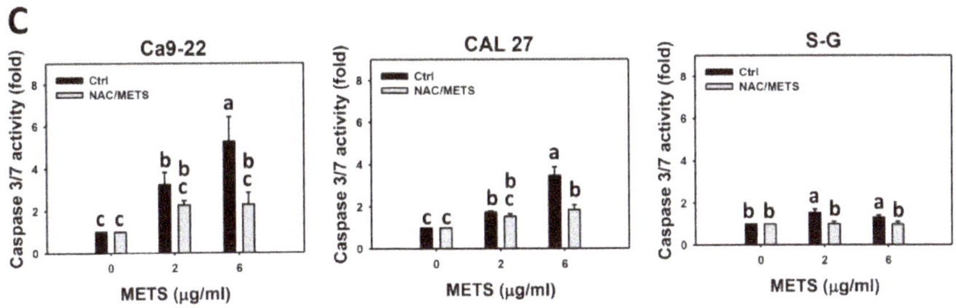

Figure 5. METS leads to caspase 3 and 3/7 activations in oral cancer cells. (**A**) Caspase 3 histogram changes. Cells were treated with METS for 24 h. Oral cancer and normal cells (S-G) were included. (+) indicates a high intensity of caspase 3. (**B**,**C**). NAC response on caspase 3 and 3/7 activations. For caspase 3 assay, untreated control and NAC pretreatment were followed by posttreatment of vehicle control and METS (6 μg/mL) for 0, 12, and 24 h, i.e., control, NAC, METS, and NAC/METS, respectively. Except for 24 h, other conditions of caspase 3/7 assays were the same as caspase 3 assay. Data = mean ± SD (n = 3). Data showing non-overlapping lower-case letters have significant results ($p < 0.05$) as analyzed by the one-way ANOVA analysis and post hoc test.

3.6. METS Causes More Caspases 8 and 9 Activations to Oral Cancer Cells than Normal Cells

Caspase 8 and 9 activations were monitored by flow cytometry. The caspase 8 and 9 (+) events of histograms were used to measure the apoptosis. Caspase 8 and 9 (+) events were dose- and time-dependently increased by METS in oral cancer cells (Ca9-22 and CAL 27) (Figure 6A,C). However, it showed lower caspase 8 and 9 (+) events in normal S-G cells than oral cancer cells by METS treatment. Accordingly, METS exert more caspase 8 and 9 activations in oral cancer cells than in normal cells. Additionally, NAC inhibited caspase 8 and 9 activations of METS-treated oral cancer cells (Figure 6B,D), revealing that oxidative stress modulated METS-triggered apoptotic signaling.

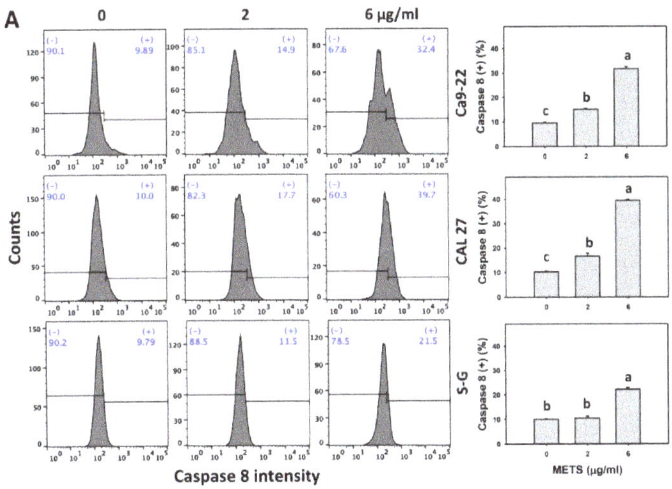

Figure 6. *Cont.*

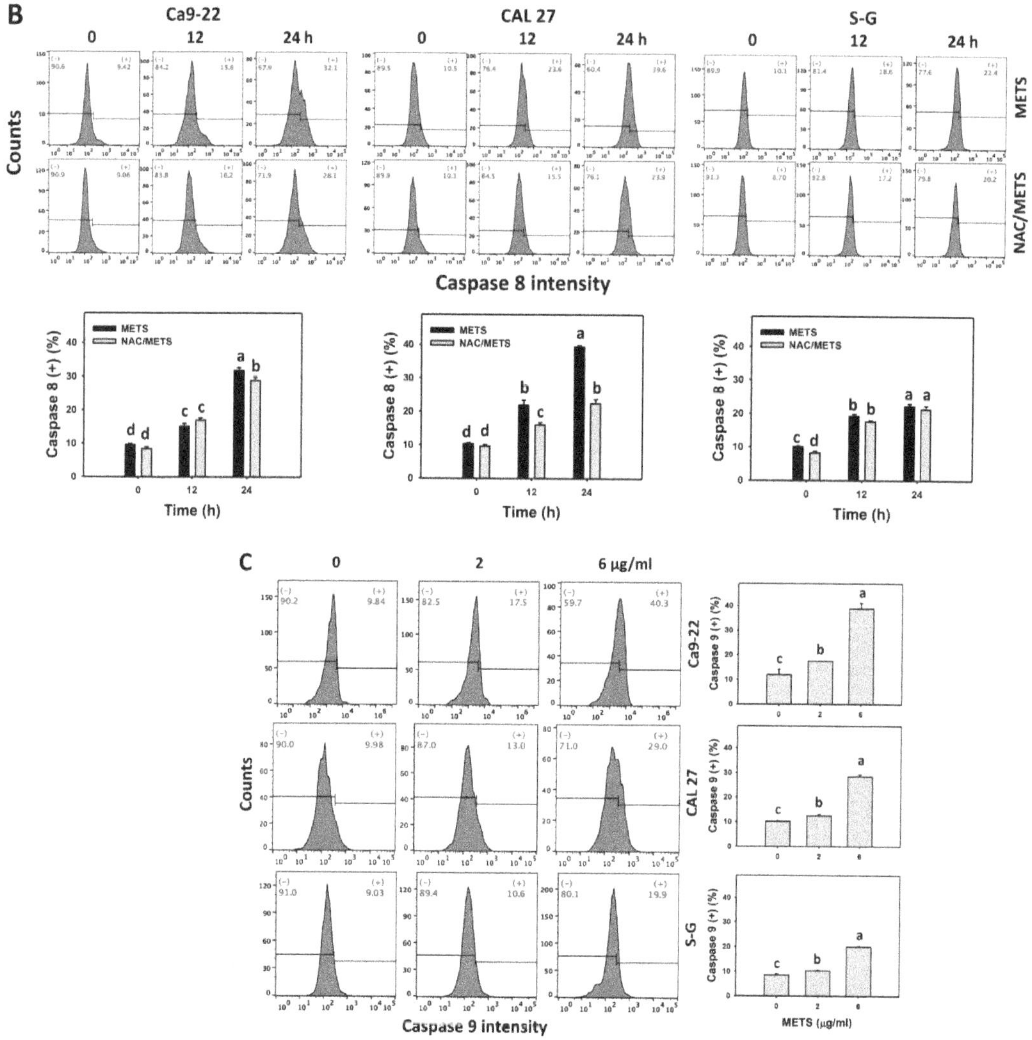

Figure 6. *Cont.*

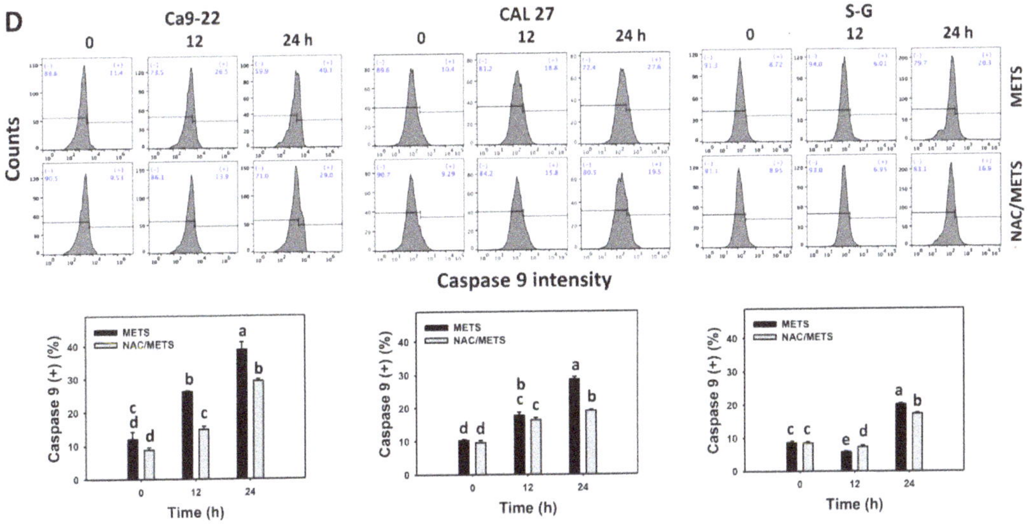

Figure 6. METS leads to Cas 8 and 9 activations in oral cancer cells. (**A**,**C**) Caspases 8 and 9 histogram changes. Cells were treated with METS for 24 h. Oral cancer (Ca9-22 and CAL 27) and normal cells (S-G) were included. (+) indicates a high intensity of caspases 8 and 9. (**B**,**D**). NAC response on caspases 8 and 9 activations. Untreated control and NAC pretreatment were followed by posttreatment of vehicle control and METS (2 and 6 μg/mL) for 0, 12, and 24 h, i.e., control, NAC, METS, and NAC/METS, respectively. Data = mean ± SD ($n = 3$). Data showing non-overlapping lower-case letters show significant results ($p < 0.05$) as analyzed by the one-way ANOVA analysis and post hoc test.

3.7. METS Causes More ROS and MitoSOX but Less GSH Generations to Oral Cancer Cells than Normal Cells

The ROS and MitoSOX (+) events of histograms were used to measure oxidative stress. In general, ROS and MitoSOX (+) events were dose- and time-dependently increased by METS in oral cancer cells (Ca9-22 and CAL 27) (Figure 7). However, it showed lower ROS and MitoSOX (+) events in normal S-G cells than in oral cancer cells by METS treatment. Accordingly, METS exerts more ROS and MitoSOX in oral cancer cells than in normal cells. Additionally, NAC inhibited ROS and MitoSOX increment of METS-treated oral cancer cells (Figure 7B), revealing that oxidative stress participated in METS-triggered ROS and MitoSOX generation.

Moreover, GSH (+) events were decreased by METS in oral cancer cells (Ca9-22 and CAL 27) in dose and time course experiments (Figure 8). However, it showed higher GSH (+) events in normal S-G cells than in oral cancer cells by METS treatment. Accordingly, METS exerts less GSH (+) in oral cancer cells than in normal cells. Additionally, NAC inhibited GSH (+) depletion of METS-treated oral cancer cells (Figure 8B), revealing that oxidative stress participated in METS-triggered GSH depletion.

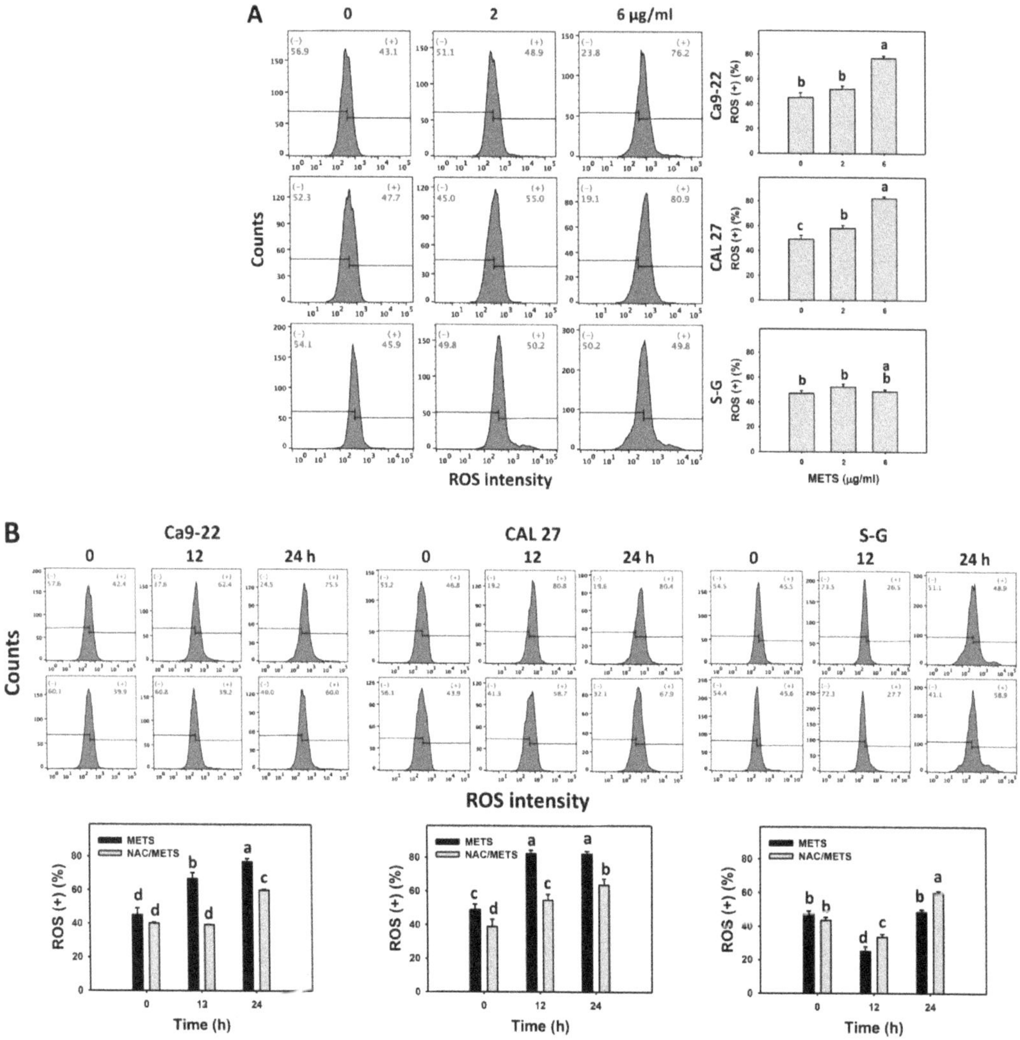

Figure 7. *Cont.*

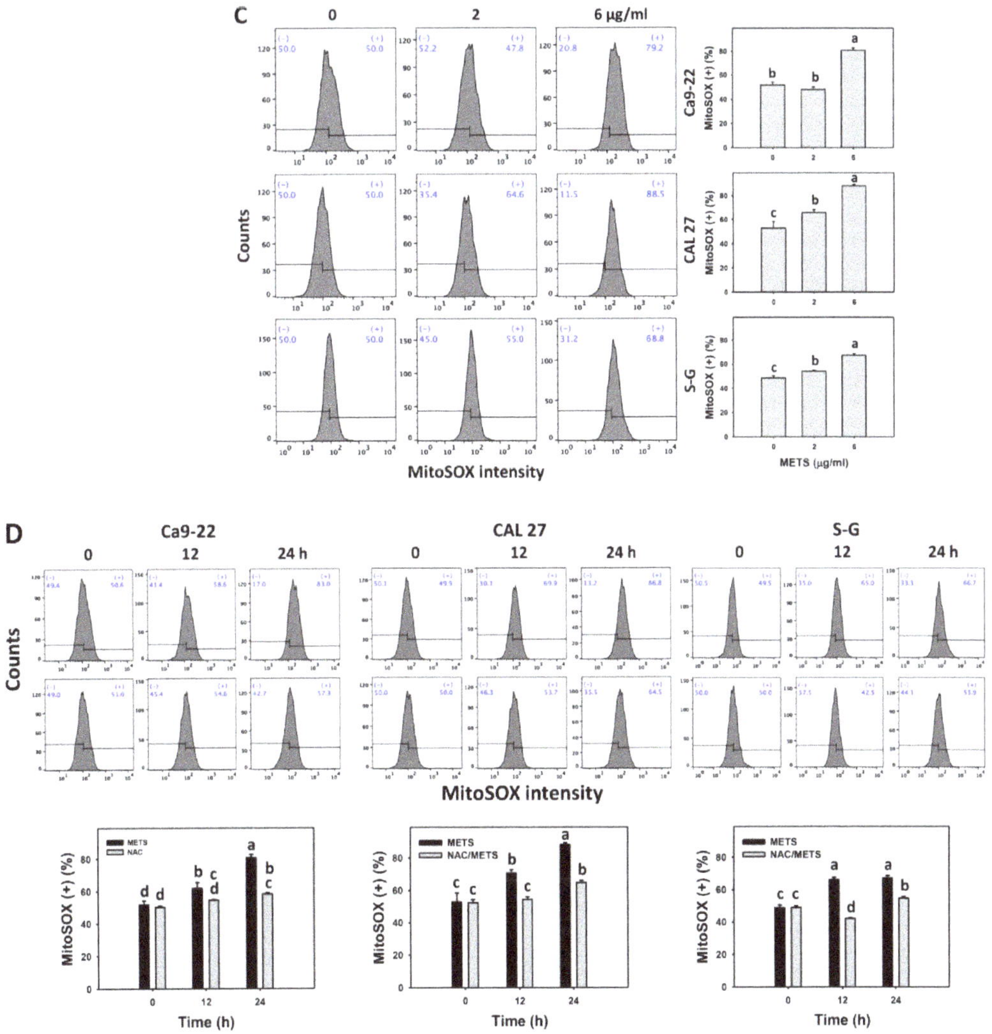

Figure 7. METS improves ROS and MitoSOX levels of oral cancer cells. (**A**,**C**) ROS and MitoSOX histogram changes. Cells were treated with METS for 24 h. Oral cancer and normal cells (S-G) were included. (+) indicates the high intensity of ROS and MitoSOX. (**B**,**D**) NAC response on ROS and MitoSOX levels. Untreated control and NAC pretreatment were followed by posttreatment of vehicle control and METS (6 μg/mL) for 0, 12, and 24 h, i.e., control, NAC, METS, and NAC/METS, respectively. Data = mean ± SD ($n = 3$). Data showing non-overlapping lower-case letters provide significant results ($p < 0.05$) as analyzed by the one-way ANOVA analysis and post hoc test.

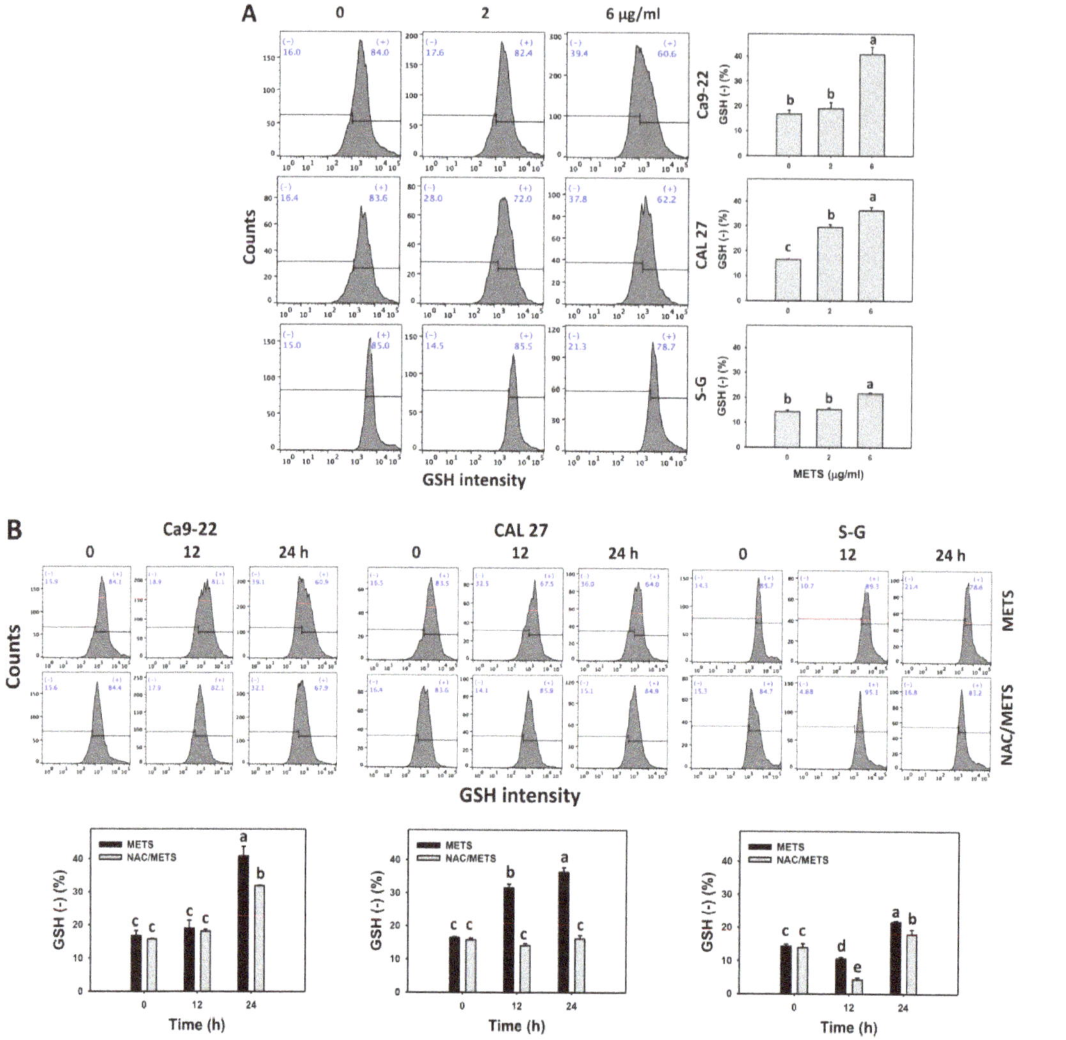

Figure 8. METS depletes the GSH level of oral cancer cells. (**A**) GSH histogram changes. Cells were treated with METS for 24 h. Oral cancer and normal cells (S-G) were included. (+) indicates the high intensity of GSH. (**B**) NAC response on GSH levels. Untreated control and NAC pretreatment were followed by posttreatment of vehicle control and METS (6 μg/mL) for 0, 12, and 24 h, i.e., control, NAC, METS, and NAC/METS, respectively. Data = mean ± SD (n = 3). Data marked showing non-overlapping lower-case letters have significant results ($p < 0.05$) as analyzed by the one-way ANOVA analysis and post hoc test.

3.8. METS Causes More DNA Damage to Oral Cancer Cells than Normal Cells

The γH2AX and 8-OHdG (+) events of histograms were used to measure DNA damage. γH2AX and 8-OHdG (+) events were dose- and time-dependently increased by METS in oral cancer cells (Ca9-22 and CAL 27) (Figures 9A and 10A). However, it showed lower γH2AX and 8-OHdG (+) events in normal S-G cells than in oral cancer cells by METS treatment. Accordingly, METS exerts more γH2AX and 8-OHdG in oral cancer cells than in normal cells. Additionally, NAC inhibited γH2AX and 8-OHdG increment of METS-treated oral cancer cells (Figures 9B and 10B), particularly for 24 h. These results reveal that oxidative stress was involved in METS-triggered γH2AX and 8-OHdG generation.

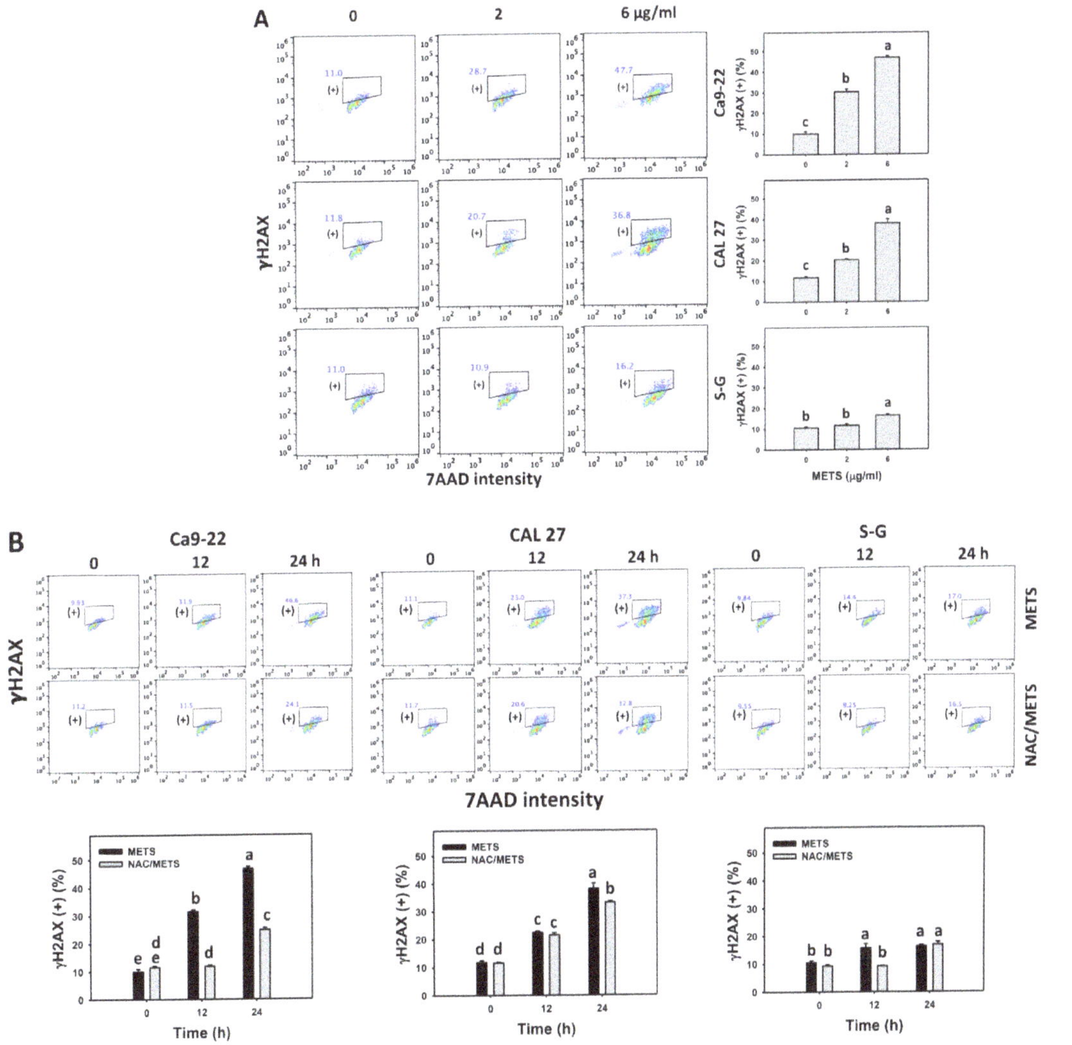

Figure 9. METS enhances the γH2AX level of oral cancer cells. (**A**) γH2AX histogram changes. Cells were treated with METS for 24 h. Oral cancer and normal cells (S-G) were included. (+) indicates the high intensity of γH2AX. (**B**) NAC response on γH2AX levels. Untreated control and NAC pretreatment were followed by posttreatment of vehicle control and METS (6 μg/mL) for 0, 12, and 24 h, i.e., control, NAC, METS, and NAC/METS, respectively. Data = mean ± SD ($n = 3$). Data showing non-overlapping lower-case letters have significant results ($p < 0.05$) as analyzed by the one-way ANOVA analysis and post hoc test.

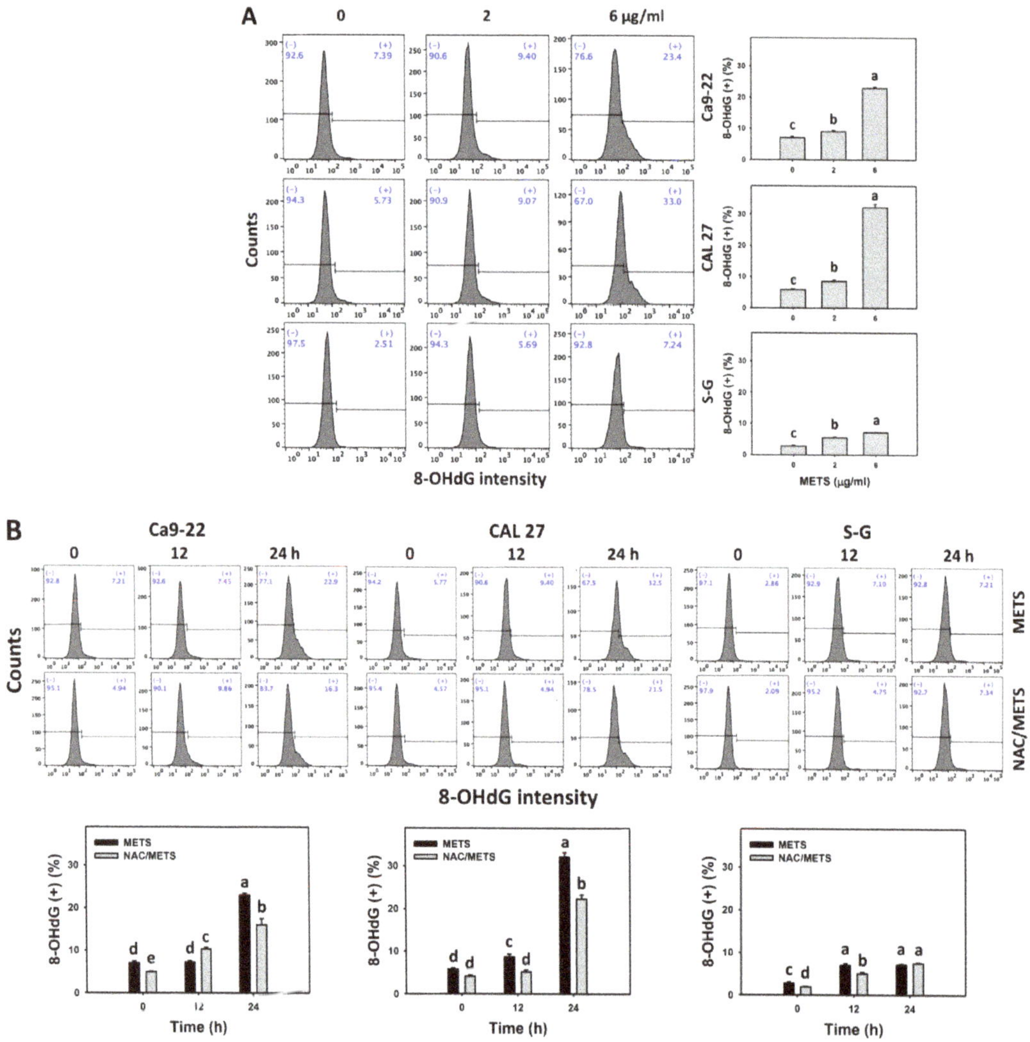

Figure 10. METS enhances the 8-OHdG level of oral cancer cells. (**A**) 8-OHdG histogram changes. Cells were treated with METS for 24 h. Oral cancer and normal cells (S-G) were included. (+) indicates the high intensity of 8-OHdG. (**B**) NAC response on 8-OHdG levels. Untreated control and NAC pretreatment were followed by posttreatment of vehicle control and METS (6 μg/mL) for 0, 12, and 24 h, i.e., control, NAC, METS, and NAC/METS, respectively. Data = mean ± SD ($n = 3$). Data showing non-overlapping lower-case letters have significant results ($p < 0.05$) as analyzed by the one-way ANOVA analysis and post hoc test.

4. Discussion

Several kinds of marine sponge extracts have demonstrated anticancer [43] effects. However, the impacts of antiproliferation of the marine sponge *Theonella* extract (METS) on oral cancer cells were rarely investigated. Moreover, most marine sponge extract studies reported only on the cytotoxicity of some cancer cells but did not consider the drug response to normal cells. These studies for marine sponge extracts lack a detailed investigation of their acting mechanism. The present study demonstrated the preferential antiproliferation to oral cancer cells but only revealed the low cytotoxic effect on normal cells. The exact impacts of METS treatments on oral cancer cells were discussed.

4.1. Comparison of Antiproliferation Effects of Several Marine Sponge Extracts on Cancer Cells

Several marine sponge extracts reported antiproliferation against cancer cells. For example, methanol/CH_2Cl extracts of two kinds of *Theonella* sp. show IC_{50} values of 4 and 10 μg/mL for liver cancer cells (Hep3B) at the 24 h CCK-8 assay [44]. Methanol/CH_2Cl extracts of *Theonella* sp. and *Haliclona* sp. show IC_{50} values of 0.5 and 1 μg/mL for colon cancer cells (HT29) at the 72 h CCK-8 assay [44]. The 70% ethanol extract of *Dysidea avara* shows IC_{50} values of 7.17, 9.55, 20.88, and 25.15 μg/mL for leukemia (K562), myeloma (KMS-12PE and A375), and lung (K549) cancer cells at a 48 h MTT assay [45].

In the present assessment, METS shows IC_{50} values of 4.5 and 5 μg/mL in oral cancer cells (Ca9-22 and CAL 27) at a 24 h MTS assay. In contrast, normal cells (S-G) only show no change in viability. Moreover, this antiproliferation was alleviated by ROS inhibitor (NAC) pretreatment. Accordingly, these results suggest that METS possesses ROS-dependent preferential antiproliferation to oral cancer cells but causes little cell death in normal cells. Since METS is the crude extract from the marine sponge, it shows a high drug sensitivity (IC_{50} 4.5 μg/mL) for oral cancer treatment; however, the in vivo antioral cancer effects of METS were not investigated. The in vivo effect of METS warrants a detailed evaluation that may improve the drug progression of its clinical treatment to oral cancer. Moreover, combining natural products with clinical drugs can improve oral cancer therapy [46,47]. Combined treatment may sensitize cancer cells to clinical medications and reduce their potential adverse effects. It warrants a detailed assessment of future combined treatment, including METS and other clinical drugs.

4.2. METS Exhibits Preferential Generation of Oxidative Stress to Oral Cancer Cells

Cancer cells exhibit a higher extent of basal oxidative stress than normal cells [42,48–51]. Elevated oxidative stress may overload cancer cells and cause cell death but tolerate normal cells to maintain cell survival. Similarly, METS also demonstrated oxidative stress inductions, such as upregulation of ROS and MitoSOX in oral cancer cells more than in normal cells (Figure 7). These oxidative stress inductions were suppressed by NAC, suggesting that MET is an oxidative modulating agent. Accordingly, METS showed preferential induction of oxidative stress in oral cancer but not in normal cells.

Moreover, redox homeostasis is balanced by cellular antioxidants and oxidative stress [52,53]. GSH can alleviate cellular oxidative stress [54]. Some drugs suppress endogenous antioxidants and induce oxidative stress. For instance, amygdalin decreases the GSH level and is associated with oxidative stress induction of breast cancer cells [55]. Similarly, METS downregulate GSH levels in oral cancer cells (Figure 8), accompanied by oxidative stress generation (Figure 7). After METS treatment, the GSH level of oral cancer cells is lower than normal cells. Therefore, METS-induced oxidative stress may partly be attributed to GSH depletion.

4.3. METS Preferentially Provokes Apoptosis in Oral Cancer Cells

Oxidative stress elevation is an anticancer strategy promoting apoptosis [49,56]. Marine sponge extracts promote apoptosis in cancer cells [57]. For example, petroleum ether extract of *Negombata magnifica* induces apoptosis in terms of DNA ladder assay in liver cancer cells [57]. Ethyl acetate extract of *Hyalella cribriformis* activates caspase 3 in rhabdomyosarcoma cells [58].

Similarly, METS causes a dramatic elevation of subG1 (Figure 3), and it also was associated with apoptosis inductions as detected by annexin V staining (Figure 4). Moreover, apoptosis-related caspase signaling, such as caspases 8, 9, and 3, were activated by METS of oral cancer cells but displayed a small activation in normal cells. These results indicate that METS preferentially causes apoptosis in oral cancer cells compared to normal cells. Moreover, it also suggests that METS activates extrinsic and intrinsic apoptosis in oral cancer cells since caspases 8 and 9 are activated. Furthermore, these apoptosis inductions were suppressed by NAC (Figure 6). Hence, METS promotes apoptosis of oral cancer cells relying on oxidative stress.

4.4. METS Preferentially Provokes DNA Damage to Oral Cancer Cells

Several natural products use the anticancer strategy by targeting DNA damage responses, such as γH2AX expression [59]. 8-OHdG is a typical oxidative-stress-dependent DNA damage [60]. Since METS induces oxidative stress, the DNA damage status warrants a detailed assessment. Using flow cytometry, the γH2AX and 8-OHdG DNA damage were caused after METS treatment for oral cancer, but they showed lower levels in normal cells (Figures 9 and 10). Moreover, these DNA damage inductions were suppressed by NAC. Accordingly, METS preferentially provokes ROS-dependent DNA damage to oral cancer cells, solidly impacting preferential antiproliferation.

4.5. METS Preferentially Arrests the Cell Cycle in Oral Cancer Cells

Marine sponge extracts disturb cell cycle progression in cancer cells. For example, petroleum ether extract of *Negombata magnifica* induces G1 arrest in liver cancer cells [57]. Methanol extract of *Crambe crambe* causes G2/M arrest in prostate cancer cells [61]. In addition to subG1 accumulation, METS caused G1 blocking of oral cancer Ca9-22 cells and G2/M blocking of oral cancer CAL 27 and normal S-G cells. Consequently, METS induced different responses to cell cycle disturbance in different oral cancer cell lines. Therefore, different marine sponge extracts may block the cell cycle progression at different phases in different cancer cells.

5. Conclusions

This study confirms that the antiproliferation effects of METS are effective in oral cancer cells but do not affect normal cells. This METS-induced preferential antiproliferation effects is associated with higher expressions of cellular and mitochondrial oxidative stress (ROS and mitoSOX), apoptosis (subG1 accumulation, annexin V enhancement, and upregulation of the extrinsic and intrinsic caspase signaling), and DNA damage (DNA double-strand breaks and oxidative damage) in breast cancer cells than in normal cells. Utilizing NAC pretreatment demonstrates that preferential antiproliferation function and mechanism of METS provides an oxidative-stress-mediated regulation. In conclusion, METS is a promising marine-sponge-derived natural product for antioral cancer treatment.

Author Contributions: Conceptualization, J.-P.S., Y.-T.C., Y.-B.C. and H.-W.C.; data curation, Y.-T.C.; formal analysis, Y.-T.C.; methodology, J.-P.S., J.-Y.T., S.-R.C., M.-F.H. and J.-H.J.; supervision, Y.-B.C. and H.-W.C.; writing—original draft, J.-P.S. and H.-W.C.; writing—review and editing, Y.-B.C. and H.-W.C. All authors have read and agreed to the published version of the manuscript.

Funding: This work was partly supported by funds from the Ministry of Science and Technology (MOST 111-2320-B-037-015-MY3; MOST-108-2320-B-110-009-MY3; MOST 110-2314-B-037-074-MY3), the Kaohsiung Medical University (KMU-DK(A)111008), and the Kaohsiung Medical University Research Center (KMU-TC108A04).

Institutional Review Board Statement: Not applicable.

Informed Consent Statement: Not applicable.

Data Availability Statement: Data are contained within the article.

Acknowledgments: The authors thank our colleague Hans-Uwe Dahms for editing the manuscript.

Conflicts of Interest: The authors declare that there are no conflict of interest among them.

References

1. Petersen, P.E. Oral cancer prevention and control–the approach of the World Health Organization. *Oral Oncol.* **2009**, *45*, 454–460. [CrossRef] [PubMed]
2. Montero, P.H.; Patel, S.G. Cancer of the oral cavity. *Surg. Oncol. Clin. North Am.* **2015**, *24*, 491–508. [CrossRef]

3. Silverman, S., Jr. Oral cancer: Complications of therapy. *Oral Surg. Oral Med. Oral Pathol. Oral Radiol. Endod.* **1999**, *88*, 122–126. [CrossRef]
4. Singh, R.; Sharma, M.; Joshi, P.; Rawat, D.S. Clinical status of anti-cancer agents derived from marine sources. *Anticancer. Agents Med. Chem.* **2008**, *8*, 603–617. [CrossRef] [PubMed]
5. Sithranga Boopathy, N.; Kathiresan, K. Anticancer drugs from marine flora: An overview. *J. Oncol.* **2010**, *2010*, 214186. [CrossRef] [PubMed]
6. Farooqi, A.A.; Fayyaz, S.; Hou, M.F.; Li, K.T.; Tang, J.Y.; Chang, H.W. Reactive oxygen species and autophagy modulation in non-marine drugs and marine drugs. *Mar. Drugs* **2014**, *12*, 5408–5424. [CrossRef]
7. Lee, M.G.; Liu, Y.C.; Lee, Y.L.; El-Shazly, M.; Lai, K.H.; Shih, S.P.; Ke, S.C.; Hong, M.C.; Du, Y.C.; Yang, J.C.; et al. Heteronemin, a marine sesterterpenoid-type metabolite, induces apoptosis in prostate LNcap cells via oxidative and ER stress combined with the inhibition of topoisomerase II and Hsp90. *Mar. Drugs* **2018**, *16*, 204. [CrossRef] [PubMed]
8. Mehbub, M.F.; Lei, J.; Franco, C.; Zhang, W. Marine sponge derived natural products between 2001 and 2010: Trends and opportunities for discovery of bioactives. *Mar. Drugs* **2014**, *12*, 4539–4577. [CrossRef]
9. Mehbub, M.F.; Perkins, M.V.; Zhang, W.; Franco, C.M.M. New marine natural products from sponges (Porifera) of the order Dictyoceratida (2001 to 2012); a promising source for drug discovery, exploration and future prospects. *Biotechnol. Adv.* **2016**, *34*, 473–491. [CrossRef] [PubMed]
10. Calcabrini, C.; Catanzaro, E.; Bishayee, A.; Turrini, E.; Fimognari, C. Marine sponge natural products with anticancer potential: An updated review. *Mar. Drugs* **2017**, *15*, 310. [CrossRef] [PubMed]
11. Varijakzhan, D.; Loh, J.Y.; Yap, W.S.; Yusoff, K.; Seboussi, R.; Lim, S.E.; Lai, K.S.; Chong, C.M. Bioactive compounds from marine sponges: Fundamentals and applications. *Mar. Drugs* **2021**, *19*, 246. [CrossRef]
12. Perdicaris, S.; Vlachogianni, T.; Valavanidis, A. Bioactive natural substances from marine sponges: New developments and prospects for future pharmaceuticals. *Nat. Prod. Chem. Res.* **2013**, *1*, 114. [CrossRef]
13. Warsidah, M.; Sofiana, M.S.J.; Safitri, I.; Sapar, A.; Aritonang, A.B.; Muttalib, Y.; Fadly, D. Protein isolation from sponge *Niphates* sp. as an antibacterial and antioxidant. *Syst. Rev. Pharm.* **2020**, *11*, 518–521.
14. Kumar, M.S.; Vijayalaxmi, K.; Pal, A. Antiinflamatuvar and antioxidant properties of *Spongosorites halichondriodes*, a marine sponge. *Turk. J. Pharm. Sci.* **2014**, *11*, 285–294.
15. De Marino, S.; Ummarino, R.; D'Auria, M.V.; Chini, M.G.; Bifulco, G.; Renga, B.; D'Amore, C.; Fiorucci, S.; Debitus, C.; Zampella, A. Theonellasterols and conicasterols from *Theonella swinhoei*. Novel marine natural ligands for human nuclear receptors. *J. Med. Chem.* **2011**, *54*, 3065–3075. [CrossRef] [PubMed]
16. Oiki, S.; Muramatsu, I.; Matsunaga, S.; Fusetani, N. A channel-forming peptide toxin: Polytheonamide from marine sponge (*Theonella swinhoei*). *Nihon Yakurigaku Zasshi* **1997**, *110* (Suppl. 1), 195P–198P. [CrossRef]
17. De Marino, S.; Festa, C.; D'Auria, M.V.; Cresteil, T.; Debitus, C.; Zampella, A. Swinholide J, a potent cytotoxin from the marine sponge *Theonella swinhoei*. *Mar. Drugs* **2011**, *9*, 1133–1141. [CrossRef]
18. Sinisi, A.; Calcinai, B.; Cerrano, C.; Dien, H.A.; Zampella, A.; D'Amore, C.; Renga, B.; Fiorucci, S.; Taglialatela-Scafati, O. New tridecapeptides of the theonellapeptolide family from the Indonesian sponge *Theonella swinhoei*. *Beilstein J. Org. Chem.* **2013**, *9*, 1643–1651. [CrossRef]
19. Youssef, D.T.; Shaala, L.A.; Mohamed, G.A.; Badr, J.M.; Bamanie, F.H.; Ibrahim, S.R. Theonellamide G, a potent antifungal and cytotoxic bicyclic glycopeptide from the Red Sea marine sponge *Theonella swinhoei*. *Mar. Drugs* **2014**, *12*, 1911–1923. [CrossRef] [PubMed]
20. Fukuhara, K.; Takada, K.; Okada, S.; Matsunaga, S. Nazumazoles A-C, cyclic pentapeptides dimerized through a disulfide bond from the marine sponge *Theonella swinhoei*. *Org. Lett.* **2015**, *17*, 2646–2648. [CrossRef] [PubMed]
21. Issac, M.; Aknin, M.; Gauvin-Bialecki, A.; De Voogd, N.; Ledoux, A.; Frederich, M.; Kashman, Y.; Carmeli, S. Cyclotheonellazoles A-C, potent protease inhibitors from the marine sponge *Theonella* aff. *swinhoei*. *J. Nat. Prod.* **2017**, *80*, 1110–1116. [CrossRef] [PubMed]
22. Yang, F.; Li, Y.Y.; Tang, J.; Sun, F.; Lin, H.W. New 4-methylidene sterols from the marine sponge *Theonella swinhoei*. *Fitoterapia* **2018**, *127*, 279–285. [CrossRef] [PubMed]
23. Lai, K.H.; Peng, B.R.; Su, C.H.; El-Shazly, M.; Sun, Y.L.; Shih, M.C.; Huang, Y.T.; Yen, P.T.; Wang, L.S.; Su, J.H. Anti-proliferative potential of secondary metabolites from the marine sponge *Theonella* sp.: Moving from correlation toward causation. *Metabolites* **2021**, *11*, 532. [CrossRef] [PubMed]
24. Kumar, M.S.; Adki, K.M. Marine natural products for multi-targeted cancer treatment: A future insight. *Biomed. Pharmacother.* **2018**, *105*, 233–245. [CrossRef] [PubMed]
25. Koeberle, A.; Werz, O. Multi-target approach for natural products in inflammation. *Drug Discov. Today* **2014**, *19*, 1871–1882. [CrossRef]
26. Chamberlin, S.R.; Blucher, A.; Wu, G.; Shinto, L.; Choonoo, G.; Kulesz-Martin, M.; McWeeney, S. Natural product target network reveals potential for cancer combination therapies. *Front. Pharmacol.* **2019**, *10*, 557. [CrossRef]
27. Mioso, R.; Marante, F.J.; Bezerra, R.S.; Borges, F.V.; Santos, B.V.; Laguna, I.H. Cytotoxic compounds derived from marine sponges. A review (2010–2012). *Molecules* **2017**, *22*, 208. [CrossRef] [PubMed]
28. Kasten, F.H.; Pineda, L.F.; Schneider, P.E.; Rawls, H.R.; Foster, T.A. Biocompatibility testing of an experimental fluoride releasing resin using human gingival epithelial cells *in vitro*. *Vitro Cell Dev. Biol.* **1989**, *25*, 57–62. [CrossRef]

29. Kasten, F.H.; Soileau, K.; Meffert, R.M. Quantitative evaluation of human gingival epithelial cell attachment to implant surfaces *in vitro*. *Int. J. Periodontics Restor. Dent.* **1990**, *10*, 68–79.
30. Hsieh, P.L.; Liao, Y.W.; Hsieh, C.W.; Chen, P.N.; Yu, C.C. Soy isoflavone genistein impedes cancer stemness and mesenchymal transition in head and neck cancer through activating miR-34a/RTCB axis. *Nutrients* **2020**, *12*, 1924. [CrossRef] [PubMed]
31. Wang, H.R.; Tang, J.Y.; Wang, Y.Y.; Farooqi, A.A.; Yen, C.Y.; Yuan, S.F.; Huang, H.W.; Chang, H.W. Manoalide preferentially provides antiproliferation of oral cancer cells by oxidative stress-mediated apoptosis and DNA damage. *Cancers* **2019**, *11*, 1303. [CrossRef]
32. Hung, J.H.; Chen, C.Y.; Omar, H.A.; Huang, K.Y.; Tsao, C.C.; Chiu, C.C.; Chen, Y.L.; Chen, P.H.; Teng, Y.N. Reactive oxygen species mediate Terbufos-induced apoptosis in mouse testicular cell lines via the modulation of cell cycle and pro-apoptotic proteins. *Environ. Toxicol.* **2016**, *31*, 1888–1898. [CrossRef]
33. Huang, C.H.; Yeh, J.M.; Chan, W.H. Hazardous impacts of silver nanoparticles on mouse oocyte maturation and fertilization and fetal development through induction of apoptotic processes. *Environ. Toxicol.* **2018**, *33*, 1039–1049. [CrossRef] [PubMed]
34. Wu, C.F.; Lee, M.G.; El-Shazly, M.; Lai, K.H.; Ke, S.C.; Su, C.W.; Shih, S.P.; Sung, P.J.; Hong, M.C.; Wen, Z.H.; et al. Isoaaptamine induces T-47D cells apoptosis and autophagy via oxidative stress. *Mar. Drugs* **2018**, *16*, 18. [CrossRef] [PubMed]
35. Yeh, C.C.; Tseng, C.N.; Yang, J.I.; Huang, H.W.; Fang, Y.; Tang, J.Y.; Chang, F.R.; Chang, H.W. Antiproliferation and induction of apoptosis in Ca9-22 oral cancer cells by ethanolic extract of *Gracilaria tenuistipitata*. *Molecules* **2012**, *17*, 10916–10927. [CrossRef] [PubMed]
36. Vignon, C.; Debeissat, C.; Georget, M.T.; Bouscary, D.; Gyan, E.; Rosset, P.; Herault, O. Flow cytometric quantification of all phases of the cell cycle and apoptosis in a two-color fluorescence plot. *PLoS ONE* **2013**, *8*, e68425. [CrossRef]
37. Fan, H.C.; Hsieh, Y.C.; Li, L.H.; Chang, C.C.; Janouskova, K.; Ramani, M.V.; Subbaraju, G.V.; Cheng, K.T.; Chang, C.C. Dehydroxyhispolon methyl ether, a hispolon derivative, inhibits WNT/beta-catenin signaling to elicit human colorectal carcinoma cell apoptosis. *Int. J. Mol. Sci.* **2020**, *21*, 8839. [CrossRef]
38. Liu, W.; Lin, L.C.; Wang, P.J.; Chen, Y.N.; Wang, S.C.; Chuang, Y.T.; Tsai, I.H.; Yu, S.Y.; Chang, F.R.; Cheng, Y.B.; et al. Nepenthes ethyl acetate extract provides oxidative stress-dependent anti-leukemia effects. *Antioxidants* **2021**, *10*, 1410. [CrossRef]
39. Lin, C.H.; Chan, H.S.; Tsay, H.S.; Funayama, S.; Kuo, C.L.; Chung, J.G. Ethyl acetate fraction from methanol extraction of *Vitis thunbergii* var. taiwaniana induced G0/G1 phase arrest via inhibition of cyclins D and E and induction of apoptosis through caspase-dependent and -independent pathways in human prostate carcinoma DU145 cells. *Environ. Toxicol.* **2018**, *33*, 41–51.
40. Liu, S.L.; Yang, K.H.; Yang, C.W.; Lee, M.Y.; Chuang, Y.T.; Chen, Y.N.; Chang, F.R.; Chen, C.Y.; Chang, H.W. Burmannic acid inhibits proliferation and induces oxidative stress response of oral cancer cells. *Antioxidants* **2021**, *10*, 1588. [CrossRef]
41. Shiau, J.P.; Chuang, Y.T.; Yang, K.H.; Chang, F.R.; Sheu, J.H.; Hou, M.F.; Jeng, J.H.; Tang, J.Y.; Chang, H.W. Brown algae-derived fucoidan exerts oxidative stress-dependent antiproliferation on oral cancer cells. *Antioxidants* **2022**, *11*, 841. [CrossRef] [PubMed]
42. Chiu, C.C.; Huang, J.W.; Chang, F.R.; Huang, K.J.; Huang, H.M.; Huang, H.W.; Chou, C.K.; Wu, Y.C.; Chang, H.W. Golden berry-derived 4beta-hydroxywithanolide E for selectively killing oral cancer cells by generating ROS, DNA damage, and apoptotic pathways. *PLoS ONE* **2013**, *8*, e64739. [CrossRef] [PubMed]
43. El-Damhougy, K.; El-Naggar, H.A.; Ibrahim, H.; Bashar, M.A.; Abou-Senna, F.M. Biological activities of some marine sponge extracts from Aqaba Gulf, Red Sea, Egypt. *Int. J. Fish. Aquat. Stud.* **2017**, *5*, 652–659.
44. Choi, C.; Son, A.; Lee, H.S.; Lee, Y.J.; Park, H.C. Radiosensitization by marine sponge *Agelas* sp. extracts in hepatocellular carcinoma cells with autophagy induction. *Sci. Rep.* **2018**, *8*, 6317. [CrossRef] [PubMed]
45. Ciftci, H.I.; Can, M.; Ellakwa, D.E.; Suner, S.C.; Ibrahim, M.A.; Oral, A.; Sekeroglu, N.; Ozalp, B.; Otsuka, M.; Fujita, M.; et al. Anticancer activity of Turkish marine extracts: A purple sponge extract induces apoptosis with multitarget kinase inhibition activity. *Invest New Drugs* **2020**, *38*, 1326–1333. [CrossRef]
46. Lin, A. Radiation therapy for oral cavity and oropharyngeal cancers. *Dent. Clin. North Am.* **2018**, *62*, 99–109. [CrossRef]
47. Hartner, L. Chemotherapy for oral cancer. *Dent. Clin. North Am.* **2018**, *62*, 87–97. [CrossRef]
48. Acharya, A.; Das, I.; Chandhok, D.; Saha, T. Redox regulation in cancer: A double-edged sword with therapeutic potential. *Oxid. Med. Cell Longev.* **2010**, *3*, 23–34. [CrossRef]
49. Tang, J.Y.; Ou-Yang, F.; Hou, M.F.; Huang, H.W.; Wang, H.R.; Li, K.T.; Fayyaz, S.; Shu, C.W.; Chang, H.W. Oxidative stress-modulating drugs have preferential anticancer effects—Involving the regulation of apoptosis, DNA damage, endoplasmic reticulum stress, autophagy, metabolism, and migration. *Semin. Cancer Biol.* **2019**, *58*, 109–117. [CrossRef]
50. Gorrini, C.; Harris, I.S.; Mak, T.W. Modulation of oxidative stress as an anticancer strategy. *Nat. Rev. Drug Discov.* **2013**, *12*, 931–947. [CrossRef]
51. Kim, S.J.; Kim, H.S.; Seo, Y.R. Understanding of ROS-inducing strategy in anticancer therapy. *Oxid Med. Cell Longev.* **2019**, *2019*, 5381692. [CrossRef] [PubMed]
52. Sies, H. Oxidative stress: Eustress and distress in redox homeostasis. In *Stress: Physiology, Biochemistry, and Pathology*; Elsevier: Amsterdam, The Netherlands, 2019; pp. 153–163.
53. Ahmad, T.; Suzuki, Y.J. Juglone in oxidative stress and cell signaling. *Antioxidants* **2019**, *8*, 91. [CrossRef] [PubMed]
54. Li, R.; Huang, C.; Ho, J.C.H.; Leung, C.C.T.; Kong, R.Y.C.; Li, Y.; Liang, X.; Lai, K.P.; Tse, W.K.F. The use of glutathione to reduce oxidative stress status and its potential for modifying the extracellular matrix organization in cleft lip. *Free Radic. Biol. Med.* **2021**, *164*, 130–138. [CrossRef] [PubMed]

55. Abboud, M.M.; Al Awaida, W.; Alkhateeb, H.H.; Abu-Ayyad, A.N. Antitumor action of amygdalin on human breast cancer cells by selective sensitization to oxidative stress. *Nutr. Cancer* **2019**, *71*, 483–490. [CrossRef]
56. Zou, Z.; Chang, H.; Li, H.; Wang, S. Induction of reactive oxygen species: An emerging approach for cancer therapy. *Apoptosis* **2017**, *22*, 1321–1335. [CrossRef]
57. Rady, H.M.; Hassan, A.Z.; Salem, S.M.; Mohamed, T.K.; Esmaiel, N.N.; Ez-El-Arab, M.A.; Ibrahim, M.A.; Fouda, F.K. Induction of apoptosis and cell cycle arrest by *Negombata magnifica* sponge in hepatocellular carcinoma. *Med. Chem. Res.* **2016**, *25*, 456–465. [CrossRef]
58. Annamalai, P.; Thayman, M.; Rajan, S.; Raman, L.S.; Ramasubbu, S.; Perumal, P. Ethyl acetate extract from marine sponge *Hyattella cribriformis* exhibit potent anticancer activity by promoting tubulin polymerization as evidenced mitotic arrest and induction of apoptosis. *Pharmacogn. Mag.* **2015**, *11*, 345.
59. Van Stuijvenberg, J.; Proksch, P.; Fritz, G. Targeting the DNA damage response (DDR) by natural compounds. *Bioorg. Med. Chem.* **2020**, *28*, 115279. [CrossRef]
60. Arfin, S.; Jha, N.K.; Jha, S.K.; Kesari, K.K.; Ruokolainen, J.; Roychoudhury, S.; Rathi, B.; Kumar, D. Oxidative stress in cancer cell metabolism. *Antioxidants* **2021**, *10*, 642. [CrossRef]
61. Ottinger, S.; Kloppel, A.; Rausch, V.; Liu, L.; Kallifatidis, G.; Gross, W.; Gebhard, M.M.; Brummer, F.; Herr, I. Targeting of pancreatic and prostate cancer stem cell characteristics by *Crambe crambe* marine sponge extract. *Int. J. Cancer* **2012**, *130*, 1671–1681. [CrossRef]

antioxidants

MDPI

Article

Genetic Polymorphisms of MnSOD Modify the Impacts of Environmental Melamine on Oxidative Stress and Early Kidney Injury in Calcium Urolithiasis Patients

Chia-Chu Liu [1,2,3,4], Chia-Fang Wu [1,5], Yung-Chin Lee [2,3,6], Tsung-Yi Huang [2], Shih-Ting Huang [1,7], Hsun-Shuan Wang [6], Jhen-Hao Jhan [6], Shu-Pin Huang [2,3], Ching-Chia Li [2,3], Yung-Shun Juan [2,3], Tusty-Jiuan Hsieh [1,7], Yi-Chun Tsai [1,8,9], Chu-Chih Chen [1,10] and Ming-Tsang Wu [1,11,12,13,*]

1 Research Center for Environmental Medicine, Kaohsiung Medical University, Kaohsiung City 807, Taiwan; ccliu0204@gmail.com (C.-C.L.); cfwu27@nuu.edu.tw (C.-F.W.); u107800006@kmu.edu.tw (S.-T.H.); hsiehjun@kmu.edu.tw (T.-J.H.); 920254@kmuh.org.tw (Y.-C.T.); ccchen@nhri.edu.tw (C.-C.C.)
2 Department of Urology, Kaohsiung Medical University Hospital, Kaohsiung Medical University, Kaohsiung City 807, Taiwan; 890197@kmuh.org.tw (Y.-C.L.); 970417@kmuh.org.tw (T.-Y.H.); shpihu@kmu.edu.tw (S.-P.H.); 850144@kmuh.org.tw (C.-C.L.); 840066@kmuh.org.tw (Y.-S.J.)
3 Department of Urology, School of Medicine, College of Medicine, Kaohsiung Medical University, Kaohsiung City 807, Taiwan
4 Department of Urology, Pingtung Hospital, Ministry of Health and Welfare, Pingtung City 900, Taiwan
5 International Master Program of Translational Medicine, National United University, Miaoli 360, Taiwan
6 Department of Urology, Kaohsiung Municipal Siaogang Hospital, Kaohsiung City 812, Taiwan; 940199@kmuh.org.tw (H.-S.W.); 1030398@kmuh.org.tw (J.-H.J.)
7 Graduate Institute of Medicine, College of Medicine, Kaohsiung Medical University, Kaohsiung City 807, Taiwan
8 Department of Internal Medicine, Divisions of Nephrology and General Medicine, Kaohsiung Medical University Hospital, Kaohsiung Medical University, Kaohsiung City 807, Taiwan
9 Department of Internal Medicine, School of Medicine, College of Medicine, Kaohsiung Medical University, Kaohsiung City 807, Taiwan
10 Division of Biostatistics and Bioinformatics, Institute of Population Health Sciences, National Health Research Institutes, Miaoli 350, Taiwan
11 Environmental and Occupational Medicine and Graduate Institute of Clinical Medicine, Kaohsiung Medical University, Kaohsiung City 807, Taiwan
12 Department of Family Medicine, Kaohsiung Medical University Hospital, Kaohsiung Medical University, Kaohsiung City 807, Taiwan
13 Department of Public Health, College of Health Sciences, Kaohsiung Medical University, Kaohsiung City 807, Taiwan
* Correspondence: 960021@cc.kmuh.org.tw; Tel.: +886-7-3121101 (ext. 2315)

Citation: Liu, C.-C.; Wu, C.-F.; Lee, Y.-C.; Huang, T.-Y.; Huang, S.-T.; Wang, H.-S.; Jhan, J.-H.; Huang, S.-P.; Li, C.-C.; Juan, Y.-S.; et al. Genetic Polymorphisms of MnSOD Modify the Impacts of Environmental Melamine on Oxidative Stress and Early Kidney Injury in Calcium Urolithiasis Patients. *Antioxidants* **2022**, *11*, 152. https://doi.org/10.3390/antiox11010152

Academic Editor: Tim Hofer

Received: 15 December 2021
Accepted: 11 January 2022
Published: 13 January 2022

Abstract: Environmental melamine exposure increases the risks of oxidative stress and early kidney injury. Manganese superoxide dismutase (MnSOD), glutathione peroxidase, and catalase can protect the kidneys against oxidative stress and maintain normal function. We evaluated whether their single-nucleotide polymorphisms (SNPs) could modify melamine's effects. A total of 302 patients diagnosed with calcium urolithiasis were enrolled. All patients provided one-spot overnight urine samples to measure their melamine levels, urinary biomarkers of oxidative stress and renal tubular injury. Median values were used to dichotomize levels into high and low. Subjects carrying the T allele of rs4880 and high melamine levels had 3.60 times greater risk of high malondialdehyde levels than those carrying the C allele of rs4880 and low melamine levels after adjustment. Subjects carrying the G allele of rs5746136 and high melamine levels had 1.73 times greater risk of high *N*-Acetyl-β-D-glucosaminidase levels than those carrying the A allele of rs5746136 and low melamine levels. In conclusion, the SNPs of MnSOD, rs4880 and rs5746136, influence the risk of oxidative stress and renal tubular injury, respectively, in calcium urolithiasis patients. In the context of high urinary melamine levels, their effects on oxidative stress and renal tubular injury were further increased.

Keywords: manganese superoxide dismutase; genetic polymorphism; melamine; kidney injury; oxidative stress; calcium urolithiasis

1. Introduction

Melamine is a synthetic chemical used in manufacturing a variety of commercial daily life products including housewares, countertops, fabrics, glues, and flame retardants [1,2]. Because it has a high nitrogen content, it has been misused in animal feed and milk to deceptively elevate the protein content [1,2]. The adverse effects of melamine exposure drew worldwide attention after its scandalous addition to animal feeds, leading to the deaths of thousands of pet animals in the US in 2007, as well as to infant formula, resulting in urolithiasis in more than 50,000 children and six deaths in China in 2008 [1,2]. To date, melamine remains ubiquitously present in our environment. Several studies have detected melamine in water, soil, crops, daily food products, and animal tissues [3–5], and others detected it in most urine samples of several general populations from different countries [6–9].

After melamine intake, ninety percent of its original form will be excreted in urine within 24 h, so the kidneys may be more susceptible to melamine [1]. In addition to the effects of high doses of melamine on acute nephrotoxicity in children in 2008, long-term low-dose exposure to melamine has been linked to risk of kidney complications, including stone formation and deterioration of renal function in adults [10–12]. One probable mechanism that chronic low-dose melamine exposure could lead to early kidney injury and stone formation is its adverse effect on renal tubules, as was found by two of our human studies in melamine tableware workers [13] and adult patients with calcium urolithiasis [14]. We also conducted an in vitro study by using human renal proximal tubular HK-2 cells and found that melamine could induce renal tubular damage by increasing oxidative stress [15]. In two recent human studies, we also found that exposure to melamine increased the urinary biomarkers of oxidative stress, malondialdehyde (MDA), and 8-oxo-2′-deoxyguanosine (8-OHdG), with MDA mediating 36–53% of the total effect of melamine on a biomarker of renal tubular injury, *N*-Acetyl-β-D-glucosaminidase (NAG) [16]. Those findings suggest that exposure to low-dose environmental melamine might increase oxidative stress and further the risk of early kidney injury in humans.

Oxidative stress occurs when the generation of pro-oxidants or reactive oxygen species (ROS) exceeds the endogenous antioxidant capacity. Our kidneys are particularly sensitive to oxidative stress, which is thought to be an important factor in the initiation, development, and progression of most kidney diseases [17,18]. The cause of oxidative stress in kidneys may be associated with the interactions between medical diseases and environmental exposure to chemicals such as melamine [16]. Endogenous antioxidant systems including manganese superoxide dismutase (MnSOD), glutathione peroxidase (GPX1), and catalase (CAT) have been found to protect the kidneys against oxidative stress and subsequently help maintain normal function [18]. MnSOD can decompose toxic ROS, superoxide anions, to hydrogen peroxide, which is then converted to non-toxic water and oxygen by GPX1 and CAT in the mitochondria [18]. Several single-nucleotide polymorphisms (SNPs) of those antioxidant enzyme genes have been associated with various diseases [19], including kidney diseases [20–22].

Since environmental toxicant-associated kidney damage might be influenced by genetic factors [23], it is possible that the SNPs of antioxidant enzyme genes (e.g., *MnSOD*, *GPX1*, *CAT*) could modify the effects of environmental melamine exposure on the risk of oxidative stress and renal tubular injury in humans. To find out, we selected five candidate SNPs of antioxidant enzyme genes (*MnSOD:* rs4880 and rs5746136, *GPX1:* rs1800668, *CAT:* rs1001179 and rs769217), most of which have been linked to kidney diseases [20–22], to study each of their effects when combined with environmental melamine exposure on 8-OHdG and MDA, two biomarkers of oxidative stress, and NAG, a biomarker of renal tubular injury, in urine.

2. Materials and Methods

2.1. Subjects

In total, 309 patients diagnosed with upper urinary tract calcium urolithiasis were enrolled between November 2010 and January 2015. The detailed study designs and protocols for their inclusion were described previously [14,16]. In brief, all patients were from Kaohsiung Medical University-affiliated hospitals in southwestern Taiwan. The eligible patients were individuals aged $\geq$ 20 years who had been diagnosed with urolithiasis in the upper urinary tract by radiography or ultrasonography, and who had provided stone specimens confirmed to have calcium components by infrared spectroscopy analysis (Spectrum RX I Fourier Transform-Infrared System, PerkinElmer, Waltham, MA, USA). None of the participants was found by X-ray to have radiolucent stones or by clinical evaluation to have uric acid or cystine stones.

Patient subjects were excluded if they had a history of chronic urinary tract infection, chronic diarrhea, gout, hyperparathyroidism, renal tubular acidosis, renal failure, or cancer. Any subjects who had regularly taken vitamin D, calcium supplements, diuretics, or potassium citrate more than once per week within six months prior to the diagnosis of urolithiasis or interview were also excluded. The study protocol was approved by the Institutional Review Board of KMUH and all eligible patients provided signed informed written consent forms. This study followed the guidelines of STREGA [24] (see Supplementary Materials Table S1).

2.2. Collection of Clinical Data and Biological Samples

Upon admission, all participants provided blood and one-spot first-void morning urine samples after overnight fasting and before any treatment of urolithiasis for biochemical and genetic analyses. They also answered a structured questionnaire to collect their demographic data, medical history, and substance use (cigarette, alcohol, and betel quid) [14,16]. Participants were defined as cigarette smokers, alcohol drinkers, or betel quid chewers if they had regularly smoked $\geq$ 10 cigarettes per week, had consumed any alcoholic beverage $\geq$ 1 time per week, or had chewed $\geq$ 7 betel quids per week, respectively, for at least six months. Current users were those still practicing the above habits within one year before diagnosis of urolithiasis or the interview [11,25].

Clinical information, including stone location, stone number, stone diameter, and stone episodes, was also collected by questionnaire and further reviewed using the patients' medical charts by one urologist (C.-C.L.) who was unaware of the exposure of interest, including melamine and biomarkers of oxidative stress (8-OHdG and MDA) and renal tubular injury (NAG) in urine and genotyping of antioxidant enzyme genes.

2.3. Analyses of Melamine, Biomarkers of Oxidative Stress (MDA and 8-OHdG), and Renal Tubular Injury (NAG) in Urine

Urinary melamine was measured using an isotopic liquid chromatography–tandem mass spectrometry method (LC-MS/MS) (API4000Q, Applied Biosystems/MDS SCIEX, Concord, Vaughan, ON, Canada) [13]. Urinary MDA was measured using high-performance liquid chromatography with fluorescence (HPLC-FL) detection in a reversed-phase column (Luna C18, 250 $\times$ 4.6 nm) [26]. Urinary 8-OHdG was measured using a validated method for online solid-phase extraction (SPE) LC-MS/MS [27]. The detailed methods were described previously [14,16]. The limits of detection (LOD) for urinary melamine and biomarkers of oxidative stress were 0.4 ng/ml for melamine, 0.02 μmol/L for MDA, and 0.01 ng/mL for 8-OHdG. All urinary measurements of MDA and 8-OHdG were detectable. In contrast, 31 (10.0%) of the 309 urinary melamine measurements were below the LOD and were substituted as $LOD/\sqrt{2}$.

Urinary NAG was measured using an NAG assay kit (Diazyme Laboratory, Poway, CA, USA) [13,14]. All urinary NAG measurements were detectable. Urinary creatinine was analyzed by spectrophotometry (U-2000; Hitachi, Tokyo, Japan) set at a wavelength of 520 nm to measure the creatinine–picrate reaction [14]. The measurement of all biochemical parameters above were performed by two different laboratory technicians blinded to each other's findings, study design, and participant information.

2.4. Genotyping of Five SNPs

DNA was extracted from peripheral whole blood using a Puregene DNA Isolation Kit (Gentra Systems Inc., Minneapolis, MN, USA). The five SNPs (*MnSOD*-rs4880, *MnSOD*-rs5746136, *GPX1*-rs1800668, *CAT*-rs1001179, and *CAT*-rs769217) were analyzed using an assay-on-demand SNP genotyping kit for a TaqMan 5′ allelic discrimination assay (Applied Biosystems, Foster City, CA, USA). Briefly, SNP amplification assays including 10 ng of sample DNA in 25 μL of reaction solution contained 12.5 μL of the 2× TaqMan®Universal PCR Mix (Applied Biosystems), while 1.25 μL of pre-developed assay reagent from the SNP genotyping product (Applied Biosystems) contained two primers. Two MCB-Taqman probes were performed following the instructions of manufacturer. Polymerase chain reactions (PCRs) were performed using an ABI Prism 7500 Sequence Detection System (Applied Biosystems) [28,29]. These genotypes were confirmed by direct sequencing after full-scale genotyping. Randomly, ~10% of the study samples (30 cases) were repeated for quality control, with the results showing 100% accuracy.

2.5. Statistical Analyses

Quantitative data were expressed as means ± standard deviations (SD) or medians with interquartile ranges (IQR), and categorical data were presented as numbers (*n*) and percentages. The Hardy–Weinberg equilibrium for the distribution of genotypes was checked using the chi-square test.

Urinary melamine levels and urinary biomarkers (MDA, 8-OHdG, NAG) were corrected by urinary creatinine values before further analyses. After correction, urinary biomarkers of oxidative stress (MDA and 8-OHdG) and renal tubular injury (NAG) were dichotomized into high or low using their median values [14,16]. Simple logistic regression models were first used to evaluate the associations between genotypes of antioxidant enzyme genes and urinary biomarkers of oxidative stress and renal tubular injury. If a significant relationship was noted in the initial analysis, the combined effects between genotypes of antioxidant enzyme genes and urinary melamine levels dichotomized by median values were further tested by multiple logistic regression analyses after adjusting for covariates, such as age, sex, BMI, educational level, personal habits, stone number, stone size, stone location, and comorbidities. All statistical analyses were performed using the SAS statistical package. All *p*-values were two-sided and considered significant if <0.05.

2.6. Sensitivity Analysis

To examine the robustness of our findings, we performed sensitivity analyses to compare the results when urinary melamine levels were divided into tertiles. Furthermore, we also compare the findings using a new method of covariate-adjusted standardization plus creatinine adjustment to correct for urine concentration and complicated confounding structures [14,30].

3. Results

Of the 309 study patients, 302 had data for genetic polymorphisms. There were 226 males and 76 females, with a mean age of 54.5 ± 12.9 years. In total, 156 participants (51.7%) presented first stone episodes and 194 (64.2%) presented single stones (Table 1). The median levels of the biomarkers in urine were 1.26 μg/mmol Cr for melamine, 0.24 μmol/mmol Cr for MDA, 5.78 mg/g Cr for 8-OHdG, and 0.86 IU/mmol Cr for NAG, which were the values used to dichotomize the levels into high and low for further analyses (Table 1).

Table 1. Characteristics and laboratory data in 302 calcium urolithiasis patients.

Variables	*N* (%)
Gender	
Male	226 (74.8)
Female	76 (25.2)
Years of education	
≤9	143 (47.4)
>9	159 (52.6)
Personal habits	
Current smokers	104 (34.4)
Current betel quid chewers	21 (7.0)
Current drinkers	41(13.6)
Diabetes mellitus	52 (17.2)
Hypertension	105 (34.8)
Dyslipidemia	21 (7.0)
Stone episodes	
1	156 (51.7)
≥2	146 (48.3)
Stone location	
Kidney	69 (22.8)
Ureter	181 (59.9)
Kidney and Ureter	52 (17.2)
Stone number	
Single	194 (64.2)
Multiple	108(35.8)
Maximal stone diameter	
<1 cm	184 (60.9)
1–2 cm	55 (18.2)
>2 cm	63 (20.9)
	Mean ± SD (median, IQR)
Age (years)	54.5 ± 12.9 (55.0, 46.8–63.0)
BMI (kg/m^2)	26.3 ± 4.0 (25.9, 23.8–28.4)
Urinary biomarkers	
Melamine (ng/mL)	12.47 ± 29.51 (4.86, 1.94–12.67)
MDA (μmol/L)	1.16 ± 0.86 (0.93, 0.52–1.60)
8-OHdG (ng/mL)	3.30 ± 2.46 (2.59, 1.55–4.46)
NAG (IU/L)	5.23 ± 5.62 (3.77, 1.90–6.33)
Urine biomarkers corrected by urine creatinine	
Melamine (μg/mmol Cr)	3.27 ± 6.68 (1.26, 0.48–3.34)
MDA (μmol/mmol Cr)	0.31 ± 0.28 (0.24, 0.15–0.37)
8-OHdG (mg/g Cr)	7.82 ± 7.41 (5.78, 4.08–9.11)
NAG (IU/mmol Cr)	1.36 ± 1.52 (0.86, 0.55–1.60)
Urinary biomarkers corrected by method of covariate-adjusted standardization plus covariate adjustment	
Melamine (ng/mL)	12.59 ± 26.10 (4.89, 1.94–12.76)
MDA (μmol/L)	1.18 ± 1.05 (0.91, 0.58–1.37)
8-OHdG (ng/mL)	3.37 ± 3.18 (2.57, 1.70–3.81)
NAG (IU/L)	5.12 ± 5.80 (3.14, 2.09–6.06)

Abbreviations: *N* = number; SD = standard deviation; IQR = interquartile range; BMI = Body mass index; MDA = malondialdehyde; 8-OHdG = 8-oxo-2′-deoxyguanosine; NAG = *N*-acetyl b-D-glucosaminidase; Cr = creatinine.

Five SNPs of antioxidant enzyme genes (*MnSOD*, *GPX1*, and *CAT*) are shown in Table 2, all of which conformed to the Hardy–Weinberg equilibrium. Because the rare homozygotes of both GPX1-rs1800668 and CAT-rs1001179 had only one case, they were combined with heterozygotes for the subsequent analyses (Tables 3, 4 and S2–S5).

Table 2. Distribution of SNPs of antioxidant enzyme genes (*MnSOD, GPX1,* and *CAT*) in 302 calcium urolithiasis patients.

	Alleles	N (%)	Genotypes	N (%)
MnSOD-rs4880	*C*	80 (13.2)	*CC*	8 (2.6)
	T	524 (86.8)	*TC*	64 (21.2)
			TT	230 (76.2)
MnSOD-rs5746136	*A*	238 (39.4)	*AA*	48 (15.9)
	G	366 (60.6)	*AG*	142 (47.0)
			GG	112 (37.1)
GPX1-rs1800668	*T*	21 (3.5)	*TT*	1 (0.3)
	C	583 (96.5)	*TC*	19 (6.3)
			CC	282 (93.4)
CAT-rs1001179	*T*	17 (2.8)	*TT*	1 (0.3)
	C	587 (97.2)	*TC*	15 (5.0)
			CC	286 (94.7)
CAT-rs769217	*T*	289 (47.8)	*TT*	65 (21.5)
	C	315 (52.2)	*TC*	159 (52.6)
			CC	78 (25.9)

Abbreviations: SNP = single nucleotide polymorphism; *MnSOD* = manganese superoxide dismutase; *GPX1* = glutathione peroxidase; *CAT* = catalase; *N* = number.

Table 3. Relationships of SNPs of antioxidant enzyme genes with a urinary biomarker of oxidative stress, MDA, in 302 calcium urolithiasis patients.

		MDA, *N* (%)			
***MnSOD*-rs4880**		**Low < 50%**	**High ≥ 50%**	**Crude OR (95% CI)**	***p* Value**
Alleles	*C*	50 (16.6)	30 (9.9)	Ref	
	T	252 (83.4)	272 (90.1)	1.80 (1.11–2.92)	0.017
Genotypes	*CC*	5 (3.3)	3 (2.0)	Ref	
	CT	40(26.5)	24 (15.9)	1.00 (0.22-4.56)	1
	TT	106 (70.2)	124 (82.1)	1.95 (0.46-8.35)	0.368
	p for trend				0.023
***MnSOD*-rs5746136**					
Alleles	*A*	118 (39.1)	120 (39.7)	Ref	
	G	184 (60.9)	182 (60.3)	0.97 (0.73–1.35)	0.868
Genotypes	*AA*	27 (17.9)	21 (13.9)	1	
	AG	64 (42.4)	78 (51.7)	1.57 (0.81–3.03)	0.182
	GG	60 (39.7)	52 (34.4)	1.11 (0.56–2.20)	0.755
	p for trend				0.869
***GPX1*-rs1800668**					
Alleles	*T*	7 (2.3)	14 (4.6)	Ref	
	C	295 (97.7)	288 (95.4)	0.49 (0.19–1/23)	0.127
Genotypes	*TT* + *TC*	7 (4.6)	13 (8.6)	Ref	
	CC	144 (95.4)	138 (91.4)	0.52 (0.20–1.33)	0.171
***CAT*-rs1001179**					
Alleles	*T*	7 (2.3)	10 (3.3)	Ref	
	C	295 (97.7)	292 (96.7)	0.69 (0.26–1.85)	0.463
Genotypes	*TT* + *TC*	6 (4.0)	10 (6.6)	Ref	
	CC	145 (96.0)	141 (93.4)	0.58 (0.21–1.65)	0.309
***CAT*-rs769217**					
Alleles	*T*	146 (48.3)	143 (47.4)	Ref	
	C	156 (51.7)	159 (52.6)	1.04 (0.76–1.63)	0.807
Genotypes	*TT*	35 (23.2)	30 (19.9)	Ref	
	TC	76 (50.3)	83 (55.0)	1.27 (0.71–2.27)	0.412
	CC	40 (26.5)	38 (25.1)	1.11 (0.57–2.14)	0.76
	p for trend				0.802

Abbreviations: SNP = single nucleotide polymorphism; MDA = malondialdehyde; *N* = number; *MnSOD* = manganese superoxide dismutase; *GPX1* = glutathione peroxidase; *CAT* = catalase.

Table 4. Relationships of SNPs of antioxidant enzyme genes with a urinary biomarker of renal tubular injury, NAG, in 302 calcium urolithiasis patients.

		NAG, *N* (%)			
***MnSOD*-rs4880**		**Low < 50%**	**High ≥ 50%**	**Crude OR (95% CI)**	***p* Value**
Alleles	*C*	42 (13.9)	38 (12.6)	Ref	
	T	260 (86.1)	264 (87.4)	1.12 (0.70–1.80)	0.631
Genotypes	*CC*	4 (2.6)	4 (2.6)	Ref	
	CT	34 (22.5)	30 (19.9)	0.88 (0.20–3.84)	0.87
	TT	113 (74.9)	117 (77.5)	1.04 (0.25–4.24)	0.96
	p for trend				0.898
***MnSOD*-rs5746136**					
Alleles	*A*	130 (43.0)	108 (35.8)	Ref	
	G	172 (57.0)	194 (64.2)	1.36 (0.98–1.88)	0.067
Genotypes	*AA*	30(19.8)	18 (11.9)	Ref	
	AG	70 (46.4)	72 (47.7)	1.71 (0.88–3.35)	0.115
	GG	51 (33.8)	61 (40.4)	1.99 (1.00–3.99)	0.051
	p for trend				0.07
***GPX1*-rs1800668**					
Alleles	*T*	11 (3.6)	10 (3.3)	Ref	
	C	291 (96.4)	292 (96.7)	1.10 (0.46–2.64)	0.824
Genotypes	*TT* + *TC*	11 (7.3)	9 (6.0)	Ref	
	CC	140 (92.7)	142 (94.0)	1.24 (0.50–3.08)	0.64
***CAT*-rs1001179**					
Alleles	*T*	9 (3.0)	8 (2.6)	Ref	
	C	293 (97.0)	294 (97.4)	1.13 (0.43–2.97)	0.806
Genotypes	*TT* + *TC*	8 (5.3)	8 (5.3)	Ref	
	CC	143 (94.7)	143 (94.7)	1 (0.37–2.74)	1
***CAT*-rs769217**					
Alleles	*T*	137 (45.4)	152 (50.3)	Ref	
	C	165 (54.6)	150 (49.7)	0.82(0.60–1.13)	0.222
Genotypes	*TT*	34 (22.5)	31 (20.5)	Ref	
	TC	69 (45.7)	90 (59.6)	1.43(0.80–2.55)	0.225
	CC	48 (31.8)	30 (19.9)	0.69(0.35–1.34)	0.267
	p for trend				0.21

Abbreviations: SNP = single-nucleotide polymorphism; NAG = *N*-acetyl b-D-glucosaminidase; *N* = number; *MnSOD* = manganese superoxide dismutase; *GPX1* = glutathione peroxidase; *CAT* = catalase.

In *MnSOD*-rs4880, subjects who carried the *T* allele had a significantly higher risk of having high MDA levels than those carrying the *C* allele (odds ratio (OR) = 1.80, 95% CI = 1.11–2.92) (Table 3). In *MnSOD*-rs5746136, subjects who carried *GG* genotype had 1.99-fold higher risk of having high NAG levels (95% CI = 1.00–3.99) compared to those with *AA* genotype (Table 4). There were no significant relationships between five SNPs of antioxidant enzyme genes and 8-OHdG (Table S2). Sensitivity analyses using a new method of covariate-adjusted standardization plus creatinine adjustment found similar results (Tables S3–S5).

Table 5 shows the effects of *MnSOD*-rs4880 combined with high and low urine melamine levels on the risk of high MDA levels (≥50% vs. <50%). Subjects who carried the *T* allele and had high melamine levels had a significantly higher risk of high MDA levels than those who carried the C allele and had low melamine levels (adjusted OR = 3.60, 95% CI = 1.79–7.22), after adjusting for age, sex, BMI, educational level, personal habits, clinical stone characteristics, and comorbidities. Dichotomized urinary MDA levels to <66.6% and ≥66.6%, the results remained the same (Table 5). The results also remained similar when we used a new method of covariate-adjusted standardization plus creatinine adjustment (Table S6).

Table 5. Combined effects of *MnSOD* genetic polymorphism (rs4880) and urine melamine levels on the risk of a high urinary marker of oxidative stress, MDA, in 302 calcium urolithiasis patients.

		MDA, *N* (%)				Model 1		Model 2	
rs4880 and Melamine	*N*	**Low <50%**	**High ≥50%**	**Crude OR (95% CI)**	*p* **Value**	**Adjusted OR (95% CI)**	*p* **Value**	**Adjusted OR (95% CI)**	*p* **Value**
C allele + Melamine < 50%	48	32 (10.6)	16 (5.3)	1		1		1	
C allele + Melamine > 50%	32	18 (6.0)	14 (4.6)	1.56 (0.62–3.91)	0.347	1.70 (0.64–4.49)	0.286	1.65 (0.62–4.38)	0.317
T allele + Melamine < 50%	254	156 (51.6)	98 (32.5)	1.26 (0.66–2.41)	0.492	1.28 (0.64–2.56)	0.484	1.28 (0.64–2.56)	0.491
T allele + Melamine > 50%	270	96 (31.8)	174 (57.6)	3.63 (1.89–6.94)	<0.001	3.64 (1.82–7.27)	<0.001	3.60 (1.79–7.22)	<0.001
p for trend					<0.001		<0.001		<0.001
		MDA, *N* (%)				**Model 1**		**Model 2**	
rs4880 and Melamine	*N*	**Low <66.6%**	**High ≥66.6%**	**Crude OR (95% CI)**	*p* **Value**	**Adjusted OR (95% CI)**	*p* **Value**	**Adjusted OR (95% CI)**	*p* **Value**
T allele + Melamine < 50%	48	37 (9.2)	11 (5.5)	1		1		1	
T allele + Melamine > 50%	32	22 (5.4)	10 (5.0)	1.53 (0.56–4.18)	0.408	1.64 (0.57–4.73)	0.365	1.61 (0.55–4.68)	0.384
C allele + Melamine < 50%	254	185 (45.8)	69 (34.5)	1.26 (0.61–2.60)	0.541	1.28 (0.59–2.74)	0.534	1.28 (0.59–2.76)	0.528
C allele + Melamine > 50%	270	160 (39.6)	110 (55.0)	2.31 (1.13–4.73)	0.022	2.19 (1.03–4.65)	0.042	2.20 (1.03–4.70)	0.042
p for trend					0.002		0.009		0.01

Abbreviations: MnSOD = manganese superoxide dismutase; MDA = malondialdehyde; *N* = number. Model 1: Adjusting for age, sex, BMI, educational level, alcohol drinking, cigarette smoking, betel quid chewing, stone number, stone location, and stone size. Model 2: Adjusting for age, sex, BMI, educational level, alcohol drinking, cigarette smoking, betel quid chewing, stone number, stone location, stone size, hypertension, diabetes mellitus, and dyslipidemia.

Table 6 shows the effects of *MnSOD*-rs5746136 combined with urine melamine levels on the risk of high NAG levels (≥50% vs. <50%). Subjects who carried the *G* allele and had high melamine levels had a significantly higher risk of high NAG levels than those who carried the *A* allele and had low melamine levels (adjusted OR = 1.73, 95% CI = 1.04–2.89) in the fully adjusted model. Dichotomized urinary NAG levels to <66.6% and ≥66.6%, the results remained consistent (Table 6). The results also remained similar when we used a new method of covariate-adjusted standardization plus creatinine adjustment (Table S7).

Table 6. Combined effects of *MnSOD* genetic polymorphism (rs5746136) and urine melamine levels on the risk of a high urinary marker of renal tubular injury, NAG, in 302 calcium urolithiasis patients.

		NAG, *N* (%)				Model 1		Model 2	
rs5746136 and Melamine	*N*	**Low < 50%**	**High ≥50%**	**Crude OR (95% CI)**	*p* **Value**	**Adjusted OR (95% CI)**	*p* **Value**	**Adjusted OR (95% CI)**	*p* **Value**
A allele + Melamine < 50%	110	62 (20.5)	48 (15.9)	1		1		1	
A allele + Melamine > 50%	126	66 (21.9)	60 (19.9)	1.14 (0.68–1.90)	0.617	1.07 (0.62–1.83)	0.808	1.02 (0.59–1.77)	0.931
G allele + Melamine < 50%	192	100 (33.1)	92 (30.4)	1.21 (0.76–1.94)	0.424	1.11 (0.68–1.82)	0.68	1.11 (0.67–1.82)	0.689
G allele + Melamine > 50%	176	74 (24.5)	102 (33.8)	1.78 (1.10–2.88)	0.019	1.78 (1.08–2.95)	0.025	1.73 (1.04–2.89)	0.036
p for trend					0.019		0.024		0.029
		NAG, *N* (%)				**Model 1**		**Model 2**	
rs5746136 and Melamine	*N*	**Low < 66.6%**	**High ≥66.6%**	**Crude OR (95% CI)**	*p* **Value**	**Adjusted OR (95% CI)**	*p* **Value**	**Adjusted OR (95% CI)**	*p* **Value**
A allele + Melamine < 50%	110	80 (19.9)	30 (14.9)	1		1		1	
A allele + Melamine > 50%	126	88 (21.9)	38 (18.8)	1.13 (0.64–1.98)	0.681	1.05 (0.58–1.89)	0.877	1.01 (0.55–1.82)	0.991
G allele + Melamine < 50%	192	132 (32.8)	60 (29.7)	1.23 (0.73–2.07)	0.433	1.12 (0.66–1.92)	0.67	1.12 (0.65–1.93)	0.678
G allele + Melamine > 50%	176	102 (25.4)	74 (36.6)	1.94 (1.16–3.24)	0.012	1.86 (1.09–3.17)	0.024	1.78 (1.03–3.07)	0.038
p for trend					0.008		0.016		0.023

Abbreviations: *MnSOD* = manganese superoxide dismutase; NAG = *N*-acetyl b-D-glucosaminidase; *N* = number. Model 1: Adjusting for age, sex, BMI, educational level, alcohol drinking, cigarette smoking, betel quid chewing, stone number, stone location, and stone size. Model 2: Adjusting for age, sex, BMI, educational level, alcohol drinking, cigarette smoking, betel quid chewing, stone number, stone location, stone size, hypertension, diabetes mellitus, and dyslipidemia.

4. Discussion

Increased risk of oxidative stress and renal tubular injury was previously found in adult patients with calcium urolithiasis exposed to environmental melamine [14,16]. In that same population, the current study found that two SNPs of the antioxidant enzyme gene, *MnSOD*, modified the effect of melamine exposure on those risks. In conjunction with high urinary melamine levels (≥50%), subjects carrying *T* allele of *MnSOD*-rs4880 were at even greater risk of high urinary MDA levels, a biomarker of oxidative stress, compared to its *C* allele carriers (adjusted OR = 3.60, 95% CI = 1.79–7.22 vs. 1.65, 95% CI = 0.62–4.37) (Table 5). Furthermore, subjects carrying *G* allele of *MnSOD*-rs5746136 were also at even greater risk of high urinary NAG levels, a biomarker of renal tubular injury, than its *A* allele carriers (adjusted OR = 1.73, 95% CI = 1.04–2.89 vs. 1.02, 95% CI = 0.59–1.77) (Table 6).

Melamine is widely used in the production of laminates, plastics, glues, adhesives, and coatings, and its derivatives are also used in flame retardants and in insulation [31]. Because of its resilience, light weight, and low cost, melamine is also widely used as a substitute for porcelain in the production of tableware and food utensils [31]. Our early studies found that melamine-made tableware could leach substantial amounts of melamine, especially in highly acidic soups or soups served at high temperatures, making them a major source of environmental melamine exposure [32,33]. In addition to our own in vitro study using human renal proximal tubular HK-2 cells, several studies have found that melamine could increase ROS and induce injury in other kidney cell lines, including a rat kidney epithelial cell line (NRK-52e cells) [34] and a human embryonic kidney cell line [35]. Furthermore, several extrinsic antioxidants, such as catechin [36] and bee honey [37], have been found to reverse melamine-induced ROS and prevent further nephrotoxicity in animal models.

The current study found that two SNPs of *MnSOD*, rs4880 and rs5746136, modified the effects of melamine exposure on the risk of oxidative stress and renal tubular injury, respectively. MnSOD is a key enzyme found in our antioxidant defense systems and the most important member of the superoxide dismutase family known to play a crucial role in controlling mitochondrial superoxide radicals [18]. Oxidative stress in the mitochondria has been found to play a vital role in the pathogenesis of kidney disease and injury [18]. *MnSOD*-rs 4880 is present in exon 2 and substitutes a $C > T$ at position 2734, changing the amino acid from alanine (Ala) to valine (Val) at position 16 [19,20]. The presence of the *T* (Val) allele leads to the production of instable mRNA and reduces the transport of the enzyme into the mitochondrial matrix, reducing its antioxidant function [19]. This study found that subjects who carried the *T* allele had a significantly higher risk of high urinary MDA, a biomarker of oxidative stress, than those who carried the *C* allele. In a recent case–control study of 256 patients with end-stage renal disease (ESRD) underlying hemodialysis and 374 controls, subjects with the *TT* genotype were found to have significantly higher serum MDA levels than those with the *CC* genotype (2.57 ± 0.79 vs 2.17 ± 0.78 mmol/L, $p < 0.05$) [38]. That study also found the *TT* genotype to be an independent risk factor for the development of ESRD [38]. In a prospective cohort study of 185 patients with chronic kidney disease followed for up to 12 months, subjects with *TT* and *CT* genotypes were found to have a greater decline in kidney function than those with the *CC* genotype [20]. In another prospective cohort study (SURGENE study) of 340 patients with type 1 diabetes followed for up to ten years, the *T* allele was associated with a higher incidence of renal events (new cases of incipient nephropathy or the progression to a more severe stage of nephropathy) and with a decline in estimated glomerular filtration rate during follow-up [39].

MnSOD-rs5746136 is located at 65 bp downstream from the poly(A) site and near the SP1 and the NF-kB transcription element sequences, which may play a role in the regulation of *MnSOD* gene expression [40]. The current study found that subjects who carried the *G* allele of *MnSOD*-rs5746136 had a significantly higher risk of high urinary NAG, a biomarker of renal tubular injury, than those carrying the *A* allele. This SNP has also been linked to several diseases associated with oxidative stress. In a recent case–control study of 100 Chinese children with Kawasaki disease and 102 healthy controls, subjects with the *A* allele had a 0.558-times lower risk (95% CI = 0.371–0.838) of Kawasaki disease than those with the *G* allele. In another case–control study of 164 Turkish patients with polycystic ovary syndrome and 148 healthy controls, subjects with *GG* and *AG* genotypes had a 2.95-fold increase in risk (95% Cl: 1.2–3.1) of having polycystic ovary syndrome compared those with the AA genotype [41]. However, in the SURGENE study, which followed 340 Caucasian patients with type 1 diabetes for up to ten years, subjects with the

G allele were reported to have a significantly lower risk of having established or advanced nephropathy compared to those with the *A* allele (OR:0.30, 95% GI = 0.11–0.72) at the end of follow-up [39]. Further studies are needed to elucidate the role of *MnSOD*-rs5746136 on the risk of kidney diseases, including urolithiasis formation and the deterioration of renal function in populations of different clinical and ethnic backgrounds.

Traditionally, melamine was considered a relatively safe chemical. After the toxic milk and food scandals, the WHO and US FDA lowered their recommended tolerable daily intake (TDI) of melamine to 200 and 63 μg per kg body weight per day (μg/kg_bw/day), respectively [31]. Recently, we evaluated the one-sided 95% lower bound of benchmark dose (BMDL) of melamine exposure, taking into consideration a benchmark response of 0.10 in two vulnerable human populations, calcium urolithiasis patients and early chronic kidney disease patients [31,42]. It was concluded that both BMDLs should be lowered to 4.89 μg/kg_bw/day and 0.74 to 2.03 μg/kg_bw/day, respectively, to protect calcium urolithiasis patients and early chronic kidney disease patients from further deterioration of renal function [31,42]. Both simulated BMDLs for melamine exposure threshold in those susceptible populations were much lower than the current WHO and the US FDA recommended TDIs. These findings were supported by a toxicological analysis reporting that a lower TDI of melamine to 8.1 μg/kg_bw/day should be considered for general population [43]. This study found that two SNPs of *MnSOD*, rs4880 and rs5746136, modify the risks of oxidative stress and renal tubular injury, respectively. Furthermore, they strengthen the risks of melamine on high urinary biomarkers of oxidative stress (MDA) and renal tubular injury (NAG), respectively. Our findings support the hypothesis that genetic factors can influence melamine-induced oxidative stress and kidney damage. Therefore, in the future, genetic factors may need to be taken into consideration when assessing the human TDI, especially in vulnerable populations. In addition, it is important to prevent environmental melamine exposure, for example by avoiding the use of melamine-made tableware, especially when containing highly acidic or high-temperature soups [32,33]. A behavior intervention to decrease the use of melamine-made tableware was found to effectively prevent melamine exposure from the environment [44]. Other studies have also reported that natural products such as catechin [36] and bee honey [37] could reverse melamine-induced ROS and prevent further nephrotoxicity. The effects of other natural product or plant extract supplementation are warranted for further investigation [45,46].

This study has some limitations. First, we only measured melamine levels based on one-spot first-void morning urine samples. Although we previously found good correlations between melamine levels in one-spot first-void morning urine and total melamine excretion in previous 24 h urine [7,11], these measurements might not completely represent the long-term melamine exposure of participants. Second, we did not evaluate melamine analogues, especially cyanuric acid, or other environmental nephrotoxins, such as lead and cadmium. Further studies might need to elucidate their possible synergistic effects on the adverse outcomes of kidney. Third, we did not evaluate SNPs of other *SOD* genes such as Cu/ZnSOD and their impacts on mitochondrial function. Further studies might need to elucidate their possible relationships.

5. Conclusions

Two SNPs, rs4880 and rs5746135, of *MnSOD* were found to modify the risks of oxidative stress and renal tubular injury, respectively, in adult patients with calcium urolithiasis. Their presence increased the risk already posed by high urinary melamine levels in the same population, supporting the hypothesis that genetic factors can influence melamine-induced oxidative stress and kidney damage. Future assessments may want to factor in these SNPs by setting a TDI for melamine, especially in vulnerable populations. Further studies of different ethnic populations are needed to confirm our preliminary results.

Supplementary Materials: The following are available online at https://www.mdpi.com/article/10.3390/antiox11010152/s1.

Author Contributions: Conceptualization, C.-C.L. (Chia-Chu Liu), C.-F.W. and M.-T.W.; data curation, Y.-C.L., T.-Y.H., S.-T.H., H.-S.W., J.-H.J., S.-P.H., C.-C.L. (Ching-Chia Li) and Y.-S.J.; formal analysis, C.-C.L. (Chia-Chu Liu), C.-F.W., and M.-T.W.; funding acquisition, C.-C.L. (Chia-Chu Liu) and M.-T.W.; investigation, Y.-C.L., T.-Y.H., S.-T.H., H.-S.W., J.-H.J., S.-P.H., C.-C.L. (Ching-Chia Li), Y.-S.J., T.-J.H., Y.-C.T. and C.-C.C.; methodology, C.-C.L. (Chia-Chu Liu)., C.-F.W. and M.-T.W.; resources, C.-C.L. (Chia-Chu Liu) and M.-T.W.; supervision, M.-T.W.; Validation, C.-C.L. (Chia-Chu Liu), C.-F.W. and M.-T.W.; writing—original draft, C.-C.L. (Chia-Chu Liu); writing—review and editing, M.-T.W. All authors have read and agreed to the published version of the manuscript.

Funding: This work was financially supported by grants from Kaohsiung Medical University Hospital (KMUH-DK(C)110003; KMUH-DK109008; KMUH110-0R56; KMUH109-9R62; KMUH108-8R54); from The Featured Areas Research Center Program within the framework of the Higher Education Sprout Project by the Ministry of Education (MOE), and Kaohsiung Medical University Research Center Grant (KMU-TC109A01); from the Ministry of Science and Technology (MOST 109-2314-B-037-104-MY3; MOST 110-2314-B-037-056-; MOST 110-2314-B-037-133-; MOST 110-2314-B-037-047-MY3); from the National Health Research Institutes (NHRI-EX110-10703PI); and from Kaohsiung Municipal Siaogang Hospital (S-108-017), all in Taiwan.

Institutional Review Board Statement: This study was conducted according to the guidelines of the Declaration of Helsinki, and approved by the Institutional Review Board of Kaohsiung Medical University Hospital (protocol code: KMUH-IRB-960437, date of approval: 19 February 2008; protocol code: KMUH-IRB-G(I)-20170018, date of approval: 8 December 2017).

Informed Consent Statement: Informed consent was obtained from all subjects involved in the study.

Data Availability Statement: Data are contained in the article and Supplementary Materials.

Acknowledgments: We thank James F. Steed for language editing.

Conflicts of Interest: The authors declare no conflict of interest. The funders had no role in the design of the study; in the collection, analyses, or interpretation of data; in the writing of the manuscript, or in the decision to publish the results.

References

1. Hau, A.K.; Kwan, T.H.; Li, P.K. Melamine toxicity and the kidney. *J. Am. Soc. Nephrol.* **2009**, *20*, 245–250. [CrossRef]
2. Bhalla, V.; Grimm, P.C.; Chertow, G.M.; Pao, A.C. Melamine nephrotoxicity: An emerging epidemic in an era of globalization. *Kidney Int.* **2009**, *75*, 774–779. [CrossRef]
3. Qin, Y.; Lv, X.; Li, J.; Qi, G.; Diao, Q.; Liu, G.; Xue, M.; Wang, J.; Tong, J.; Zhang, L.; et al. Assessment of melamine contamination in crop, soil and water in China and risks of melamine accumulation in animal tissues and products. *Environ. Int.* **2010**, *36*, 446–452. [CrossRef]
4. Zhu, H.; Kannan, K. Occurrence and distribution of melamine and its derivatives in surface water, drinking water, precipitation, wastewater, and swimming pool water. *Environ. Pollut.* **2020**, *258*, 113743. [CrossRef]
5. Zhu, H.; Kannan, K. Melamine and cyanuric acid in foodstuffs from the United States and their implications for human exposure. *Environ. Int.* **2019**, *130*, 104950. [CrossRef]
6. Panuwet, P.; Nguyen, J.V.; Wade, E.L.; D'Souza, P.E.; Ryan, P.B.; Barr, D.B. Quantification of melamine in human urine using cation-exchange based high performance liquid chromatography tandem mass spectrometry. *J. Chromatogr. B Anal. Technol. Biomed Life Sci.* **2012**, *887–888*, 48–54. [CrossRef]
7. Lin, Y.-T.; Tsai, M.-T.; Chen, Y.-L.; Cheng, C.-M.; Hung, C.-C.; Wu, C.-F.; Liu, C.-C.; Hsieh, T.-J.; Shiea, J.; Chen, B.H.; et al. Can melamine levels in 1-spot overnight urine specimens predict the total previous 24-hour melamine excretion level in school children? *Clin. Chim. Acta* **2013**, *420*, 128–133. [CrossRef]
8. Sathyanarayana, S.; Flynn, J.T.; Messito, M.J.; Gross, R.; Whitlock, K.B.; Kannan, K.; Karthikraj, R.; Morrison, D.; Huie, M.; Christakis, D.; et al. Melamine and cyanuric acid exposure and kidney injury in US children. *Environ. Res.* **2019**, *171*, 18–23. [CrossRef]
9. Shi, X.; Dong, R.; Chen, J.; Yuan, Y.; Long, Q.; Guo, J.; Li, S.; Chen, B. An assessment of melamine exposure in Shanghai adults and its association with food consumption. *Environ. Int.* **2020**, *135*, 105363. [CrossRef]
10. Wu, C.-F.; Liu, C.-C.; Chen, B.-H.; Huang, S.-P.; Lee, H.-H.; Chou, Y.-H.; Wu, W.-J.; Wu, M.-T. Urinary melamine and adult urolithiasis in Taiwan. *Clin. Chim. Acta* **2010**, *411*, 184–189. [CrossRef]
11. Liu, C.-C.; Wu, C.-F.; Chen, B.-H.; Huang, S.-P.; Goggins, W.; Lee, H.-H.; Chou, Y.-H.; Wu, W.-J.; Huang, C.-H.; Shiea, J.; et al. Low exposure to melamine increases the risk of urolithiasis in adults. *Kidney Int.* **2011**, *80*, 746–752. [CrossRef]
12. Tsai, Y.-C.; Wu, C.-F.; Liu, C.-C.; Hsieh, T.-J.; Lin, Y.-T.; Chiu, Y.-W.; Hwang, S.-J.; Chen, H.-C.; Wu, M.-T. Urinary Melamine Levels and Progression of CKD. *Clin. J. Am. Soc. Nephrol. CJASN* **2019**, *14*, 1133–1141. [CrossRef]

13. Wu, C.-F.; Peng, C.-Y.; Liu, C.-C.; Lin, W.-Y.; Pan, C.-H.; Cheng, C.-M.; Hsieh, H.-M.; Hsieh, T.-J.; Chen, B.H.; Wu, M.-T. Ambient Melamine Exposure and Urinary Biomarkers of Early Renal Injury. *J. Am. Soc. Nephrol.* **2015**, *26*, 2821–2829. [CrossRef] [PubMed]
14. Liu, C.-C.; Hsieh, T.-J.; Wu, C.-F.; Tsai, Y.-C.; Huang, S.-P.; Lee, Y.-C.; Huang, T.-Y.; Shen, J.-T.; Chou, Y.-H.; Huang, C.-N.; et al. Urinary melamine excretion and increased markers of renal tubular injury in patients with calcium urolithiasis: A cross-sectional study. *Environ. Pollut.* **2017**, *231*, 1284–1290. [CrossRef]
15. Hsieh, T.J.; Hsieh, P.C.; Tsai, Y.H.; Wu, C.; Liu, C.; Lin, M.; Wu, M. Melamine induces human renal proximal tubular cell injury via transforming growth factor-beta and oxidative stress. *Toxicol. Sci.* **2012**, *130*, 17–32. [CrossRef]
16. Liu, C.-C.; Hsieh, T.-J.; Wu, C.-F.; Lee, C.-H.; Tsai, Y.-C.; Huang, T.-Y.; Wen, S.-C.; Lee, C.-H.; Chien, T.-M.; Lee, Y.-C.; et al. Interrelationship of environmental melamine exposure, biomarkers of oxidative stress and early kidney injury. *J. Hazard Mater.* **2020**, *396*, 122726. [CrossRef] [PubMed]
17. Gorin, Y. The Kidney: An Organ in the Front Line of Oxidative Stress-Associated Pathologies. *Antioxid. Redox Signal.* **2016**, *25*, 639–641. [CrossRef]
18. Ratliff, B.B.; Abdulmahdi, W.; Pawar, R.; Wolin, M.S. Oxidant Mechanisms in Renal Injury and Disease. *Antioxid. Redox Signal.* **2016**, *25*, 119–146. [CrossRef]
19. Crawford, A.; Fassett, R.G.; Geraghty, D.P.; Kunde, D.A.; Ball, M.J.; Robertson, I.K.; Coombes, J.S. Relationships between single nucleotide polymorphisms of antioxidant enzymes and disease. *Gene* **2012**, *501*, 89–103. [CrossRef] [PubMed]
20. Crawford, A.; Fassett, R.G.; Coombes, J.; Kunde, D.; Ahuja, K.; Robertson, I.K.; Ball, M.J.; Geraghty, D. Glutathione peroxidase, superoxide dismutase and catalase genotypes and activities and the progression of chronic kidney disease. *Nephrol. Dial. Transplant.* **2011**, *26*, 2806–2813. [CrossRef]
21. Kidir, V.; Uz, E.; Yigit, A.; Altuntas, A.; Yigit, B.; Inal, S.; Uz, E.; Sezer, M.T.; Yilmaz, H.R. Manganese superoxide dismutase, glutathione peroxidase and catalase gene polymorphisms and clinical outcomes in acute kidney injury. *Ren. Fail.* **2016**, *38*, 372–377. [CrossRef] [PubMed]
22. Ewens, K.G.; George, R.A.; Sharma, K.; Ziyadeh, F.N.; Spielman, R.S. Assessment of 115 candidate genes for diabetic nephropathy by transmission/disequilibrium test. *Diabetes* **2005**, *54*, 3305–3318. [CrossRef]
23. Lash, L.H. Environmental and Genetic Factors Influencing Kidney Toxicity. *Semin. Nephrol.* **2019**, *39*, 132–140. [CrossRef]
24. Little, J.; Higgins, J.P.T.; Ioannidis, J.P.A.; Moher, D.; Gagnon, F.; von Elm, E.; Khoury, M.J.; Cohen, B.; Davey-Smith, G.; Grimshaw, J.; et al. Strengthening the reporting of genetic association studies (STREGA): An extension of the STROBE Statement. *Hum. Genet.* **2009**, *125*, 131–151. [CrossRef]
25. Liu, C.-C.; Huang, S.-P.; Wu, W.J.; Chou, Y.-H.; Juo, S.; Tsai, L.-Y.; Huang, C.-H.; Wu, M.-T. The impact of cigarette smoking, alcohol drinking and betel quid chewing on the risk of calcium urolithiasis. *Ann. Epidemiol.* **2009**, *19*, 539–545. [CrossRef] [PubMed]
26. Hsu, K.C.; Hsu, P.F.; Chen, Y.C.; Lin, H.; Chen, H.P.; Huang, Y. Oxidative stress during bacterial growth characterized through microdialysis sampling coupled with HPLC/fluorescence detection of malondialdehyde. *J. Chromatogr. B Anal. Technol. Biomed Life Sci.* **2016**, *1019*, 112–116. [CrossRef] [PubMed]
27. Hu, C.W.; Chao, M.R.; Sie, C.H. Urinary analysis of 8-oxo-7,8-dihydroguanine and 8-oxo-7,8-dihydro-2′-deoxyguanosine by isotope-dilution LC-MS/MS with automated solid-phase extraction: Study of 8-oxo-7,8-dihydroguanine stability. *Free Radic. Biol. Med.* **2010**, *48*, 89–97. [CrossRef]
28. Liu, C.-C.; Lee, Y.-C.; Huang, S.-P.; Cheng, K.-H.; Hsieh, T.-J.; Huang, T.-Y.; Lee, C.-H.; Geng, J.-H.; Li, C.-C.; Wu, W.J. Hepatocyte Nuclear Factor-4alpha P2 Promoter Variants Are Associated With the Risk of Metabolic Syndrome and Testosterone Deficiency in Aging Taiwanese Men. *J. Sex Med.* **2018**, *15*, 1527–1536. [CrossRef]
29. Liu, C.-C.; Huang, S.-P.; Tsai, L.-Y.; Wu, W.J.; Juo, S.-H.H.; Chou, Y.-H.; Huang, C.-H.; Wu, M.-T. The impact of osteopontin promoter polymorphisms on the risk of calcium urolithiasis. *Clin. Chim. Acta* **2010**, *411*, 739–743. [CrossRef]
30. O'Brien, K.M.; Upson, K.; Cook, N.R.; Weinberg, C.R. Environmental Chemicals in Urine and Blood: Improving Methods for Creatinine and Lipid Adjustment. *Environ. Health Perspect.* **2016**, *124*, 220–227. [CrossRef]
31. Chen, C.C.; Tsai, Y.C.; Wang, Y.H.; Wu, C.; Chiu, Y.; Hwang, S.; Liu, C.; Hsieh, T.; Wu, M. Melamine exposure threshold in early chronic kidney disease patients-A benchmark dose approach. *Environ. Int.* **2021**, *156*, 106652. [CrossRef]
32. Chien, C.-Y.; Wu, C.-F.; Liu, C.-C.; Chen, B.H.; Huang, S.-P.; Chou, Y.-H.; Chang, A.-W.; Lee, H.-H.; Pan, C.-H.; Wu, W.J.; et al. High melamine migration in daily-use melamine-made tableware. *J. Hazard Mater.* **2011**, *188*, 350–356. [CrossRef] [PubMed]
33. Wu, C.F.; Hsieh, T.J.; Chen, B.H.; Liu, C.C.; Wu, M.T. A Crossover Study of Noodle Soup Consumption in Melamine Bowls and Total Melamine Excretion in Urine. *JAMA Intern. Med.* **2013**, *173*, 317–319. [CrossRef] [PubMed]
34. Guo, C.; Yuan, H.; He, Z. Melamine causes apoptosis of rat kidney epithelial cell line (NRK-52e cells) via excessive intracellular ROS (reactive oxygen species) and the activation of p38 MAPK pathway. *Cell Biol. Int.* **2012**, *36*, 383–389. [CrossRef]
35. Kuo, F.C.; Tseng, Y.T.; Wu, S.R.; Wu, M.T.; Lo, Y.C. Melamine activates NFκB/COX-2/PGE2 pathway and increases NADPH oxidase-dependent ROS production in macrophages and human embryonic kidney cells. *Toxicol. In Vitro* **2013**, *27*, 1603–1611. [CrossRef] [PubMed]
36. Li, X.; Wu, G.; Shang, P.; Bao, J.; Lu, J.; Yue, Z. Anti-nephrolithic potential of catechin in melamine-related urolithiasis via the inhibition of ROS, apoptosis, phospho-p38, and osteopontin in male Sprague-Dawley rats. *Free Radic. Res.* **2015**, *49*, 1249–1258. [CrossRef] [PubMed]

37. Al-Seeni, M.N.; El Rabey, H.A.; Al-Solamy, S.M. The protective role of bee honey against the toxic effect of melamine in the male rat kidney. *Toxicol. Ind. Health* **2015**, *31*, 485–493. [CrossRef]
38. Jerotic, D.; Matic, M.; Suvakov, S.; Vucicevic, K.; Damjanovic, T.; Savic-Radojevic, A.; Pljesa-Ercegovac, M.; Coric, V.; Stefanovic, A.; Ivanisevic, J.; et al. Association of Nrf2, SOD2 and GPX1 Polymorphisms with Biomarkers of Oxidative Distress and Survival in End-Stage Renal Disease Patients. *Toxins* **2019**, *11*, 431. [CrossRef]
39. Mohammedi, K.; Bellili-Muñoz, N.; Driss, F.; Roussel, R.; Seta, N.; Fumeron, F.; Hadjadj, S.; Marre, M.; Velho, G. Manganese superoxide dismutase (SOD2) polymorphisms, plasma advanced oxidation protein products (AOPP) concentration and risk of kidney complications in subjects with type 1 diabetes. *PLoS ONE* **2014**, *9*, e96916. [CrossRef]
40. Wan, X.S.; Devalaraja, M.N.; St Clair, D.K. Molecular structure and organization of the human manganese superoxide dismutase gene. *DNA Cell Biol.* **1994**, *13*, 1127–1136. [CrossRef] [PubMed]
41. Polat, S.; Şimşek, Y. Five variants of the superoxide dismutase genes in Turkish women with polycystic ovary syndrome. *Free Radic. Res.* **2020**, *54*, 467–476. [CrossRef] [PubMed]
42. Wang, Y.-H.; Wu, C.-F.; Liu, C.-C.; Hsieh, T.-J.; Tsai, Y.-C.; Wu, M.-T.; Chen, C.-C. A probabilistic approach for benchmark dose of melamine exposure for a marker of early renal dysfunction in patients with calcium urolithiasis. *Ecotoxicol. Environ. Saf.* **2020**, *200*, 110741. [CrossRef]
43. Hsieh, D.P.; Chiang, C.F.; Chiang, P.H.; Wen, C.P. Toxicological analysis points to a lower tolerable daily intake of melamine in food. *Regul. Toxicol. Pharmacol.* **2009**, *55*, 13–16. [CrossRef] [PubMed]
44. Wu, M.T.; Wu, C.F.; Chen, B.H. Behavioral Intervention and Decreased Daily Melamine Exposure from Melamine Tableware. *Environ. Sci. Technol.* **2015**, *49*, 9964–9970. [CrossRef] [PubMed]
45. Elhelaly, A.E.; Albasher, G.; Alfarraj, S.; Almeer, R.; Bahbah, E.; Fouda, M.M.A.; Bungau, S.; Aleya, L.; Abdel-Daim, M.M. Protective effects of hesperidin and diosmin against acrylamide-induced liver, kidney, and brain oxidative damage in rats. *Environ. Sci. Pollut. Res. Int.* **2019**, *26*, 35151–35162. [CrossRef] [PubMed]
46. Mirkov, I.; Stojković, D.; Aleksandrov, A.P.; Ivanov, M.; Kostić, M.; Glamočlija, J.; Soković, M. Plant Extracts and Isolated Compounds Reduce Parameters of Oxidative Stress Induced by Heavy Metals: An up-to-Date Review on Animal Studies. *Curr. Pharm. Des.* **2020**, *26*, 1799–1815. [CrossRef] [PubMed]

antioxidants

Article

Serum 5-Hydroxyindoleacetic Acid and Ratio of 5-Hydroxyindoleacetic Acid to Serotonin as Metabolomics Indicators for Acute Oxidative Stress and Inflammation in Vancomycin-Associated Acute Kidney Injury

Hyun-Seung Lee [1], Sang-Mi Kim [1], Ja-Hyun Jang [1], Hyung-Doo Park [1] and Soo-Youn Lee [1,2,3,*]

1 Department of Laboratory Medicine and Genetics, Samsung Medical Center, Sungkyunkwan University School of Medicine, Seoul 06351, Korea; hyunseung1011.lee@samsung.com (H.-S.L.); jeehee0520@gmail.com (S.-M.K.); jahyun.jang@samsung.com (J.-H.J.); nayadoo@hanmail.net (H.-D.P.)
2 Department of Clinical Pharmacology & Therapeutics, Samsung Medical Center, Sungkyunkwan University School of Medicine, Seoul 06351, Korea
3 Department of Health Science and Technology, Samsung Advanced Institute of Health Science and Technology, Sungkyunkwan University, Seoul 06351, Korea
* Correspondence: suddenbz@skku.edu; Tel.: +82-2-3410-1834; Fax: +82-2-3410-2719

Abstract: The incidence of vancomycin-associated acute kidney injury (VAKI) varies from 5–43%, and early detection of VAKI is important in deciding whether to discontinue nephrotoxic agents. Oxidative stress is the main mechanism of VAKI, and serotonin (5-HT) and its metabolite 5-hydroxyindoleacetic acid (5-HIAA) have been examined with respect to their involvement in ischemia/reperfusion damage in experimental animal models. In the current study, we assessed 5-HT and 5-HIAA as novel biomarkers for detecting VAKI in patients who have infections or compromised renal function, using a mass spectrometry–based metabolomics approach. We conducted amino acid profiling analysis and measurements of 5-HT and 5-HIAA using serum from subjects with VAKI (n = 28) and non-VAKI control subjects (n = 69), consisting of the infection subgroup (n = 23), CKD subgroup (n = 23), and healthy controls (HCs, n = 23). 5-HT was significantly lower in the VAKI group than in the non-VAKI groups, and the concentration of 5-HIAA and the ratio of 5-HIAA to 5-HT (5-HIAA/5-HT) showed higher values in the VAKI group. The infection subgroup presented a significantly greater 5-HIAA/5-HT ratio compared with the HC subgroup. Our study revealed that increased 5-HIAA/5-HT ratio has the potential to act as a VAKI surrogate marker, reflecting acute oxidative stress and inflammation.

Keywords: tryptophan; serotonin; 5-hydroxyindoleacetic acid; vancomycin; nephrotoxicity; oxidative stress; inflammation

Citation: Lee, H.-S.; Kim, S.-M.; Jang, J.-H.; Park, H.-D.; Lee, S.-Y. Serum 5-Hydroxyindoleacetic Acid and Ratio of 5-Hydroxyindoleacetic Acid to Serotonin as Metabolomics Indicators for Acute Oxidative Stress and Inflammation in Vancomycin-Associated Acute Kidney Injury. *Antioxidants* **2021**, *10*, 895. https://doi.org/10.3390/antiox10060895

Academic Editor: Tim Hofer

Received: 28 April 2021
Accepted: 27 May 2021
Published: 2 June 2021

1. Introduction

Acute kidney injury (AKI) is a serious complication with high associated rates of mortality and morbidity among hospitalized patients [1]. Sepsis is the leading cause of AKI, and 45% to 70% of all AKI cases can be linked to sepsis [2]. For sepsis management, it is often necessary to use antibiotic agents with the potential to cause AKI, although nephrotoxic drugs were contributing factors in 16% to 25% of cases of severe AKI [3]. Vancomycin, one nephrotoxic drug [4–6], is a widely used glycopeptide antimicrobial against Gram-positive organisms, such as methicillin-resistant Staphylococcus aureus (MRSA), in clinical practice [6,7]. As a subtype of AKI, vancomycin-associated AKI (VAKI) is characterized by dose-dependent acute tubular necrosis, which is associated with oxidative stress in proximal tubule cells [5]. VAKI is correlated closely with elderly, compromised renal function, concomitant nephrotoxic therapy, longer duration of vancomycin administration, high trough level of vancomycin, and critically ill status in the intensive care unit [4–6]. The incidence of VAKI has been reported to range from 5% to 43% [6], and therapeutic drug monitoring and customized dosing of vancomycin have been studied for the prevention of VAKI in a high-risk population [6].

The rapid and accurate diagnosis of VAKI is important for decision-making about MRSA treatment [8]. Earlier detection of VAKI with a kidney-specific biomarker can result in discontinuation of nephrotoxic agents and the earlier initiation of alternate therapeutics [8,9]. An increase in serum creatinine concentration (SCr) and a reduction in urine output are the current diagnostic markers for VAKI [5,6,10]; however, these traditional markers of kidney function might not be sensitive to kidney damage, and their changes lag several days behind actual changes in glomerular filtration rate (GFR) [11]. Recent studies have shown that novel biomarkers of kidney damage, including kidney injury molecule–1, neutrophil gelatinase-associated lipocalin, liver-type fatty acid–binding protein, and interleukin-18, increase before SCr [12–16]. However, it is unclear whether these biomarkers could reveal the different AKI-inducing mechanisms and conditions, such as toxicity, nephron obstruction, ischemia, and inflammation [10,11]. A viable novel marker for VAKI must be suitable for detecting kidney damage in septic patients with or without compromised renal function [17].

Recent studies have attempted the prediction of drug-induced nephrotoxicity using a metabolomics approach [18,19]. This approach could be used to identify novel kidney biomarkers and might elucidate biological mechanisms of disease [20]. Zhang et al. reported that branched-chain amino acids could be predictors for individual differences of cisplatin-induced nephrotoxicity [21]. Oxidative stress is the main mechanism of VAKI, similar to cisplatin-induced nephrotoxicity [6,10]. Therefore, we hypothesized that changes in amino acid concentrations in the blood could be distinguished between patients with VAKI and those with infection or compromised renal function. Furthermore, specific metabolites of amino acids and their ratios have the potential to act as novel biomarkers of VAKI.

In this study, we conducted an exploratory metabolome profiling analysis and amino acids profiling analysis from clinical residual samples obtained from VAKI patients and non-VAKI patients, including those with infection and chronic kidney disease (CKD). Using this approach, we aimed to analyze alterations in the metabolomics profile in acute and chronic oxidative stress conditions and to discover a novel biomarker of VAKI.

2. Materials and Methods

2.1. Study Cohort

This study enrolled a total of 97 subjects. A total of 3370 samples with supratherapeutic level of vancomycin (vancomycin trough concentration (C_{min}) > 20 μg/mL) [6] were screened from 23,182 samples referred for vancomycin concentration measurements from January 2018 to December 2019. Of the 3370 samples, we finally selected 28 patients with confirmed VAKI (VAKI group) by medical records review. VAKI was defined as follows: (1) an increase in SCr level of at least 0.5 mg/dL or (2) an increase from the pre-vancomycin treatment SCr of at least 50% from baseline on two consecutive days in the absence of an alternative explanation [22]. The control group consisted of three subgroups of 23 patients without VAKI who were suspected to have bacterial infections and who received vancomycin treatment (infection subgroup); 23 patients with CKD who were outpatients of the nephrology department with GFRs of less than 60 mL/min/1.73 m^2 and not receiving vancomycin treatment (CKD subgroup); and 23 subjects referred for routine health examination without abnormal medical findings (healthy control (HC) subgroup). Each subgroup was matched by age and sex to the VAKI group. All subjects met the following criteria: (1) age of at least 18 years; (2) no antidepressant drug therapy, including selective serotonin reuptake inhibitors; (3) no hemodialysis; and (4) no current pregnancy.

The study was conducted according to the guidelines of the Declaration of Helsinki and was approved by the Institutional Review Board of Samsung Medical Center Hospital

(IRB file no. 2017-12-038). A waiver of consent was obtained given the nature of the project. The study design and cohort subjects are described in Figure 1.

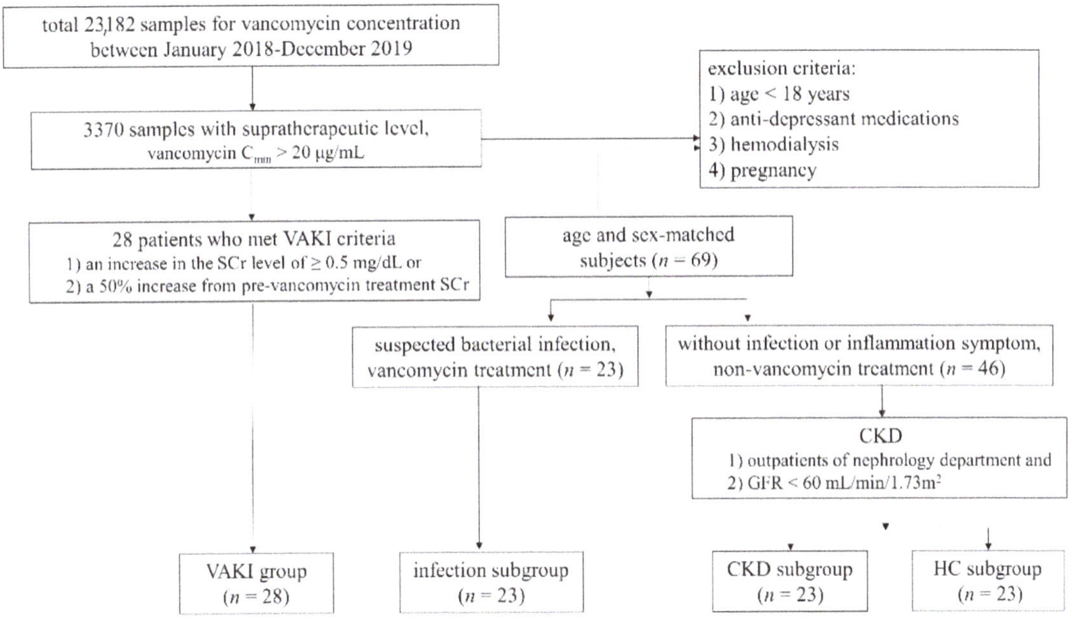

Figure 1. Schematic study flow and cohort population. Abbreviations: VAKI, vancomycin-associated acute kidney injury; CKD, chronic kidney disease; HC, healthy control; C_{min}, trough concentration; GFR, glomerular filtration rate; SCr, serum creatinine concentration.

2.2. Blood Sampling

Clinical samples referred for SCr and vancomycin concentration from subjects with vancomycin treatment (VAKI group and infection subgroup) or referred for SCr from subjects without vancomycin treatment (CKD subgroup and HC subgroup) were collected, and residual serum samples were used for our metabolomics analyses. Peripheral venous blood samples were collected by venipuncture into a serum separator tube and centrifuged for 10 min at 2270× *g*, and the serum was separated and stored in an Eppendorf tube at −70 °C until analysis.

2.3. The Exploratory Metabolome Profiling Analysis

Among 97 subjects, exploratory metabolome profiling analysis was performed using seven serum samples, including from two subjects in the VAKI group, two subjects in the infection subgroup, and three subjects in the HC subgroup (Table S1). To avoid interference with metabolomics profiling, we selectively included subjects without CKD history, previous malignancy, or chemotherapy by medical record review. The analysis was performed by Human Metabolome Technologies, Inc. (Yamagata, Japan) using capillary electrophoresis time-of-flight mass spectrometry with a fused silica capillary inner diameter of 50 μm × 80 cm (Agilent Technologies, Santa Clara, CA, USA) in two modes for cationic and anionic metabolites, as previously described [23]. Comparative analysis and hierarchical cluster analysis were performed using the relative peak area of putative metabolites. The Visualization and Analysis of Networks Containing Experimental Data (VANTED) software (VANTED version 2.5, www.vanted.org, accessed on 28 April 2021) was used for creating metabolic pathway maps based on peak profiles of putative metabolites.

2.4. The Amino Acids Profiling Analysis

Serum concentrations of 43 amino acids and amino acid derivatives were measured using the Agilent 6460 Triple Quadrupole MS/MS (Agilent Technologies Inc.), as previously described [24]. Briefly, protein precipitation was performed aTRAQ™ kits for Amino Acid Analysis in Physiological Fluids (AB Sciex, Foster City, CA, USA) according to the manufacturer's instructions. The Agilent 1260 Infinity LC system (Agilent Technologies Inc.) with a reverse-phase C18 (5 mm × 4.6 mm × 9150 mm) column at 50 °C was used to separate amino acids.

2.5. Measurement of 5-HT, and 5-HIAA

Serotonin (5-HT) and 5-hydroxyindoleacetic acid (5-HIAA) concentrations were measured by NeuroVIS (Cheonan, Korea), using AB SCIEX Triple Quadrupole 6500+ (AB Sciex, Framingham, MA, USA), in ExionLC™ Series UHPLC (AB Sciex, Framingham, MA, USA) with ACQUITY UPLC HSS T3 (2.1 × 100 mm, 1.8 µm, Waters, Milford, MA, USA) column at 30 °C was used to separate 5-HT and 5-HIAA by a gradient mobile phase of 0.1% formic acid in water and 5 mM of ammonium formate in methanol at a flow rate of 0.3 mL/min. These substances were monitored using positive electrospray ionization in multiple-reaction monitoring (MRM) mode (Q1 mass (m/z)/ Q3 mass (m/z); 177.1/160.1 5-HT; 192.0/146.1 5-HIAA, respectively).

2.6. Measurement of Serum Creatinine and Vancomycin Concentration

SCr was measured by an IDMS traceable Jaffe and enzymatic assay with Cobas c702 module (Roche Diagnostics Corp., Indianapolis, IN, USA). The GFR was calculated with the Cockcroft–Gault equation. Samples for measurement of vancomycin concentration were collected within one hour of regularly scheduled vancomycin infusion and were measured by the Cobas c702 analyzer using VANC3 (Roche Diagnostics GmbH, Indianapolis, IN, USA).

2.7. Statistical Analysis

Fisher's exact test or the chi-square test was used for categorical data. The Kruskal–Wallis test or Mann–Whitney U test with Bonferroni correction was used for continuous data. For discriminating metabolite concentrations between the VAKI group and non-VAKI subgroups, receiver operating characteristic (ROC) curves were drawn, and area under curve (AUC) analyses were performed. Data analyses were conducted using The Statistical Package for the Social Sciences version 25.0 (IBM Corporation, Armonk, NY, USA). p-values < 0.05 were considered statistically significant.

3. Results

3.1. Subject Characteristics

The clinical characteristics of this study cohort are summarized in Table 1. The age and sex distribution of a total of 97 subjects (mean age ± standard deviation (SD): 60.2 ± 13.1 years; 39% female) was balanced between the VAKI group and non-VAKI group. The body mass index (BMI) of the VAKI group was lower than that of the non-VAKI group (21.8 ± 3.2 kg/m^2 vs. 24.0 ± 3.2 kg/m^2; p = 0.002), whereas the BMI of the infection subgroup was comparable to that of the VAKI group. The SCr and GFR of the VAKI group were significantly different from those of the non-VAKI group (1.60 ± 0.96 mg/dL vs. 1.05 ± 0.51 mg/dL; p < 0.001 and 45.9 ± 12.8 mL/min/1.73 m^2 vs. 76.8 ± 32.4 mL/min/1.73 m^2; p < 0.001), whereas the SCr and GFR of the CKD subgroup were comparable to those of the VAKI group. The C-reactive protein (CRP) level of the VAKI group was higher than that of the non-VAKI group (5.78 ± 8.74 mg/dL vs. 2.39 ± 4.88 mg/dL; p < 0.001), whereas the CRP level of the infection subgroup was comparable to that of the VAKI group. The serum vancomycin C_{min} was significantly higher in the VAKI group than in the infection subgroup (27.3 ± 5.3 µg/mL vs. 14.2 ± 2.6 µg/mL; p < 0.001).

3.2. The Results of the Exploratory Metabolome Profiling Analysis

Of a total of 221 detected peaks, 192 were matched on the basis of the standard library of Human Metabolome Technologies, Inc. (Yamagata, Japan) from the exploratory metabolome profiling analysis. Hierarchical cluster analysis was performed using the relative peak area of putative metabolites (Figure 2A), and the total metabolomics pathway map was plotted with the obtained metabolome and statistic data (Figure S1). In the pathway map, amino acid metabolism as well as central carbon metabolism, urea cycle, lipid metabolism, nucleotide metabolism, and metabolism of coenzymes showed differences between the two subjects from the VAKI group and the five subjects from the non-VAKI group. When sorted by Human Metabolome Database (HMDB) classification, the concentrations of indolic compounds such as 3-indoxylsulfuric acid, IAA, and 5-MIAA were greater in the two subjects from the VAKI group compared with the five subjects from the non-VAKI group, whereas 5-HT and tryptophan (Trp) were not (Figure 2B). Similarly, when sorted by VANTED pathway cluster categorization, the concentrations of Trp metabolism compounds such as 5-MIAA, anthranilic acid, IAA, kynurenine, and quinolinic acid were greater in the two subjects from the VAKI group compared with the five subjects from the non-VAKI group, whereas 5-HT and Trp were not (Figure 2B).

3.3. The Results of the Amino Acids Profiling Analysis

A total of 43 amino acids and amino acid derivatives were measured by the amino acids profiling analysis. The comparisons of these concentrations between the VAKI group and non-VAKI group were summarized in Table 2. Beta-alanine, anserine, gamma-aminobutyric acid, aspartic acid, ethanolamine, glutamic acid, methionine, O-phosphoserine, taurine, argininosuccinic acid, and all-isoleucine showed reduced serum concentrations in the VAKI group compared to non-VAKI group, whereas beta-aminoisobutyric acid, citrulline, 3-methylhistidine, and homocitrulline showed elevated serum concentrations in the VAKI group compared to the non-VAKI group, respectively. Among these compounds, gamma-aminobutyric acid and O-phosphoserine showed reduced serum concentrations in the VAKI group compared to all of the non-VAKI subgroups, whereas citrulline and homocitrulline showed significantly elevated serum concentrations in the VAKI group compared to all of the non-VAKI subgroups, respectively. Trp showed no significant difference between the VAKI group and the non-VAKI group, while Trp showed an elevated serum concentration in the VAKI group compared to the HC subgroup.

Table 1. Clinical characteristics of subjects in this study.

Clinical Characteristics	Overall	VAKI	Non-VAKI							
			Total Non-VAKI	p-Value [a]	Infection	p-Value [a]	CKD	p-Value [a]	HC	p-Value [a]
Number of Subjects	97	28	69		23		23		23	
Age, yr (mean ± SD)	60.2 ± 13.1	60.4 ± 14.5	60.2 ± 13.1	N.S	58.4 ± 12.2	N.S	63.4 ± 12.1	N.S	58.4 ± 13.1	N.S
range	20–85	20–85	21–82		37–82		21–72		24–81	
Female (%)	38 (39%)	10 (36%)	28 (41%)	N.S	9 (39%)	N.S	9 (39%)	N.S	10 (43%)	N.S
BMI (kg/m^2)	23.4 ± 3.4	21.8 ± 3.2	24.0 ± 3.2	0.002	23.2 ± 3.5	N.S	24.3 ± 3.4	0.018	24.6 ± 2.7	0.02
SCr (mg/dL)	1.16 ± 0.60	1.60 ± 0.96	1.05 ± 0.51	<0.001	0.62 ± 0.21	<0.001	1.61 ± 0.29	N.S	0.85 ± 0.15	<0.001
GFR (mL/min/1.73 m^2)	67.8 ± 31.4	45.9 ± 12.8	76.8 ± 32.4	<0.001	107.8 ± 23.9	<0.001	39.5 ± 4.3	N.S	84.6 ± 11.8	<0.001
CRP (mg/dL)	0.26 ± 6.78	5.78 ± 8.74	2.39 ± 4.88	<0.001	7.19 ± 6.31	N.S	0.07 ± 0.94	<0.001	0.04 ± 0.71	<0.001
Vancomycin C_{min} (μg/mL)		27.3 ± 5.3			14.2 ± 2.6	<0.001				

Notes: a Each p-value was calculated relative to the VAKI group. Abbreviations: VAKI, vancomycin-associated acute kidney injury; CKD, chronic kidney disease; HC, healthy control; SD, standard deviation; BMI, body mass index; SCr, serum creatinine; CRP, C reactive protein; C_{min}, trough concentration.

Table 2. Comparison of serum amino acids and amino acid derivatives concentration between VAKI group and non-VAKI group.

Parameters (μmol/mL)	VAKI	Non-VAKI							
		Total Non-VAKI	p-Value [a]	Infection	p-Value [a]	CKD	p-Value [a]	HC	p-Value [a]
Alanine	417.3 (324.0–487.3)	296.0 (207.5–393.3)	0.002	290.6 (241.5–529.9)	N.S	201.2 (141.7–267.2)	<0.001	371.7 (339.4–420.3)	N.S
Beta-alanine	2.1 (1.4–2.9)	4.7 (2.4–7.2)	<0.001	7.3 (6.2–8.6)	<0.001	3.5 (2.4–6.5)	0.01	2.4 (1.9–4.7)	N.S
Anserine	0.0 (0.0–0.0)	0.0 (0.0–0.7)	<0.001	0.5 (0.0–0.8)	<0.001	0.0 (0.0–0.0)	N.S	0.6 (0.0–0.9)	<0.001
Aminoadipic acid	0.0 (0.0–1.0)	0.6 (0.0–1.1)	N.S	0.6 (0.0–1.0)	N.S	0.7 (0.0–1.1)	N.S	0.5 (0.0.–0.9)	N.S
Alpha-aminobutyric acid	10.2 (8.2–13.4)	13.5 (8.5–17.9)	0.05	14.9 (11.4–19.1)	0.002	6.4 (3.9–11.5)	0.037	16.6 (11.7–20.0)	<0.001
Gamma-aminobutyric acid	0.0 (0.0–0.2)	1.4 (0.9–2.0)	<0.001	1.9 (1.3–2.5)	<0.001	0.9 (0.5–1.9)	<0.001	1.3 (1.1–1.6)	<0.001
Beta-aminoisobutyric acid	2.8 (1.4–4.8)	0.7 (0.0–2.3)	0.002	0.5 (0.0–1.7)	0.001	2.6 (0.0–9.8)	N.S	0.7 (0.0–1.1)	<0.001
Arginine	72.8 (59.2–84.2)	61.9 (37.3–95.0)	N.S	62.0 (40.6–143.2)	N.S	37.3 (23.2–59.1)	<0.001	73.7 (59.4–117.9)	N.S
Asparagine	29.5 (23.7–34.9)	28.2 (19.4–36.3)	N.S	30.1 (20.4–43.0)	N.S	19.4 (13.0–33.5)	0.011	30.5 (23.2–36.5)	N.S
Aspartic acid	11.3 (8.5–16.4)	19.6 (11.3–36.0)	0.01	24.2 (19.0–41.6)	<0.001	11.8 (9.1–37.3)	N.S	15.3 (10.4–32.6)	N.S
Carnosine	0.0 (0.0–0.0)	0.0 (0.0–0.0)	0.024	0.0 (0.0–0.3)	0.017	0.0 (0.0–0.0)	N.S	0.0 (0.0–0.0)	0.039
Citrulline	35.0 (29.3–39.4)	20.9 (13.2–30.5)	<0.001	17.6 (11.2–39.5)	0.002	14.1 (9.8–22.0)	<0.001	27.4 (22.4–33.0)	0.007
Cystathionine	0.0 (0.0–0.0)	0.0 (0.0–0.8)	N.S	0.0 (0.0–0.0)	N.S	0.8 (0.0–1.6)	0.002	0.6 (0.0–0.5)	N.S
Cystine	3.7 (1.9–8.1)	5.6 (2.0–13.0)	N.S	11.4 (7.8–13.5)	<0.001	10.4 (5.2–29.4)	<0.001	1.2 (0.0–2.0)	<0.001
Ethanolamine	5.8 (4.8–6.6)	9.1 (5.5–11.9)	0.037	8.9 (7.0–11.6)	0.01	5.5 (3.4–10.3)	N.S	10.9 (6.7–15.4)	0.003

Table 2. *Cont.*

Parameters	VAKI	Non-VAKI							
(μmol/mL)		Total Non-VAKI	*p*-Value [a]	Infection	*p*-Value [a]	CKD	*p*-Value [a]	HC	*p*-Value [a]
Glutamic acid	54.1 (39.8–76.9)	82.2 (54.1–114.5)	0.016	100.4 (72.5–158.6)	0.002	58.9 (35.6–113.1)	N.S	72.0 (56.4–99.6)	0.034
Glutamine	313 (275.1–370.9)	257.5 (145.1–378.5)	N.S	232.6 (170.5–326.9)	0.022	140.5 (97.4–244.7)	<0.001	382.3 (298.5–570.1)	0.028
Glycine	136 (117.2–198.9)	142.7 (119.8–266.4)	N.S	149.7 (125.4–267.9)	N.S	110.3 (98.4–244.7)	0.043	165.9 (133.3–288.5)	N.S
Histidine	34.9 (32.1–38.6)	40.2 (29.4–57.7)	N.S	37.1 (28.2–54.3)	N.S	35.7 (17.3–46.6)	N.S	47.0 (37.7–80.2)	0.001
Homocysteine	0.0 (0.0–0.0)	0.0 (0.0–0.0)	N.S	0.0 (0.0–0.0)	N.S	0.0 (0.0–0.0)	N.S	0.0 (0.0–0.0)	N.S
Dehydroxylysine	0.0 (0.0–0.0)	0.0 (0.0–0.0)	N.S	0.0 (0.0–0.0)	N.S	0.0 (0.0–0.0)	N.S	0.0 (0.0–0.0)	N.S
Hydroxyproline	7.6 (5.4–10.4)	6.1 (3.9–9.4)	N.S	6.6 (3.9–9.4)	N.S	6.1 (3.1–14.9)	N.S	5.2 (4.4–8.8)	N.S
Isoleucine)	32.0 (25.8–39.0)	35.2 (25.5–51.5)	N.S	41.0 (31.1–59.0)	0.041	24.4 (16.4–40.7)	N.S	37.9 (29.2–53.1)	N.S
Leucine	80.2 (59.3–99.1)	93.8 (72.1–137.3)	N.S	103.1 (87.4–140.8)	0.015	53.1 (34.5–105.9)	N.S	106.7 (81.6–147.4)	0.006
Lysine	99.2 (83.1–114.8)	122.2 (78.2–157.5)	N.S	122.2 (83.7–170.2)	N.S	73.1 (46.0–127.1)	N.S	131.1 (105.5–185.6)	0.001
Methionine	6.0 (5.1–8.9)	16.1 (7.2–29.7)	0.002	9.3 (5.9–15.9)	N.S	8.0 (4.6–15.9)	N.S	34.7 (26.9–40.0)	<0.001
1-methylhistidine	7.4 (5.9–9.9)	9.2 (6.2–12.6)	N.S	13.6 (10.1–14.9)	<0.001	6.3 (4.1–10.1)	N.S	8.8 (6.2–10.3)	N.S
3-methylhistidine	4.8 (4.0–7.2)	3.4 (1.6–6.8)	0.024	1.7 (0.9–10.7)	0.018	5.7 (4.2–12.3)	N.S	2.1 (1.6–4.3)	<0.001
Ornithine	48.9 (38.5–77.1)	53.7 (40.9–92.6)	N.S	88.4 (51.3–117.3)	0.039	51.9 (35.7–90.3)	N.S	49.8 (42.5–69.9)	N.S
Phenylalanine	51.5 (46.2–67.7)	64.6 (47.0–87.6)	N.S	76.4 (55.5–100.3)	0.003	65.4 (40.9–104.7)	N.S	54.9 (37.2–82.4)	N.S
O-phosphoethanolamine	0.0 (0.0–0.0)	0.0 (0.0–0.0)	N.S	0.0 (0.0–0.0)	N.S	0.0 (0.0–0.0)	N.S	0.0 (0.0–0.0)	N.S
O-phosphoserine	1.6 (1.2–2.0)	3.0 (1.7–4.6)	<0.001	3.8 (1.9–.5.1)	0.002	2.4 (1.8–3.5)	0.009	3.4 (1.6–8.3)	0.002
Proline	97.7 (76.9–127.9)	83.3 (49.3–129.9)	N.S	89.5 (86.7–141.9)	N.S	53.2 (44.4–106.1)	0.013	74.9 (57.1–169.0)	N.S
Sarcosine	0.0 (0.0–0.6)	0.6 (0.0–0.8)	N.S	0.6 (0.0–1.2)	N.S	0.0 (0.0–0.8)	N.S	0.6 (0.0–0.8)	N.S
Serine	78.2 (69.2–93.9)	88.0 (70.1–138.7)	N.S	100.3 (86.7–141.9)	0.01	63.3 (40.8–82.7)	0.021	111.0 (75.1–1843.4)	0.009
Taurine	40.6 (32.6–50.4)	65.2 (34.7–99.3)	0.02	69.8 (39.9–109.3)	N.S	34.7 (29.1–73.9)	N.S	81.0 (60.8–127.1)	<0.001
Threonine	66.2 (52.9–80.8)	59.0 (40.7–89.4)	N.S	64.6 (46.0–114.2)	N.S	38.1 (30.4–63.9)	0.001	59.8 (50.3–118.3)	N.S
Tryptophan	21.0 (17.6–24.6)	25.0 (13.0–36.9)	N.S	18.9 (12.2–37.0)	N.S	14.3 (10.8–27.2)	N.S	35.4 (25.0–47.8)	<0.001
Tyrosine	34.1 (26.7–37.2)	39.1 (26.9–57.9)	N.S	36.4 (25.2–37.0)	N.S	37.2 (18.4–54.4)	N.S	41.7 (25.0–48.6)	0.011
Valine	129.3 (114.9–160.5)	142.1 (101.3–220.3)	N.S	178.1 (138.0–222.2)	0.034	91.9 (51.4–140.8)	0.009	156.9 (125.5–256.8)	0.014
Argininosuccinic acid	0.0 (0.0–0.0)	0.0 (0.0–1.1)	<0.001	0.6 (0.0–1.3)	<0.001	0.0 (0.0–0.6)	0.002	0.0 (0.0–2.0)	<0.001
Homocitrulline	1.1 (0.7–1.5)	0.0 (0.0–0.7)	<0.001	0.0 (0.0–0.4)	<0.001	0.0 (0.0–0.8)	0.001	0.0 (0.0–0.7)	<0.001
All-isoleucine	0.0 (0.0–0.0)	0.0 (0.0–0.8)	0.003	0.0 (0.0–0.0)	N.S	0.0 (0.0–0.9)	0.007	0.8 (0.0–1.4)	<0.001

Notes: a Each *p* value was calculated using Mann–Whitey U test, compared to VAKI group. Abbreviations: VAKI, vancomycin associated acute kidney injury; CKD, chronic kidney disease; HC, healthy control.

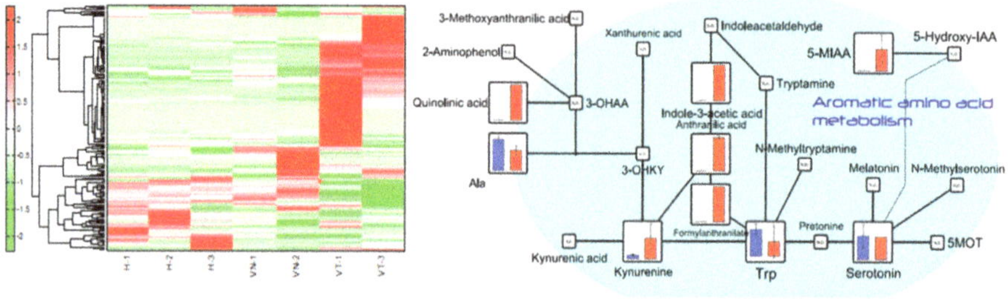

(A) (B)

Figure 2. Serotonin pathway map and kynurenine pathway map with the obtained metabolome data from exploratory metabolome profiling analysis (cropped). Notes: Hierarchical cluster analysis was performed using the relative peak area of putative metabolites from seven serum samples, including from two subjects in the VAKI group (VT-1 and VT-3), two subjects in the infection subgroup (VN-1 and VN-2), and three HCs (H-1, H-2, and H-3) (**A**), and the metabolomics pathway map was created using the VANTED software (**B**). The blue and red boxes represent the mean relative peak area of the component in non-VAKI subjects and VAKI subjects, respectively. The cropped serotonin pathway map and kynurenine pathway map showed that VAKI subjects had significantly increased concentrations of indolic compounds such as 3-indoxylsulfuric acid, IAA, and 5-MIAA compared with non-VAKI subjects. Abbreviations: 3-OHAA, 3-hydroxyanthranilic acid; 3-OHKY, 3-hydroxykynurenine; 5-MOT, 5- methoxytryptamine; 5-hydroxy-IAA, 5-hydroxyindoleacetic acid; 5-MIAA, 5-methoxyindole-3-acetic acid; Ala, alanine; Trp, tryptophan; VANTED, Visualization and Analysis of Networks Containing Experimental Data.

3.4. The Results of Measurement of 5-HT and 5-HIAA

The comparisons of the serum concentrations of Trp, 5-HT, 5-HIAA, and the ratios of 5-HT/Trp and 5-HIAA/5-HT between the VAKI group and the non-VAKI group are summarized in Table 3. Trp concentrations showed no significant difference between the VAKI group and the total non-VAKI group (median, 21.0 μmol/mL vs. 25.0 μmol/mL, p = not significant), while Trp showed a reduced concentration in the VAKI group compared with the HC subgroup (21.0 μmol/mL vs. 35.4 μmol/mL, $p = 0.017$) (Figure 3A). 5-HT and 5-HT/Trp showed a reduced concentration (47.8 ng/mL vs. 251.2 ng/mL, $p < 0.001$) (Figure 3B) and ratio (2.0 vs. 9.8, $p < 0.001$) (Figure 3D) in the VAKI group compared to the non-VAKI group, whereas 5HIAA and 5-HIAA/5-HT showed elevated concentrations (249.4 ng/mL vs. 84.5 ng/mL, $p < 0.001$) (Figure 3C) and ratio (6.3 vs. 0.4, $p < 0.001$) (Figure 3E). Among the non-VAKI subgroups, the 5-HT concentration of the infection subgroup was lower than that of the CKD subgroup (87.5 ng/mL vs. 337.6 ng/mL, $p = 0.011$) or HC subgroup (87.5 ng/mL vs. 345.2 ng/mL, $p = 0.015$) (Figure 3B), while the 5-HIAA/5-HT ratio in the infection subgroup was higher than that of HC subgroup (0.9 vs. 0.3, $p = 0.016$) (Table S2 and Figure 3E). The concentration of 5-HIAA showed no significant differences between the infection subgroup, CKD subgroup, and HC subgroup (Figure 3C).

Table 3. Comparison of serum tryptophan, serotonin, and 5-HIAA concentration between the VAKI group and non-VAKI groups.

Parameters	VAKI	Non-VAKI							
		Total Non-VAKI	*p*-Value [a]	Infection	*p*-Value [a]	CKD	*p*-Value [a]	HC	*p*-Value [a]
Trp (μmol/mL)	21.0 (17.6–24.6)	25.0 (13.0–36.9)	N.S	18.9 (12.2–37.0)	N.S	14.3 (10.8–27.2)	N.S	35.4 (25.0–47.8)	0.004
5-HT (ng/mL)	47.8 (16.6–135.5)	251.2 (114.5–389.3)	<0.001	87.5 (50.6–250.2)	N.S	337.6 (195.8–520.0)	<0.001	345.2 (218.4–386.4)	<0.001
5-HIAA (ng/mL)	249.4 (113.0–442.2)	84.5(36.3–154.5)	<0.001	80.8 (35.2–258.7)	0.006	85.7 (70.4–111.8)	0.003	38.5 (27.7–171.2)	<0.001
5-HT/Trp	2.0 (0.7–6.5)	9.8(3.8–18.0)	<0.001	4.3 (1.6–16.2)	N.S	20.0 (8.6–56.9)	<0.001	9.4 (4.8–12.1)	0.017
5-HIAA/5-HT	6.3 (1.9–14.0)	0.4 (0.2–0.9)	<0.001	0.9 (0.5–1.9)	0.032	0.2 (0.2–0.5)	<0.001	0.3 (0.1–0.7)	<0.001

Notes: a Each *p*-value was calculated using the Mann–Whitey U test and adjusted by Bonferroni correction relative to the VAKI group. Abbreviations: VAKI, vancomycin-associated acute kidney injury; CKD, chronic kidney disease; HC, healthy control; Trp, tryptophan; 5-HT, serotonin; 5-HIAA, 5-hydroxyindoleacetic acid; the ratio of serotonin to tryptophan; 5-HIAA/5-HT, the ratio of 5-hydroxyindoleacetic acid to serotonin.

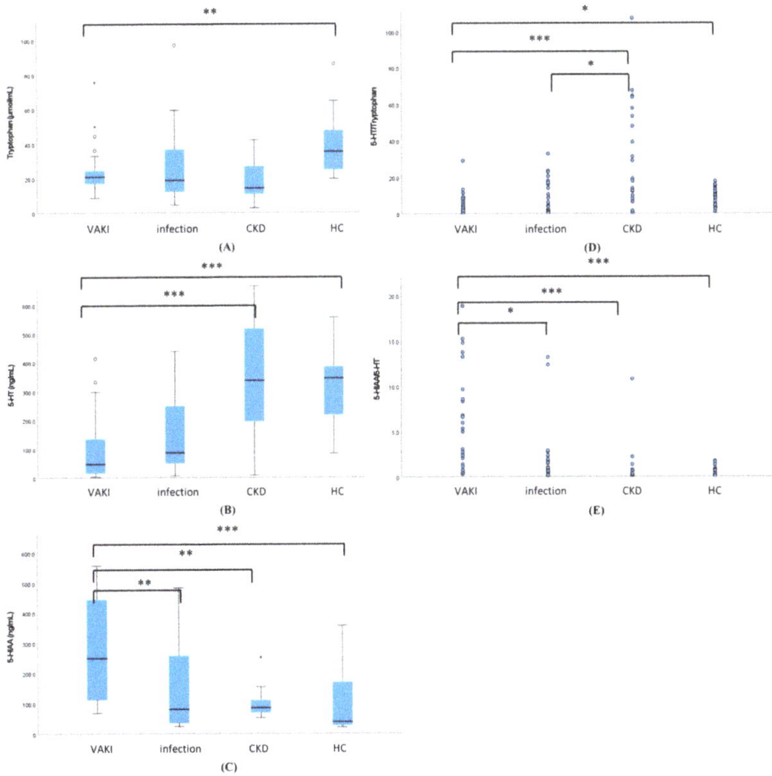

Figure 3. Comparison of serum concentration of tryptophan, 5-HT, 5-HIAA, and the ratio of 5-HT to tryptophan (5-HT/Trp) and 5-HIAA to 5-HT (5-HIAA/5-HT) between the VAKI group and non-VAKI subgroup. Notes: (**A**) Serum concentration of tryptophan (μmol/mL); (**B**) Serum concentration of 5-HT (ng/mL); (**C**) Serum concentration of 5-HIAA (ng/mL); (**D**) the ratio of 5-HT to tryptophan (5-HT/Trp); and (**E**) the ratio of 5-HIAA to 5-HT (5-HIAA/5-HT) were compared between the VAKI group and non-VAKI subgroups. Each box represents the interquartile range. Each line inside the boxes is plotted at the median. Each whisker means the 10th and 90th percentiles. * $p < 0.5$, ** $p < 0.01$, *** $p < 0.001$, respectively. Abbreviations: VAKI, vancomycin-associated acute kidney injury; CKD, chronic kidney disease; HC, healthy control; 5-HT, serotonin; 5-HIAA, 5-hydroxyindoleacetic acid; 5-HT/Trp, the ratio of serotonin to tryptophan; 5-HIAA/5-HT, the ratio of 5-hydroxyindoleacetic acid to serotonin.

3.5. The AUC-ROC Analysis of 5-HT 5-HIAA, and 5-HIAA/5-HT for VAKI

The AUC analysis of 5-HT, 5-HIAA, and 5-HIAA/5-HT for VAKI is illustrated in Figure 4. Of 97 subjects in the VAKI group and non-VAKI group, the AUCs for 5-HT, 5-HIAA, and 5-HIAA/5-HT were 0.200 (95% confidence interval, 0.105–0.296), 0.795 (0.704–0.885), and 0.884 (0.881–0.956), respectively (Figure 4A). Of 51 subjects of the VAKI group and infection subgroup, the AUCs for 5-HT, 5-HIAA, and 5-HIAA/5-HT were 0.345 (0.193–0.496), 0.747 (0.608–0.886), and 0.795 (0.671–0.919), respectively (Figure 4B). Of 51 subjects in the VAKI group and CKD subgroup, the AUCs for 5-HT, 5-HIAA, and 5-HIAA/5-HT were 0.149 (0.039–0.259), 0.848 (0.744–0.951), and 0.925 (0.852–0.999), respectively (Figure 4C). Finally, of 51 subjects in the VAKI group and HC subgroup, the AUCs for 5-HT, 5-HIAA, and 5-HIAA/5-HT were 0.107 (0.019–0.195), 0.789 (0.664–0.914), and 0.930 (0.866–0.995), respectively (Figure 4D).

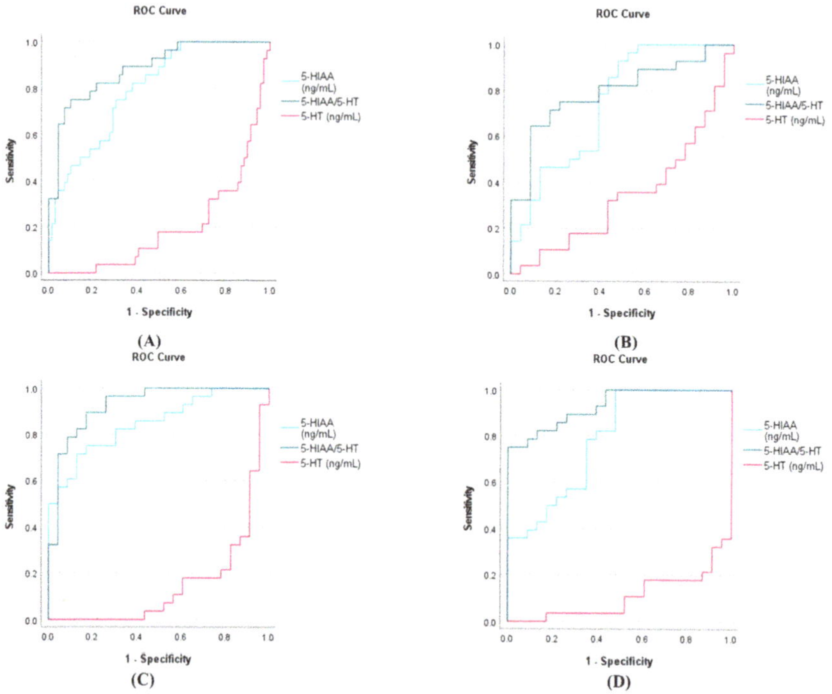

Figure 4. AUC analysis of 5-HT, 5-HIAA, and 5-HIAA/5-HT for VAKI in the VAKI group and non-VAKI subgroups. Notes: The AUC analysis of 5-HT, 5-HIAA, and 5-HIAA/5-HT was performed (**A**) for the VAKI group and all of the non-VAKI subgroups, (**B**) for the VAKI group and infection subgroup, (**C**) for the VAKI group and CKD subgroup, and (**D**) for the VAKI group and HC subgroup. Abbreviations: VAKI, vancomycin associated acute kidney injury; CKD, chronic kidney disease; HC, healthy control; 5-HT, serotonin; 5-HIAA, 5-hydroxyindoleacetic acid; 5-HT/Trp, the ratio of 5-hydroxyindoleacetic acid to serotonin.

4. Discussion

Our study is the first comparative metabolomics investigation of serum concentrations of Trp and its metabolites 5-HT, 5-HIAA, and 5-HIAA/5-HT in VAKI and non-VAKI groups, with the latter group consisting of the infection, CKD, and HC subgroups. Trp is an essential amino acid in human nutrition and is a substrate for tissue protein synthesis [25,26]. In total, 95% of Trp is metabolized via the kynurenine pathway [27], and the ratio of kynurenine to Trp (Kyn/Trp) has been studied for its impact in the settings of inflammation, neurodegeneration, and end-stage kidney disease (ESRD) [28–31]. Other metabolites of the kynurenine pathway, such as anthranilic acid and indole-3-acetic acid (IAA), have been reported as possible surrogate markers for inflammation and ESRD [32,33]. The results from our exploratory metabolome profiling analysis revealed that VAKI patients had higher serum values of 5-methoxyindoleacetic acid (5-MIAA), anthranilic acid, IAA, kynurenine, quinolinic acid, and Kyn/Trp compared with non-VAKI patients (Figure 2B), and these findings were similar to those of a previous report in AKI patients [34]. However, Kyn/Trp might not be suitable for detecting VAKI because it tends to be higher in patients with infection and CKD compared with HCs [32,33,35,36]. Furthermore, Kyn/Trp and other metabolites of the kynurenine pathway would be affected by dietary conditions, including Trp supplementation [37].

Interestingly, our data revealed a gradual increase of the 5-HT pathway components 5-HT and 5-MIAA in VAKI subjects compared with non-VAKI subjects, while 5-HIAA was not detected in the exploratory metabolome profiling analysis (Figure 2B). 5-HT pathway compounds, such as 5-HT, 5-HIAA, and 5-MIAA, are derived from 5% of Trp [27]. 5-HT is synthesized by tryptophan hydroxylase and aromatic L-amino acid decarboxylase via 5-hydroxytryptophan [38]. 5-HIAA is a primary metabolite derived from 5-HT by mitochondrial monoamine oxidase A (MAO-A) [39], and the ratio of 5-HIAA to 5-HT (5-HIAA/5-HT) is a surrogate marker for serotonergic activity or MAO-A activity [40,41]. MAO-A is an enzyme that generates hydrogen peroxide, one of the sources of oxygen free radicals [42]. An increase in oxygen free radicals is attributed to oxidative stress in living organisms. Proinflammatory oxidation, mitochondrial dysfunction, and cellular apoptosis have been studied as principal mechanisms of VAKI [6]. Both 5-HT and 5-HIAA have been considered in the occurrence of ischemia/reperfusion damage in experimental animal models [43,44]. Khovanskaya et al. reported an increase in urinary 5-HIAA concentration, consistent with a period of decrease in tissue 5-HT level, in an ischemic kidney model of rabbits [43]. Du et al. showed that increase in interstitial 5-HIAA and decrease in interstitial 5-HT occurred at a reperfusion stage rather than an ischemia stage in an ischemic myocardial rat model [44]. These phenomena are explained by the role of serotonin as an antioxidant [45] and increased MAO-A activity in oxidative stress condition [46,47]. Therefore, we focused on 5-HT and 5-HIAA as potential biomarkers for detecting VAKI among candidate compounds from exploratory metabolome profiling analysis.

BCAAs and sulfur-containing amino acids (SCAAs) are used for nutrient supplement therapy in septic patients or in liver disease [48,49], and these amino acids are considered as antioxidants to protect the oxidative stress [48–50]. Carnosine associated amino acids (CAAs) act a role as a physiological buffer in human [51,52], and the previous study has been examined with protective effects of CAAs in AKI due to septic shock in a rat model [52]. In the results from our amino acids profiling analysis, BCAAs (such as all-isoleucine, leucine and valine) and SCAAs (such as serine, O-phosphoserine, methionine, and taurine) showed reduced concentrations in the VAKI group and CKD subgroup compared to the HC subgroup, whereas urea cycle associated amino acids (such as arginine, citrulline, and homocitrulline), and cystine showed elevated concentrations in the VAKI group compared to the CKD subgroup, respectively (Table 2 and Table S3). Furthermore, CAAs (such as anserine and carnosine) showed reduced concentrations in the VAKI group and CKD subgroup compared to the HC subgroup, whereas their precursors, such as beta-alanine and 1-methyhistidine, showed elevated concentrations in the infection subgroup compared to the VAKI group (Table 2 and Table S3). These results suggest that the reduced concentrations of Trp and 5-HT were associated with oxidative stress in the VAKI group.

AKI biomarker concentrations also were high in patients with CKD, and quantitative markers and cutoff thresholds would be useful for distinguishing biomarker elevations in VAKI patients with CKD [13]. In the current study, the 5-HIAA concentration was three-fold higher and the 5-HIAA/5-HT ratio was 15-fold higher in the VAKI group than in the non-VAKI group, whereas the 5-HT concentration was one-fifth lower in the VAKI group than in the non-VAKI group (Figure 3D). The AUCs for 5-HT and 5-HIAA were comparable between the VAKI group and non-VAKI group. Meanwhile, the AUC for the 5-HIAA/5-

HT ratio was superior for identifying VAKI compared with those of 5-HT and 5-HIAA (Figure 4A). In the VAKI group and the CKD subgroup or HC subgroup, AUC-ROC for 5-HIAA/5-HT were 0.925 or 0.930, respectively (Figure 4C,D). However, when contrasting the VAKI group and the infection subgroup, the AUC for 5-HIAA/5-HT was 0.795, which was the lowest AUC value recorded among the non-VAKI subgroups (Figure 4B). When the cut-off value of 5-HIAA/5-HT ratio was set to 2.0 using the maximum Kolmogorov–Smirnov metric, specificity was lowest between the VAKI group and the infection subgroup (VAKI vs. infection, 78.3%; VAKI vs. CKD, 91.3%; VAKI vs. HC, 100%), while the sensitivity was the same at 75.0% for all non-VAKI subgroups. Our data showed that the 5-HIAA/5-HT ratios of the infection subgroup and the VAKI group were higher than that of the HC subgroup (Supplementary Table S3 and Figure 3E), and these results were similar to those of a previous study involving septic shock patients [34]. However, the CKD subgroup displayed relatively increased 5-HT and decreased Trp concentrations compared with the other groups, and the 5-HT/Trp ratio of the CKD subgroup was significantly higher than that of the VAKI group and the infection subgroup. Depending on the decrease in GFR, the 5-HT/Trp or 5-HIAA/5-HT ratio showed different alteration in the CKD subgroup or VAKI group (Figure 5A,B). These findings suggest that the 5-HIAA/5-HT ratio is a surrogate marker for acute reactive reactions such as inflammation or hypoxia rather than chronic kidney damage or dysfunction, while the 5-HT/Trp ratio can indicate chronic kidney damage or dysfunction in patients with CKD.

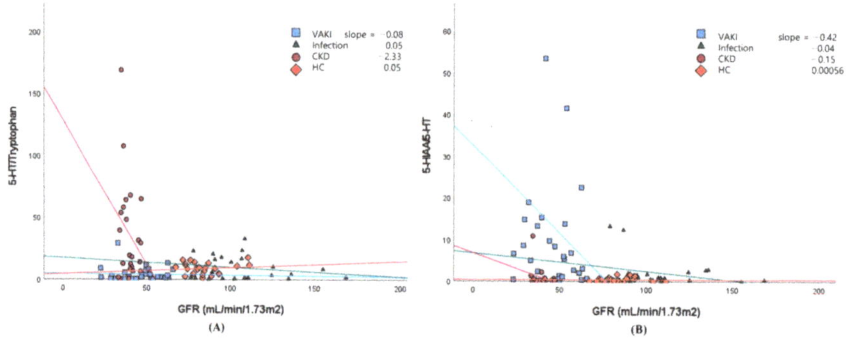

Figure 5. Scatterplots assessing the relationship between GFR and the 5-HT/Trp or 5-HIAA/5-HT ratio in the VAKI group and non-VAKI subgroups. Notes: The scatterplots assessing the relationship between GFR (mL/min/1.73 m^2) and the 5-HT/Trp ratio (**A**) or 5-HIAA/5-HT ratio (**B**) were plotted in the VAKI group, infection subgroup, CKD subgroup, and HC subgroup. The reference line of the 5-HT/Trp ratio showed a markedly negative slope for GFR in the CKD subgroup (purple), whereas that of the 5-HIAA/5-HT ratio showed a markedly negative slope for GFR in the VAKI group (blue). Abbreviations: VAKI, vancomycin-associated acute kidney injury; CKD, chronic kidney disease; HC, healthy control; GFR, glomerular filtration rate; SCr, serum creatinine concentration; Trp, tryptophan; 5-HT, serotonin; 5-HIAA, 5-hydroxyindoleacetic acid; 5-HT/Trp, the ratio of serotonin to tryptophan; 5-HIAA/5-HT, the ratio of 5-hydroxyindoleacetic acid to serotonin.

Furthermore, our study revealed that the serum 5-HT and 5-HIAA concentrations were not related to serum Trp concentration (Table 3 and Figure 3A,B), whereas the serum kynurenine concentration and Kyn/Trp ratio were affected by dietary Trp supplementation or serum Trp concentration [37,53]. These findings indicate that 5-HT and 5-HIAA would be more stable markers than serum kynurenine and Kyn/Trp despite nutrient intake and sociodemographic factors.

Our study has limitations. We collected single serum samples referred for Scr from subjects with elevated vancomycin C_{min}. An additional study with serial samples from the initial period of vancomycin treatment should be conducted to evaluate the performance of the 5-HIAA/5-HT ratio for earlier detection of VAKI.

Taken together, our results reveal that the serum concentration of 5-HT was independently decreased and the serum concentration of 5-HIAA and the 5-HIAA/5-HT ratio were increased in the VAKI group compared with all non-VAKI subgroups regardless of the serum Trp concentration.

5. Conclusions

In conclusion, our study demonstrated that increased 5-HIAA/5-HT ratio has potential to act as a novel biomarker for VAKI detection. Additionally, we showed that the 5-HIAA/5-HT ratio would be a useful indicator for oxidative stress and inflammation.

Supplementary Materials: The following are available online at https://www.mdpi.com/article/10.3390/antiox10060895/s1: Figure S1: Total pathway map and kynurenine pathway map with the obtained metabolome data from exploratory metabolome profiling analysis (full version); Table S1: Clinical history and characteristics of seven subjects in the exploratory metabolome profiling analysis; Table S2: Comparison of concentrations of serum amino acids and amino acid derivatives correlated with glomerular filtration rate between VAKI group and non-VAKI group; Table S3: Comparison of serum tryptophan, serotonin, and 5-HIAA concentrations from non-VAKI subgroups, including an infection subgroup, CKD subgroup, and healthy control subgroup.

Author Contributions: Conceptualization, S.-Y.L.; data curation, H.-S.L., S.-M.K., J.-H.J., and H.-D.P.; writing—original draft preparation, H.-S.L. All authors have read and agreed to the published version of the manuscript.

Funding: This research was funded by the Ministry of Trade, Industry and Energy (MOTIE, Korea) under Industrial Technology Innovation Program No.10080648.

Institutional Review Board Statement: The study was conducted according to the guidelines of the Declaration of Helsinki and was approved by the Institutional Review Board of Samsung Medical Center Hospital (IRB file no. 2017-12-038).

Informed Consent Statement: A waiver of consent was obtained given the nature of the project.

Data Availability Statement: Not applicable.

Conflicts of Interest: The authors declare no conflict of interest.

References

1. Nash, K.; Hafeez, A.; Hou, S. Hospital-acquired renal insufficiency. *Am. J. Kidney Dis.* **2002**, *39*, 930–936. [CrossRef]
2. Neveu, H.; Kleinknecht, D.; Brivet, F.; Loirat, P.; Landais, P. Prognostic factors in acute renal failure due to sepsis. Results of a prospective multicentre study. *Nephrol. Dial. Transplant.* **1996**, *11*, 293–299. [CrossRef] [PubMed]
3. Joyce, E.L.; Kane-Gill, S.L.; Fuhrman, D.Y.; Kellum, J.A. Drug-associated acute kidney injury: Who's at risk? *Pediatr. Nephrol.* **2017**, *32*, 59–69. [CrossRef] [PubMed]
4. Gaggl, M.; Pate, V.; Stürmer, T.; Kshirsagar, A.V.; Layton, J.B. The comparative risk of acute kidney injury of vancomycin relative to other common antibiotics. *Sci. Rep.* **2020**, *10*, 17282. [CrossRef] [PubMed]
5. Filippone, E.J.; Kraft, W.K.; Farber, J.L. The nephrotoxicity of vancomycin. *Clin. Pharmacol. Ther.* **2017**, *102*, 459–469. [CrossRef] [PubMed]
6. Bamgbola, O. Review of vancomycin-induced renal toxicity: An update. *Ther. Adv. Endocrinol. Metab.* **2016**, *7*, 136–147. [CrossRef] [PubMed]
7. Blaskovich, M.A.T.; Hansford, K.A.; Butler, M.S.; Jia, Z.; Mark, A.E.; Cooper, M.A. Developments in glycopeptide antibiotics. *ACS Infect. Dis.* **2018**, *4*, 715–735. [CrossRef]
8. Bakthavatchalam, Y.D.; Nabarro, L.E.; Veeraraghavan, B. Evolving rapid methicillin-resistant *Staphylococcus aureus* detection: Cover all the bases. *J. Glob. Infect. Dis.* **2017**, *9*, 18–22.
9. Van Meer, L.; Moerland, M.; Cohen, A.F.; Burggraaf, J. Urinary kidney biomarkers for early detection of nephrotoxicity in clinical drug development. *Br. J. Clin. Pharmacol.* **2014**, *77*, 947–957. [CrossRef]
10. Griffin, B.R.; Faubel, S.; Edelstein, C.L. Biomarkers of drug-induced kidney toxicity. *Ther. Drug Monit.* **2019**, *41*, 213–226. [CrossRef]
11. Vaidya, V.S.; Ferguson, M.A.; Bonventre, J.V. Biomarkers of acute kidney injury. *Annu. Rev. Pharmacol. Toxicol.* **2008**, *48*, 463–493. [CrossRef]
12. Holzscheiter, L.; Beck, C.; Rutz, S.; Manuilova, E.; Domke, I.; Guder, W.G.; Hofmann, W. NGAL, L-FABP, and KIM-1 in comparison to established markers of renal dysfunction. *Clin. Chem. Lab. Med.* **2014**, *52*, 537–546. [CrossRef]
13. Devarajan, P. Biomarkers for the early detection of acute kidney injury. *Curr. Opin. Pediatr.* **2011**, *23*, 194–200. [CrossRef] [PubMed]

antioxidants

MDPI

Article

Characterizing the Role of Biologically Relevant Fluid Dynamics on Silver Nanoparticle Dependent Oxidative Stress in Adherent and Suspension In Vitro Models

Katherine E. Burns, Robert F. Uhrig, Maggie E. Jewett, Madison F. Bourbon and Kristen A. Krupa *

Department of Chemical and Materials Engineering, University of Dayton, Dayton, OH 45469-0256, USA; burnsk7@udayton.edu (K.E.B.); uhrigr1@udayton.edu (R.F.U.); jewettm3@udayton.edu (M.E.J.); mbourbon1@udayton.edu (M.F.B.)

* Correspondence: kkrupa1@udayton.edu; Tel.: +1-937-229-2627

Abstract: Silver nanoparticles (AgNPs) are being employed in numerous consumer goods and applications; however, they are renowned for inducing negative cellular consequences including toxicity, oxidative stress, and an inflammatory response. Nanotoxicological outcomes are dependent on numerous factors, including physicochemical, biological, and environmental influences. Currently, NP safety evaluations are carried out in both cell-based in vitro and animal in vivo models, with poor correlation between these mechanisms. These discrepancies highlight the need for enhanced exposure environments, which retain the advantages of in vitro models but incorporate critical in vivo influences, such as fluid dynamics. This study characterized the effects of dynamic flow on AgNP behavior, cellular interactions, and oxidative stress within both adherent alveolar (A549) and suspension monocyte (U937) models. This study determined that the presence of physiologically relevant flow resulted in substantial modifications to AgNP cellular interactions and subsequent oxidative stress, as assessed via reactive oxygen species (ROS), glutathione levels, p53, NFκB, and secretion of pro-inflammatory cytokines. Within the adherent model, dynamic flow reduced AgNP deposition and oxidative stress markers by roughly 20%. However, due to increased frequency of contact, the suspension U937 cells were associated with higher NP interactions and intracellular stress under fluid flow exposure conditions. For example, the increased AgNP association resulted in a 50% increase in intracellular ROS and p53 levels. This work highlights the potential of modified in vitro systems to improve analysis of AgNP dosimetry and safety evaluations, including oxidative stress assessments.

Keywords: silver nanoparticle; reactive oxygen species; cytotoxicity; dynamic flow; p53; NFκB; cytokine secretion

Citation: Burns, K.E.; Uhrig, R.F.; Jewett, M.E.; Bourbon, M.F.; Krupa, K.A. Characterizing the Role of Biologically Relevant Fluid Dynamics on Silver Nanoparticle Dependent Oxidative Stress in Adherent and Suspension In Vitro Models. *Antioxidants* **2021**, *10*, 832. https://doi.org/10.3390/antiox10060832

Academic Editor: Tim Hofer

Received: 5 May 2021
Accepted: 21 May 2021
Published: 23 May 2021

1. Introduction

Due to their physicochemical properties, nanoparticles (NPs) possess unique behavioral traits, such as augmented reactivity, increased transport potential, and distinctive optical signatures. The utilization of these properties and behavioral characteristics has resulted in the incorporation of NPs into hundreds of consumer products and applications [1]. To date, silver NPs (AgNPs) are the most predominantly employed, with applications spanning the medical, electrical, industrial, consumer goods, and health care sectors [2,3]. Owing to their robust antibiotic and antifungal activities, AgNPs have been integrated into bandages, biomedical devices, food storage apparatuses, textiles, cosmetics, and material surface coatings [4,5]. Additionally, their plasmonic capabilities have made AgNPs attractive for biomedical applications including bioimaging, optical sensor development, drug delivery vehicles, and as a mechanism to harvest photonic energy [5,6].

As NP prevalence increases, a corresponding rise in human contact also occurs, whether intentionally via therapeutics or via unintentional introduction. Furthermore,

AgNPs have been linked to potential health concerns, with exposure resulting in significant cytotoxicity [7]. In addition to cellular death, AgNP exposure has been shown to induce numerous bioeffects including augmented oxidative stress, activation of the inflammatory and immune responses, alterations to signal transduction pathways, and genetic modifications [8–11]. For example, a study by Comfort et al. demonstrated that, even at dosages in the pg/mL range, AgNPs were able to disrupt normal cell homeostasis by augmenting cellular stress, activating Akt and Erk signal transduction pathways, stimulating pro-inflammatory cytokine secretion, and regulating gene expression [9]. In a recent study by Khan et al., 55 nm AgNPs were shown to penetrate the blood brain barrier and induce oxidative stress pathways and production of several proteins involved in neurodiseases [12]. In addition to cellular death, AgNP exposure has been linked to disruption of the human microbiota, autoimmune diseases, and environmental health concerns, highlighting the far-reaching implications of AgNP exposure [13,14] Given the potential for serious health and environmental complications, it is necessary that the cellular response to NP exposure be fully characterized and understood. However, cataloging the safety of NPs is a major challenge as bioresponses are dependent upon a unique combination of physicochemical properties, such as primary size, surface chemistry, core composition, and morphology [15,16]. This means that each AgNP set has the potential to induce a distinctive cytotoxic profile within a biological environment.

In addition to NP physicochemical properties, the biological exposure environment also plays a significant role in how cells recognize and interact with NPs. For example, the type of cell, available surface receptors, and surrounding environmental factors all contribute differently to the NP protein corona and nano-cellular interface, and the subsequent cytotoxic outcomes [17,18]. In a recent study, it was determined that two different hepatic cell models uniquely interacted and displayed differential cytotoxicity following exposure to identical gold nanoparticles [17]. Currently, NP safety assessments are carried out in either cell-based in vitro or animal-based in vivo models. While in vitro systems possess experimental flexibility, are cost-effective, and have high throughput capabilities, they significantly lack physiological relevance [19]. The discrepancy between these models has resulted in poor correlation of nanotoxicological results, with NPs inducing more diverse and greater cellular responses in vitro versus in vivo counterparts [20–24].

One approach to overcome the poor correlation between cell and animal systems is to develop enhanced biological models, which retain the in vitro advantages but incorporate key in vivo influences to produce a more realistic and relevant NP exposure scenario. One example is the inclusion of dynamic fluid movement, which can mimic the biological transport mechanisms of cardiovascular flow, and thereby include a more accurate prediction of NP transport. Initial dynamic studies identified that lateral fluid flow altered the balance between NP diffusion and sedimentation, thereby modifying dosimetry and resultant bioresponses [25,26]. A study by Fede et al. demonstrated that under static conditions gold nanoparticles had a higher rate of sedimentation and cytotoxicity than when exposure occurred within a dynamic flow environment [26]. The incorporation of fluid dynamics is critical in biological models that experience high rates of fluid motion, such as the blood brain barrier and vascular systems [27,28]. Additionally, exposure within a dynamic environment has been found to alter the protein corona, the recognized biological identity of NPs [29,30]. While several studies have examined the role of fluid dynamics on adherent cell models, to date, no work exists on how fluid dynamics influences the NP cellular association, stress, and toxicity within a suspension cell model.

The goal of this study was to elucidate and characterize the impact of dynamic flow on AgNP behavior, deposition, and biological response in both an adherent (A549) and a suspension (U937) human cell model. Lung cells, A549, were specifically chosen as inhalation is a primary form of NP exposure and A549s have been widely utilized in nanotoxicological evaluations [31]. The monocytic U937 cell line was of particular interest as monocytes are primary responders to foreign material in the body and trigger systemic oxidative and inflammatory responses [32]. Fluid flow within the cellular systems was generated using a peristaltic pump, which produced a tube-side linear velocity of 0.2 cm/s, equivalent to known capillary rates [33]. This study demonstrated that the presence of dynamic flow impacted AgNP deposition and the ensuing cellular responses for the experimental 50 nm AgNPs. In particular, several intracellular oxidative stress pathways were activated within both the A549 and U937 models. Interestingly, the adherent and suspension cells responded inversely to dynamic flow, with the A549 cells correlating to lower AgNP interactions and stress with fluid dynamics. It is hypothesized that, under the influence of fluid dynamics, the suspension U937 cells experienced augmented interactions with flowing AgNPs, thereby inducing more significant oxidative stress outcomes.

2. Materials and Methods

2.1. AgNP Characterization

The experimental 50 nm polyvinylpyrrolidone (PVP)-coated AgNPs were purchased from nanoComposix in concentrated solution form. nanoComposix high quality control guarantees that their products are endotoxin free. To minimize AgNP modifications over the duration of experimentation, the particles were stored at 4 °C in the dark. Transmission electron microscopy (TEM) analysis was performed on a Hitachi H-7600 to verify primary particle size and morphology.

For the remaining AgNP characterization assessments, the particles were diluted to a concentration of 25 μg/mL in the denoted fluid. The unique spectral signatures of these particles were visualized via ultraviolet visible (UV-Vis) spectroscopy on a Synergy 4 BioTek microplate reader. Degree of agglomeration was measured through dynamic light scattering (DLS) on an Anton Paar Litesizer 500. The Anton Paar Litesizer was also used to run zeta potential analysis in order to determine particle surface charge.

2.2. Mammalian Cell Culture

Both human cell lines utilized in this study, the lung epithelial A549 and the monocytic U937, were purchased from American Type Culture Collection. The cells were maintained in RPMI 1640 media supplemented with 10% fetal bovine serum (FBS) and 1% antibiotics. The A549 cells, which are adherent in nature, were cultured on tissue-culture-treated petri dishes. The U937s were grown in suspension in T-75 tissue culture flasks. Both cells lines were sub-cultured every 3 to 4 days, as necessary.

2.3. Establishment of Dynamic Environment

A multi-channel peristaltic pump (Ismatec–Wertheim Germany, model #ISM939D) was used to establish dynamic flow, as previously demonstrated [25]. Both inlet and outlet ends of the tubing, with a 1/16 inch inner diameter, were secured into individual wells of a 24-well plate to generate unilateral flow throughout the cellular systems. Prior to experimentation, protein-rich media were run through the system to coat the tubes, thereby minimizing unintentional AgNP binding. For A549 experimentation, the pump was primed with either untreated or AgNP-dosed media to ensure that media levels were equivalent for both static and dynamic conditions. As U937 cells are suspension cells by nature, the pump was also primed prior to experimentation, but additional cells were included in order to maintain a constant cellular density with respect to media volume. To ensure no alterations in temperature, the pump and tubing were stored within the incubator. During experimentation, the pump operated at a rate that produced a tube-side linear velocity of 0.2 cm/s, which was specifically chosen to match physiological rates within capillaries [33].

2.4. Nano-Cellular Association

To determine AgNP interaction with the biological systems, cells were seeded into 24-well plates (2×10^5 A549 cells/well or 1.5×10^5 cells/mL for U937) and incubated overnight. The denoted cells were exposed to 15 μg/mL of the 50 nm AgNPs, under static or dynamic conditions. This AgNP concentration was selected as it was large enough to provide a significant detection signal but did not induce a strong cytotoxic response. Following 24 h exposure, the AgNP-containing media was collected; the U937 cells were removed via low-speed centrifugation, which removed the cells but not the suspended AgNPs. The supernatant then underwent UV-VIS analysis to determine the final NP concentration, using previously generated concentration curves [34]. A separate calibration curve was generated for static and dynamic conditions to account for any non-specific binding that may have occurred under dynamic conditions. The difference between this final concentration and the administered 15 μg/mL dosage was the AgNP amount associated with the cells.

2.5. Cellular Viability

Cell viability was determined using the CytoTox 96 Non-Radioactive Cytotoxicity Assay (Promega), which measures the production of lactate dehydrogenase (LDH). A549 cells were plated into 24-well plates at a concentration of 2×10^5 cells per well. U937s were seeded into 24-well plates at a density of 1.5×10^5 cells/mL, as they were in suspension. The following day, the cells underwent exposure to the 50 nm AgNPs at 0, 5, or 25 μg/mL, under either static or dynamic conditions. After 24 h, the media were removed and centrifuged to remove AgNPs and suspension cells. The supernatant then underwent LDH evaluation in accordance with the manufacturer's instructions. A positive toxicity control, in which all the cells were lysed, was used to normalize data and determine degree of cytotoxicity.

2.6. Intracellular ROS Production

Intracellular stress was assessed by measuring reactive oxygen species (ROS) levels following AgNP exposure under the denoted conditions. A549 and U937 cells were seeded in 24-well plates at 2×10^5 cells per well or 1.5×10^5 cells/mL, respectively, and returned to the incubator until the following day. Next, the cells were washed and incubated with the DCFH-DA probe (Thermo Fisher Scientific) for 30 min, washed again, then dosed with the AgNPs at 0, 5, or 25 μg/mL within either a static or dynamic environment. After 24 h incubation, the ROS levels were measured via fluorescent analysis using a Synergy 4 BioTek microplate reader. Untreated cells under static conditions served as the negative control for normalization and hydrogen peroxide-dosed cells served as a positive control.

2.7. Glutathione Levels

The ratio of reduced glutathione (GSH) to oxidized glutathione (GSSG) is a metric to quantify intracellular oxidative stress [35]. The experimental cells were seeded and exposed to 0, 5, or 25 μg/mL AgNPs as described above for ROS. After 24 h, the cells were then washed, collected, and lysed in a non-denaturing lysis buffer. Intracellular GSH and GSSG levels were quantified using the GSH/GSSG Ratio Detection Assay Kit from Abcam, using a Synergy 4 BioTek microplate reader. GSH/GSSG ratios were calculated and normalized against an untreated, static control for each cellular system. Hydrogen peroxide exposure served as a positive control for known stress activation.

2.8. Activation of Intracellular Targets

A549 and U937 cells were seeded in a 24-well plate at 2×10^5 cells per well or 1.5×10^5 cells/mL, respectively. The following day the cultures were dosed with 0, 5, or 25 μg/mL AgNPs, under either static or dynamic conditions. After 24 h, the cells were washed, collected, and lysed in a non-denaturing lysis buffer. Activation of intracellular p53 and NFκB were determined by quantifying phosphorylated levels of these targets. In both cases, phosphorylated levels were normalized by the total p53 and NFκB amounts, quantified in parallel. Phospho and total levels of p53 and NFκB were assessed using protein-specific PathScan ELISA kits from Cell Signaling Technology, in accordance with the manufacturer's instructions. Experimental activation levels were normalized against untreated, static controls. Hydrogen peroxide-exposed cells served as a positive control as they are known to induce numerous intracellular stress pathways.

2.9. Inflammatory Cytokine Secretion

The inflammatory response to AgNP exposure was determined by quantifying the production of the cytokines interleukin-1β (IL-1β) and tumor necrosis factor-α (TNF-α). While a panel of cytokines are secreted within an inflammatory response, IL-1β and TNF-α were selected as representative protein markers for this assessment. A549 and U937 cells were seeded in 24-well plates at 2×10^5 cells per well or 1.5×10^5 cells/mL, respectively, incubated for 24 h, then exposed to AgNPs under the stated conditions. Hydrogen peroxide exposure served as a positive control as it is known to induce numerous intracellular stress pathways. After 24 h exposure, the media were collected, AgNPs/U937s were removed via centrifugation, and the supernatant underwent analysis for extracellular IL-1β and TNF-α levels using protein-specific ELISAs (Thermo Fisher Scientific, Waltham, MA, USA), in accordance with the manufacturer's directions.

2.10. Data Analysis

All data are presented as the mean ± the standard error of the mean. All experimentation was carried out with three independent trials. For cellular association analysis, a one-way ANOVA with Bonferroni post-test was run using GraphPad Prism to determine statistical significance between static and dynamic conditions. For the remaining experimentation, a two-way ANOVA with Bonferroni post-test was run using GraphPad Prism, with * and † indicating significance compared to the untreated control and between static and dynamic conditions, respectively.

3. Results

3.1. AgNP Characterization

Prior to cellular exposure, the AgNPs underwent numerous characterization assessments to quantify the unique physicochemical properties associated with this NP stock. As small deviations in NP properties have been correlated to differential bioresponses, it was necessary to first characterize the experimental NPs [36]. A representative TEM image of the 50 nm AgNPs is shown in Figure 1 and demonstrates a uniform, spherical morphology. Using multiple images, the primary particle size was determined to be 52.6 ± 6.9 nm. The uniformity of the AgNP stock was further verified through spectral analysis, which was comprised of a single, sharp peak (Figure 1B). When dispersed in media, there is a very slight right-shift in the spectral image, which can be attributed to the formation of a protein corona or minimal particle agglomeration.

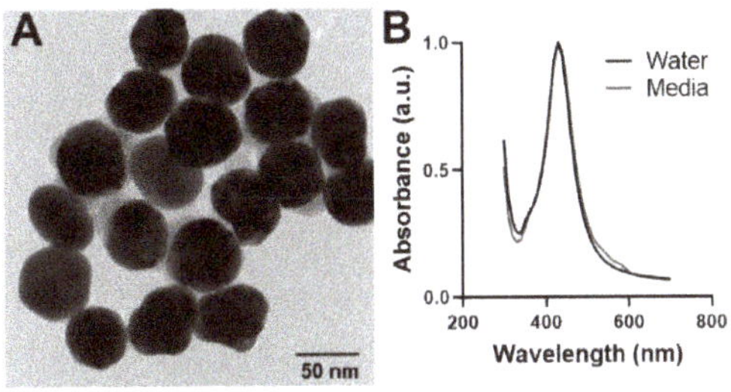

Figure 1. Characterization of the experimental 50 nm AgNP stock. (**A**) A representative TEM image of the 50 nm PVP-coated AgNPs is shown. This image verifies spherical morphology and an even particle size distribution. (**B**) The spectral profiles for the 50 nm AgNPs in both water and media demonstrate a sharp, single peak, further verifying particle uniformity.

In addition to verifying the primary particle size, the agglomerate sizes were determined in both water and media (Table 1). As all NPs will agglomerate when in solution, it is important to assess the extent of agglomeration as it can impact mechanisms of biotransport [37]. The AgNPs displayed minimal agglomeration in water with an increased effective diameter in media, due to the formation of a protein corona [38]. As the NPs had a PVP surface coating, which is known to promote particle stability in solution, the small extent of aggregation was expected [39]. Additionally, the surface charge was assessed using zeta potential measurements. The stock AgNPs displayed a negative surface charge, approximately -30 mV. Following the formation of a protein corona in media the charge shifted to -10 mV, as the outermost NP surface was covered in proteins which innately have a slight negative charge [38].

Table 1. Characterization of the AgNPs under static and dynamic conditions.

Flow Condition	Agglomerate Size (NM)		Zeta Potential (MV)	
	Water	Media	Water	Media
Static	78.6 ± 3.0	89.3 ± 2.2	−30.4 ± 1.7	−9.7 ± 0.6
Dynamic	76.7 ± 2.4	90.8 ± 2.9	−32.0 ± 2.4	−10.5 ± 1.0

A peristaltic pump was used to generate fluid dynamics within the cellular system. Pump utilization afforded the opportunity to assess the impact of fluid dynamics on both AgNP characteristics and the nano-cellular interface. The targeted pump flow rate produced a linear velocity of 0.2 cm/s within the tubing, aligning with known capillary values. This means that the velocity across the cells was orders of magnitude less, aligning with the diffusion-based transport observed within tissue. Next, the AgNPs, suspended in either water or media, were added into an acellular dynamic system, circulated for several hours, and characterized in order to determine if the shear stress impacted key physicochemical properties. Previous reports have determined that dynamic flow is capable of modifying the protein corona [29,30], which could disrupt both AgNP agglomerate size and surface charge. As seen in Table 1, the low level dynamic movement had a negligible impact on both the extent of agglomeration and zeta potential measurements.

3.2. *Dynamic Flow-Modified AgNP Cellular Interactions*

Next, the influence of dynamic flow on AgNP deposition was investigated, for both the adherent A549 and the suspension U937 cell models. As seen in Figure 2, dynamic flow significantly altered NP association efficiencies, though in opposite modalities for these culture types. For A549 cells, the presence of lateral fluid flow decreased AgNP deposition by approximately 33%. The 50 nm AgNPs would have been associated with a significant rate of sedimentation, thereby producing a high number of AgNP–A549 interactions [35]. Under dynamic flow, the lateral convective movement kept the AgNPs in suspension, resulting in the observed deposition decrease. For the suspension U937 cells, the fluid dynamics, which counterbalanced downward sedimentation effects, resulted in augmented AgNP cellular association. As shown in Figure 2, when U937 underwent AgNP exposure under dynamic conditions, there was an approximate 66% increase in NP cellular binding.

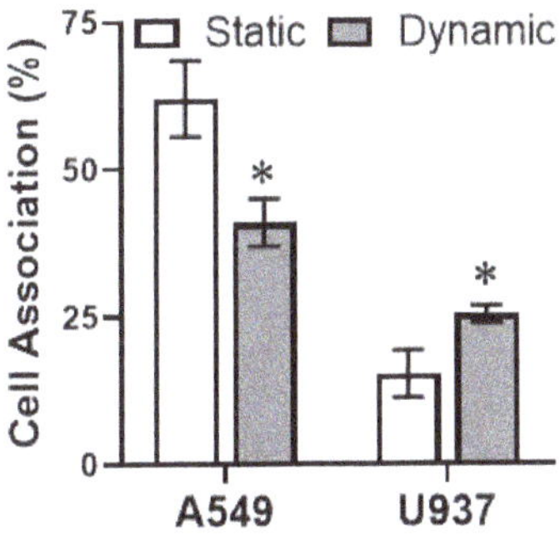

Figure 2. AgNP cellular association varies as a function of flow environment and cell model. Following 24 h exposure, AgNP interactions with the adherent A549 or suspension U937 models were measured within both static and dynamic environments. * indicates significance between static and dynamic conditions. $n = 3$, $p < 0.05$.

3.3. *AgNP-Induced Cytotoxicity Varied with Flow Condition*

As AgNPs are known to induce mammalian cytotoxicity [8–11], the first endpoint assessed was cellular viability, as a function of flow condition (Figure 3). Starting with the A549s (Figure 3A), the lower AgNP dosage produced a slight cytotoxic response. However, following 25 µg/mL exposure, there was an approximate 40% loss of cell viability under static conditions. Within a dynamic A549 model, the cells still displayed AgNP dose-dependent cytotoxicity, though to a lesser degree, indicating that cellular damage was mitigated within a dynamic environment.

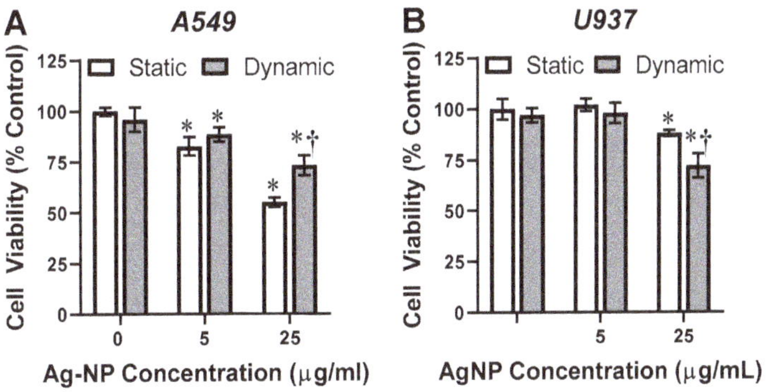

Figure 3. Cellular viability following AgNP exposure. Following 24 h AgNP exposure within either (**A**) A549 or (**B**) U937 models the cellular viability was assessed. These results indicated that within both systems viability was a function of AgNP dosage and flow condition. * and † indicate statistical significance compared to the untreated control and between static and dynamic conditions, respectively. $n = 3$, $p < 0.05$.

The U937 cells also demonstrated a dose-dependent cytotoxicity response following AgNP exposure, as expected. However, when examining the U937 dynamic flow results, opposite trends were identified (Figure 3B). Following 25 µg/mL exposure, the U937s experienced a mild 12% toxicity under static conditions. With the inclusion of dynamic flow, this toxicity rate increased to approximately 25%, presumably due to amplified AgNP–U937 interactions. These viability results are in excellent agreement with the AgNP cellular association profiles, which for A549 and U937 cells identified a respective decrease and increase in nano-cellular associations. Moreover, any cellular changes to cytotoxicity were not directly related to the presence of fluid dynamics as shown in controls.

3.4. Intracellular Oxidative Stress Levels

As the presence of dynamic flow was able to modulate AgNP-dependent cytotoxicity, the next goal was to explore if this phenomena translated to intracellular responses, which precede cell death. Based on the dose-dependent toxicity analysis, an exposure of 5 μg/mL AgNP was selected, as it did not elicit a strong degree of cellular death. Detection of ROS was used to monitor intracellular oxidative stress levels, as it is a known cellular response following NP exposure and an early indicator of apoptosis [40,41]. In the absence of AgNPs, ROS production was equivalent for static and dynamic conditions for both models (Figure 4). These controls demonstrated that dynamic flow did not impact basal A549 or U937 stress levels.

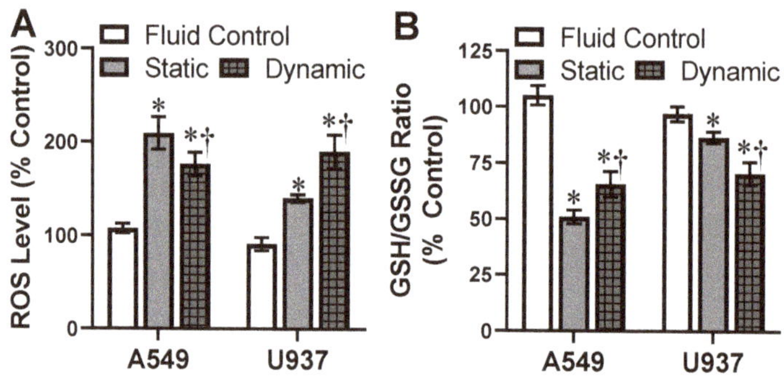

Figure 4. AgNP-induced intracellular oxidative stress levels. Both (**A**) ROS and (**B**) GSH/GSSG levels were used to monitor A549 and U937 intracellular oxidative stress following 5 μg/mL AgNP exposure for 24 h. Experimentation was carried out under both static and dynamic conditions. Dynamic flow without AgNP conditions were also included as fluid control. Oxidative stress levels were found to be a function of cell type and flow condition. * and † indicate statistical significance compared to the untreated control and between static and dynamic conditions, respectively. $n = 3$, $p < 0.05$.

As shown in Figure 4A, 50 nm AgNP exposure within a static A549 model demonstrated elevated ROS levels. In agreement with the cell viability data, under dynamic conditions the A549 stress response was significantly decreased. Following static AgNP exposure, U937 cells demonstrated a mild ROS response, indicating active intracellular oxidative stress. However, when U937 cells underwent the same AgNP dosage exposure under dynamic conditions, the resultant ROS levels were augmented by approximately 50%.

Intracellular glutathione exists in two states, reduced (GSH) and oxidized (GSSG). In normal, healthy cells, over 90% of all glutathione is in the GSH state. When cells are experiencing stress, the GSSG level rises; therefore, a decreased GSH/GSSG ratio is indicative of augmented oxidative stress [42,43]. As seen in Figure 4B, both A549 and U937 were experiencing stress following AgNP exposure under static conditions, due to GSH/GSSG ratios that were lower than the untreated control. Upon the introduction of fluid dynamics, GSH/GSSG ratios increased and decreased in the A549 and U937 cells systems, respectively, in agreement with the ROS data.

3.5. Activation of Oxidative Stress Protein Markers

Following identification of differential activation of ROS and glutathione responses within a dynamic environment, the next assessments targeted key proteins involved in the intracellular stress pathways. The goal was to elucidate whether p53 and NFκB were activated following AgNP exposure and whether the presence of dynamic flow modified the degree of this stress response. The targets of p53 and NFκB were specifically selected as these proteins have been previously shown to be induced following AgNP exposure in vitro [44,45]. As seen in Figure 5A, p53 was activated in both A549 and U937 systems following AgNP exposure. The phosphorylation of p53 mirror the ROS and glutathione results, with dynamic flow decreasing and increasing stress levels within the adherent and suspension models, respectively. When looking at NFκB activation, A549 showed both the AgNP-dependent stimulation and flow-dependent effects. However, U937 cells did not show any change in NFκB expression or activation following AgNP introduction, indicating that this pathway was not influenced by NP presence. Taken together, these results confirmed that exposure environment does influence intracellular oxidative stress proteins and pathways, though through cell-specific mechanisms.

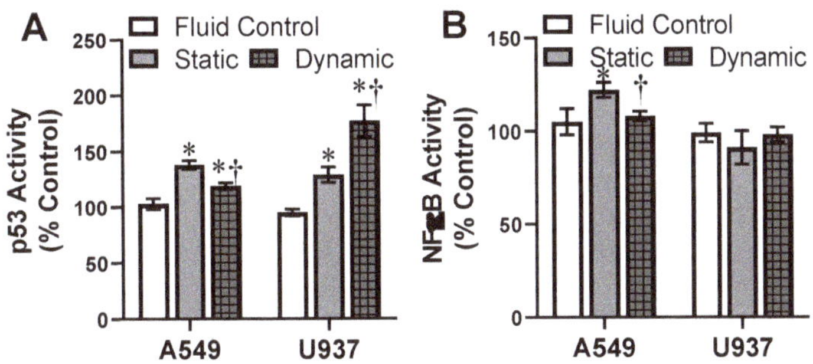

Figure 5. Activation of oxidative stress protein markers. Following exposure to 5 μg/mL AgNP, under both static and dynamic conditions, the A549 and U937 cells underwent evaluation for (**A**) p53 and (**B**) NFκB activation. Activation was determined by quantifying the phosphorylated state of target intracellular proteins, normalized by the total amounts of that protein. Activation of chosen oxidative stress pathways varied with cell type and exposure environment. Dynamic flow without AgNP conditions was also included as fluid control. * and † indicate statistical significance compared to the untreated control and between static and dynamic conditions, respectively. $n = 3$, $p < 0.05$.

3.6. Secretion of Pro-Inflammatory Cytokines Following AgNP Exposure In Vitro

AgNP-dependent acute stress induction has been linked to the activation of inflammatory responses in mammalian cells [46]. Once activated, an early inflammatory response is the production and secretion of pro-inflammatory cytokines, including IL-1β and TNF-α [47]. For untreated controls, the IL-1β and TNF-α levels were equivalent for static and dynamic conditions, indicating that fluid flow did not alter cytokine production (Figure 6). As A549 cells are alveolar epithelial, they only secrete a basal, low-level amount of cytokines. As shown in Figure 6, AgNPs did not elicit a significant inflammatory response under any condition. However, U937 monocytes, which are a part of the immune system, produced a robust array of cytokines following activation, including IL-1β and TNF-α. Regardless of exposure environment, 5 μg/mL AgNPs initiated a pro-inflammatory response, through upregulated cytokine production. Under dynamic conditions, a small increase in production levels was identified, indicating some degree of behavioral variance resulting from fluid flow conditions.

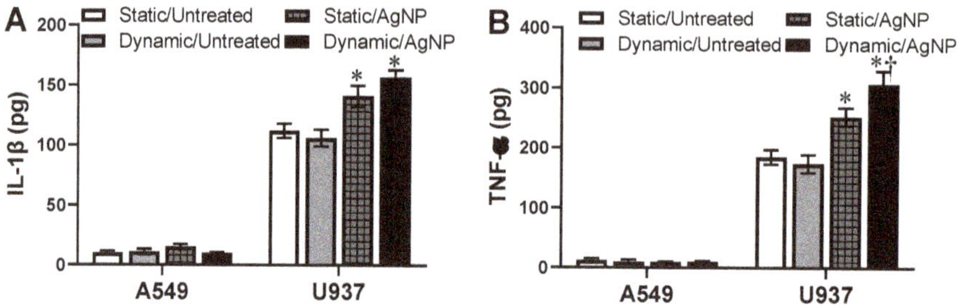

Figure 6. Secretion of pro-inflammatory cytokines following AgNP stimulation. The alveolar epithelial A549 and monocytic U937 cell lines were exposed to the denoted combinations of static/dynamic conditions and 5 μg/mL AgNPs. After 24 h the media were recovered and evaluated for levels of the pro-inflammatory cytokines of (**A**) IL-1β and (**B**) TNF-α. * and † indicate statistical significance compared to the untreated control and between static and dynamic conditions, respectively. $n = 3$, $p < 0.05$.

4. Discussion

The goal of this work was to elucidate the roles of cellular type (adhesion vs. suspension) and fluid dynamics in the AgNP cellular interactions and subsequent bioresponses, with a focus on oxidative stress markers. The results presented in this study demonstrated that both cellular and exposure characteristics significantly impacted the nano-cellular interface and the biological response. Therefore, each of these influences is discussed in greater detail below. Moreover, how this work supports the need and highlights implications for more physiologically relevant NP exposure models are included as points of discussion.

4.1. Influence of Fluid Dynamics on Different Cell Classifications

The addition of forced convective flow increased the biological relevance of the in vitro system, as all animal models are dynamic due to the presence of cardiovascular systems. By recreating a capillary-based velocity within the tubing, this model established diffusion-level transport across the A549 surface or alongside the U937s, representing tissue-level shear rates experienced in vivo [48]. This study determined that the addition of dynamic flow altered the NP transport profiles, as the lateral force counteracts the downward sedimentation and random diffusion which dominate under static conditions [49]. Firstly, when examining the impact of fluid flow, it was found that the utilized volumetric flowrate did not alter AgNP physicochemical properties (Table 1). These results suggest that all the elucidated cellular modifications were due to an altered nano-cellular interface and not the AgNPs themselves.

For both the adherent A549 and the suspension U937 cells models, the addition of dynamic flow was found to impact nano-cellular interactions, AgNP-dependent cytotoxicity, and subsequent oxidative stress responses. However, inclusion of dynamic flow in the exposure environment had opposite impacts on the A549 and U937 models, due to their differential growth behaviors. For the adherent A549 cells, dynamic flow reduced AgNP deposition, whereas nano-cellular associations were increased for the U937 model. In agreement with previous studies from the literature, the level of AgNP interaction with the surrounding biological environment dictated the degree of cytotoxicity and oxidative stress [50,51]. As such, when AgNP exposure occurred within a dynamic A549 model, the cytotoxicity and array of oxidative stress markers were assuaged. On the contrary, the higher rate of AgNP and U937 collisions when both were flowing in the dynamic model resulted in augmented toxicity and stress pathways.

Looking beyond toxicity, this study focused on numerous oxidative stress markers, including ROS, glutathione levels, p53, NFκB, and cytokine secretion. Taken together, these pathways highlight a multitude of differential intracellular stress responses. These endpoints were found to vary as a function of both exposure environment and cell model, mirroring the cytotoxicity results. This study is the first to our knowledge that has explored the role of fluid dynamics on the nano-cellular interface and oxidative stress within a suspension-based in vitro model, highlighting the variance in cellular behavior. Of particular importance is the ability of AgNPs and the exposure environment to influence the secretion of pro-inflammatory cytokines from U937 cells, which introduce the potential for a systemic inflammatory response and long-term health consequences [52].

4.2. The Need for Improved In Vitro Models

The results presented in this study highlight the necessity of developing in vitro models that are more biologically relevant and are able to better assess how NPs will behave within in vivo environments. The need for enhanced systems is further supported by the conflicting results and poor correlation between cell- and animal-based models with regards to nanotoxicity and the induction of intracellular stress and signaling pathways [20–24]. Discrepancies between these models are due in part to the fact that NP physicochemical properties and dosimetry are dependent upon environmental factors. For example, when NPs are dispersed in physiological fluids, such as interstitial and lysosomal, the rate of ionic dissolution and extent of agglomeration are significantly altered versus cell culture media [53,54]. Additionally, NP exposure within a three-dimensional in vitro model diminished the extent of cytotoxic responses versus two-dimensional models, owing to the need for NPs to translocate through the cell systems prior to internalization [55]. In agreement with this work, fluid flow has previously been found to modify cellular morphology, dosimetry, and resultant bioresponses [25–27].

In addition to further validating the need for more physiologically representative exposure scenarios, this work determined that the impact of fluid dynamics on NP dosimetry and dependent oxidative stress varied with cell system behavior. The results that emerged from this dynamic model better align with published in vivo data, in which AgNP exposure at lower dosages elicits only a minor stress response, without a large induction of cytotoxicity [56,57]. Therefore, this work indicates that the addition of dynamic flow increases the relevance of in vitro models and has the potential to improve predictive modeling capabilities.

5. Conclusions

This study explored the role of cell adhesion properties and dynamic fluid movement on NP cellular interactions, cytotoxicity, and intracellular oxidative stress pathways following exposure to AgNPs. This work revealed that fluid flow modified the nano–bio interface but through differential mechanisms for adherent and suspension cells. For adherent cells, dynamic flow reduced AgNP contact and subsequent bioresponses due to a disruption of sedimentation. In contrast, cells that grow in suspension were associated with higher AgNP contact and oxidative stress activation within a dynamic exposure environment. The model implemented in this work, which incorporated dynamic flow, demonstrates how enhanced in vitro systems capture physiological influences and can thereby produce differential biological responses. Alterations to in vitro models also allow experimental flexibility and the ability to tailor design characteristics. For example, this study focused on low, diffusion-based flow that is associated with diffusion-level circulation. However, by switching to an endothelial cell model and increasing fluid flow rates, an arterial exposure system could be recreated.

Supplementary Materials: The following are available online at https://www.mdpi.com/article/10.3390/antiox10060832/s1, Figure S1: Cytotoxicity Dynamic Controls; Figure S2: Oxidative Stress Dynamic Controls; Figure S3: Cytokine Secretion Dynamic Controls.

Author Contributions: Conceptualization, K.E.B. and K.A.K.; methodology, K.E.B., R.F.U., and K.A.K.; investigation, K.E.B., R.F.U., M.E.J., and M.F.B.; formal analysis, K.E.B. and K.A.K.; writing—original draft preparation, K.E.B. and M.E.J.; writing—review and editing, K.A.K.; visualization,

K.E.B. and K.A.K.; supervision, K.A.K.; project administration, K.A.K.; funding acquisition, K.E.B. and K.A.K. All authors have read and agreed to the published version of the manuscript.

Funding: K.E.B., K.A.K., and this work were funded in part by the Dayton Area Graduate Studies Institute, award RH-16-6. R.F.U. was funded by the University of Dayton Honors Program. M.E.J. was funded by the Claire Boothe Luce Foundation.

Institutional Review Board Statement: Not applicable.

Informed Consent Statement: Not applicable.

Data Availability Statement: All data are contained in this article and supplementary materials.

Conflicts of Interest: The authors declare no conflict of interest. The funders had no role in the design of the study; in the collection, analyses, or interpretation of data; in the writing of the manuscript, or in the decision to publish the results.

References

1. Kessler, R. Engineered nanoparticles in consumer products: Understanding a new ingredient. *Environ. Health Perspect.* **2011**, *119*, A120–A125. [CrossRef] [PubMed]
2. Yaqoob, A.A.; Umar, K. Silver nanoparticles: Various methods of synthesis, size affecting factors and their potential applications–A review. *Appl. Nanosci.* **2020**, *10*, 1369–1378. [CrossRef]
3. Zhang, X.F.; Liu, Z.G.; Shen, W.; Gurunathan, S. Silver nanoparticles: Synthesis, characterization, properties, applications, and therapeutic approaches. *Int. J. Mol. Sci.* **2016**, *17*, 1534.
4. Das, C.A.; Kumar, V.G.; Dhas, T.S.; Karthick, V.; Govindaraju, K.; Joselin, J.M.; Baalamurugan, J. Antibacterial activity of silver nanoparticles (biosynthesis): A short review on recent advances. *Biocatal. Agric. Biotechnol.* **2020**, *27*, 101593.
5. Ge, L.; Li, Q.; Wang, M.; Ouyang, J.; Li, X.; Xing, M.M.Q. Nanosilver particles in medical applications: Synthesis, performance, and toxicity. *Int. J. Nanomed.* **2014**, *9*, 2399–2407.
6. Momeni, M.M.; Shayeb, Y.; Gheibee, S. Silver nanoparticles decorated titanium dioxide-tungsten trioxide nanotube films with enhanced visible light photo catalytic activity. *Ceram. Int.* **2017**, *43*, 564–570.
7. Akter, M.; Sikder, M.T.; Rahman, M.M.; Ullah, A.A.; Hossain, K.F.B.; Banik, S.; Hosokawa, T.; Saito, T.; Kurasaki, M. A systematic review on silver nanoparticles-induced cytotoxicity: Physicochemical properties and perspectives. *J. Adv. Res.* **2018**, *9*, 1–16.
8. Breitner, E.K.; Burns, K.E.; Hussain, S.M.; Comfort, K.K. Implementation of physiological fluids to provide insight into the characterization, fate, and biological interactions of silver nanoparticles. *Nanotechnology* **2018**, *29*, 254001. [CrossRef] [PubMed]
9. Comfort, K.K.; Braydich-Stolle, L.K.; Maurer, E.I.; Hussain, S.M. Less is more: Long-term in vitro exposure to low levels of silver nanoparticles provides new insights for nanomaterial evaluation. *ACS Nano* **2014**, *8*, 3260–3271.
10. Comfort, K.K.; Maurer, E.I.; Hussain, S.M. Slow release of ions from internalized silver nanoparticles modifies the epidermal growth factor signaling response. *Colloids Surf. B Biointerfaces* **2014**, *123*, 136–142. [CrossRef] [PubMed]
11. Hadrup, N.; Lam, H.R. Oral toxicity of silver ions, silver nanoparticles and colloidal silver—A review. *Regul. Toxicol. Pharm.* **2014**, *68*, 1–7. [CrossRef] [PubMed]
12. Khan, A.M.; Korzeniowska, B.; Gorshkov, V.; Tahir, M.; Schrøder, H.; Skytte, L.; Rasmussen, K.L.; Khandige, S.; Møller-Jensen, J.; Kjeldsen, F. Silver nanoparticle-induced expression of proteins related to oxidative stress and neurodegeneration in an in vitro human blood-brain barrier model. *Nanotoxicology* **2019**, *13*, 221–239. [CrossRef] [PubMed]
13. León-Silva, S.; Fernández-Luqueño, F.; López-Valdez, F. Silver nanoparticles (AgNP) in the environment: A review of potential risks on human and environmental health. *Water Air Soil Pollut.* **2016**, *227*, 1–20. [CrossRef]
14. Li, J.; Tang, M.; Xue, Y. Review of the effects of silver nanoparticle exposure on gut bacteria. *J. Appl. Toxicol.* **2019**, *39*, 27–37.
15. Modena, M.M.; Rühle, B.; Burg, T.P.; Wuttke, S. Nanoparticle characterization: What to measure. *Adv. Mater.* **2019**, *31*, 1901556. [CrossRef]
16. Hussain, S.M.; Warheit, D.B.; Ng, S.P.; Comfort, K.K.; Grabinski, C.M.; Braydich-Stolle, L.K. At the crossroads of nanotoxicology in vitro: Past achievements and current challenges. *Toxicol. Sci.* **2015**, *147*, 5–16. [CrossRef]
17. Enea, M.; Pereira, E.; Costa, J.; Soares, M.E.; da Silva, D.D.; de Lourdes Bastos, M.; Carmo, H.F. Cellular uptake and toxicity of gold nanoparticles on two distinct hepatic cell models. *Toxicol. In Vitro* **2021**, *70*, 105046. [CrossRef]
18. Soenen, S.J.; De Cuyper, M. Assessing iron oxide nanoparticle toxicity in vitro: Current status and future prospects. *Nanomedicine* **2010**, *5*, 1261–1275. [CrossRef]
19. Clift, M.J.D.; Gehr, P.; Rothen-Rutishauser, B.R. Nanotoxicology: A perspective and discussion of whether or not in vitro testing is a valid alternative. *Arch. Toxicol.* **2011**, *85*, 723–731. [CrossRef]
20. Demokritou, P.; Gass, S.; Pyrgiotakis, G.; Cohen, J.M.; Goldsmith, W.; McKinney, W.; Frazer, D.; Ma, J.; Schwegler-Berry, D.; Brain, J.; et al. An in vivo and in vitro toxicological characterization of realistic nanoscale CeO_2 inhalation exposures. *Nanotoxicology* **2013**, *7*, 1338–1350. [CrossRef] [PubMed]
21. Frohlich, E.; Salar-Behzadi, S. Toxicological assessment of inhaled nanoparticles: Role of in vivo, ex vivo, in vitro, and in silico studies. *Int. J. Mol. Sci.* **2014**, *15*, 4795–4822. [CrossRef] [PubMed]

22. Jia, Y.P.; Ma, B.Y.; Wei, X.W.; Qian, Z.Y. The in vitro and in vivo toxicity of gold nanoparticles. *Chin. Chem. Lett.* **2017**, *28*, 691–702. [CrossRef]
23. Hulsart-Billström, G.; Dawson, J.I.; Hofmann, S.; Müller, R.; Stoddart, M.J.; Alini, M.; Redl, H.; El Haj, A.; Brown, R.; Salih, V.; et al. A surprisingly poor correlation between in vitro and in vivo testing of biomaterials for bone regeneration: Results of a multicentre analysis. *Eur. Cell Mater.* **2016**, *31*, 312–322. [CrossRef] [PubMed]
24. Kumar, V.; Sharma, N.; Maitra, S.S. In vitro and in vivo toxicity assessment of nanoparticles. *Int. Nano Lett.* **2017**, *7*, 243–256. [CrossRef]
25. Breitner, E.K.; Hussain, S.M.; Comfort, K.K. The role of biological fluid and dynamic flow in the behavior and cellular interactions of gold nanoparticles. *J. Nanobiotechnol.* **2015**, *13*, 56. [CrossRef] [PubMed]
26. Fede, C.; Fortunati, I.; Weber, V.; Rossetto, N.; Bertasi, F.; Petrelli, L.; Guidolin, D.; Signorini, R.; De Caro, R.; Albertin, G.; et al. Evaluation of gold nanoparticles toxicity toward human endothelial cells under static and flow conditions. *Microvas Res.* **2015**, *97*, 147–155. [CrossRef]
27. Ucciferri, N.; Collnot, E.M.; Gaiser, B.K.; Tirella, A.; Stone, V.; Domenici, C.; Lehr, C.M.; Ahluwalia, A. In vitro toxicology screening of nanoparticles on primary human endothelial cells and the role of flow in modulating cell response. *Nanotoxicology* **2014**, *8*, 697–708. [CrossRef]
28. Hajal, C.; Campisi, M.; Mattu, C.; Chiono, V.; Kamm, R.D. In vitro models of molecular and nano-particle transport across the blood-brain barrier. *Biomicrofluidics* **2018**, *12*, 042213. [CrossRef]
29. Braun, N.J.; DeBrosse, M.C.; Hussain, S.M.; Comfort, K.K. Modification of the protein corona-nanoparticle complex by physiological factors. *Mater. Sci. Eng. C* **2016**, *64*, 34–42. [CrossRef]
30. Abbina, S.; Takeuchi, L.E.; Anilkumar, P.; Yu, K.; Rogalski, J.C.; Shenoi, R.A.; Constantinescu, I.; Kizhakkedathu, J.N. Blood circulation of soft nanomaterials is governed by dynamic remodeling of protein opsonins at nano-biointerface. *Nat. Commun.* **2020**, *11*, 1–12. [CrossRef]
31. Brandão, F.; Fernández-Bertólez, N.; Rosário, F.; Bessa, M.J.; Fraga, S.; Pásaro, E.; Teixeira, J.P.; Laffon, B.; Valdiglesias, V.; Costa, C. Genotoxicity of tio2 nanoparticles in four different human cell lines (A549, HEPG2, A172 and SH-SY5Y). *Nanomaterials* **2020**, *10*, 412. [CrossRef] [PubMed]
32. Sahu, D.; Kannan, G.M.; Vijayaraghavan, R. Size-dependent effect of zinc oxide on toxicity and inflammatory potential of human monocytes. *J. Toxicol. Environ. Health Part A* **2014**, *77*, 177–191. [CrossRef]
33. Berger, S.A.; Goldsmith, W.; Lewis, E.R. *Introduction to Bioengineering;* Oxford University Press: Oxford, UK, 1996.
34. Zook, J.M.; Long, S.E.; Cleveland, D.; Geronimo, C.L.; MacCuspie, R.I. Measuring silver nanoparticle dissolution in complex biological and environmental matrices using UV-visible absorbance. *Anal. Bioanal. Chem.* **2011**, *401*, 1993–2002. [CrossRef] [PubMed]
35. Schulz, J.B.; Lindenau, J.; Seyfried, J.; Dichgans, J. Glutathione, oxidative stress and neurodegeneration. *Eur. J. Biochem.* **2000**, *267*, 4904–4911. [CrossRef] [PubMed]
36. Cho, E.J.; Holback, H.; Liu, K.C.; Abouelmagd, S.A.; Park, J.; Yeo, Y. Nanoparticle characterization: State of the art, challenges, and emerging technologies. *Mol. Pharm.* **2013**, *10*, 2093–2110. [CrossRef]
37. Kim, W.; Kim, W.K.; Lee, K.; Son, M.J.; Kwak, M.; Chang, W.S.; Min, J.K.; Song, N.W.; Lee, J.; Bae, K.H. A reliable approach for assessing size-dependent effects of silica nanoparticles on cellular internalization behavior and cytotoxic mechanisms. *Int. J. Nanomed.* **2019**, *14*, 7375. [CrossRef]
38. Luby, A.O.; Breitner, E.K.; Comfort, K.K. Preliminary protein corona formation stabilizes gold nanoparticles and improves deposition efficiency. *Appl. Nanosci.* **2016**, *6*, 827–836. [CrossRef]
39. Munir, T.; Mahmood, A.; Imran, M.; Sohail, A.; Fakhar-e-Alam, M.; Sharif, M.; Masood, T.; Bajwa, S.Z.; Shafiq, F.; Latif, S. Quantitative analysis of glucose by using (PVP and MA) capped silver nanoparticles for biosensing applications. *Phys. B Condens. Matter* **2021**, *602*, 412564. [CrossRef]
40. Labrador-Rached, C.J.; Browning, R.T.; Braydich-Stolle, L.K.; Comfort, K.K. Toxicological implications of platinum nanoparticle exposure: Stimulation of intracellular stress, inflammatory response, and akt signaling in vitro. *J. Toxicol.* **2018**, *2018*, 1367801. [CrossRef] [PubMed]
41. Redza-Dutordoir, M.; Averill-Bates, D.A. Activation of apoptosis signaling pathways by reactive oxygen species. *Biochem. Biophys. Acta Mol. Cell Res.* **2016**, *1863*, 2977–2992. [CrossRef]
42. Dasgupta, N.; Ranjan, S.; Mishra, D.; Ramalingam, C. Thermal co-reduction engineered silver nanoparticles induce oxidative cell damage in human colon cancer cells through inhibition of reduced glutathione and induction of mitochondria-involved apoptosis. *Chemico-Biol. Interact.* **2018**, *295*, 109–118. [CrossRef] [PubMed]
43. Matschke, V.; Theiss, C.; Matschke, J. Oxidative stress: The lowest common denominator of multiple diseases. *Neural Regen. Res.* **2019**, *14*, 238–241. [CrossRef] [PubMed]
44. Blanco, J.; Tomás-Hernández, S.; García, T.; Mulero, M.; Gómez, M.; Domingo, J.L.; Sánchez, D.J. Oral exposure to silver nanoparticles increases oxidative stress markers in the liver of male rats and deregulates the insulin signalling pathway and p53 and cleaved caspase 3 protein expression. *Food Chem. Toxicol.* **2018**, *115*, 398–404. [CrossRef] [PubMed]
45. Fehaid, A.; Fujii, R.; Sato, T.; Taniguchi, A. Silver nanoparticles affect the inflammatory response in a lung epithelial cell line. *Open Biotechnol. J.* **2020**, *14*, 113–123. [CrossRef]

46. Barbasz, A.; Oćwieja, M.; Roman, M. Toxicity of silver nanoparticles towards tumoral human cell lines U-937 and HL-60. *Coll. Surf. B Biointerfaces* **2017**, *156*, 397–404. [CrossRef]
47. Kongseng, S.; Yoovathaworn, K.; Wongprasert, K.; Chunhabundit, R.; Sukwong, P.; Pissuwan, D. Cytotoxic and inflammatory responses of TiO2 nanoparticles on human peripheral blood mononuclear cells. *J. Appl. Toxicol.* **2016**, *36*, 1364–1373. [CrossRef]
48. Roignot, J.; Peng, X.; Mostov, K. Polarity in mammalian epithelial morphogenesis. *Cold Spring Harb. Perspect. Biol.* **2013**, *5*, a013789. [CrossRef]
49. van Silfhout, A.; Erné, B. Magnetic detection of nanoparticle sedimentation in magnetized ferrofluids. *J. Magn. Magn. Mater.* **2019**, *472*, 53–58. [CrossRef]
50. De Matteis, V.; Rinaldi, R. Toxicity assessment in the nanoparticle era. *Adv. Exp. Med. Biol.* **2018**, *1048*, 1–19.
51. Savage, D.T.; Hilt, J.Z.; Dziubla, T.D. In Vitro methods for assessing nanoparticle toxicity. *Methods Mol. Biol.* **2019**, *1894*, 1–29.
52. Kany, S.; Vollrath, J.T.; Relja, B. Cytokines in inflammatory disease. *Int. J. Mol. Sci.* **2019**, *20*, 6008. [CrossRef]
53. Cathe, D.S.; Whitaker, J.N.; Breitner, E.K.; Comfort, K.K. Exposure to metal oxide nanoparticles in physiological fluid induced synergistic biological effects in a keratinocyte model. *Toxicol. Lett.* **2017**, *268*, 1–7. [CrossRef] [PubMed]
54. Cho, W.S.; Duffin, R.; Howie, S.E.M.; Scotton, C.J.; Wallace, W.A.H.; MacNee, W.; Bradley, M.; Megson, I.L.; Donaldson, K. Progressive severe lung injury by zinc oxide nanoparticles; the role of Zn2+ dissolution inside lysosomes. *Part. Fibre Toxicol.* **2011**, *8*, 27. [CrossRef] [PubMed]
55. Hoelting, L.; Scheinhardt, B.; Bondarenko, O.; Schildknecht, S.; Kapitza, M.; Tanavde, V.; Tan, B.; Lee, Q.Y.; Mecking, S.; Leist, M.; et al. A 3-dimensional embryonic stem cell (hESC)-derived model to detect developmental neurotoxicity of nanoparticles. *Arch. Toxicol.* **2015**, *87*, 721–733. [CrossRef]
56. Park, E.J.; Bae, E.; Yi, J.; Kim, Y.; Choi, K.; Lee, S.H.; Yoon, J.; Lee, B.C.; Park, K. Repeated-dose toxicity and inflammatory responses in mice by oral administration of silver nanoparticles. Environ. *Toxicol. Pharmacol.* **2010**, *30*, 162–168. [CrossRef]
57. van den Brule, S.; Ambroise, J.; Lecloux, H.; Levard, C.; Soulas, R.; de Temmerman, P.J.; Palmai-Pallag, M.; Marbaix, E.; Lison, D. Dietary silver nanoparticles can disturb the gut microbiota in mice. *Part. Fibre Toxicol.* **2016**, *13*, 38. [CrossRef] [PubMed]

antioxidants

MDPI

Article

The Ameliorative Role of *Acacia senegal* Gum against the Oxidative Stress and Genotoxicity Induced by the Radiographic Contrast Medium (Ioxitalamate) in Albino Rats

Islam El-Garawani [1,*], Sobhy Hassab El-Nabi [1], Ahmed El Kattan [2], Azza Sallam [1], Sabha Elballat [3], Shaimaa Abou-Ghanima [4], Islam H. El Azab [5], Hesham R. El-Seedi [6,7,8,*], Shaden A. M. Khalifa [6] and Sawsan El-Shamy [9]

[1] Zoology Department, Faculty of Science, Menoufia University, Shebin El-Kom, Menoufia 32511, Egypt; drsobhyhassabelnabi@science.menofia.edu.eg (S.H.E.-N.); ohoodsallam@science.menofia.edu.eg (A.S.)
[2] Medical and Radiation Department, Researcher Sector, Nuclear Materials Authority, Cairo 11381, Egypt; kattan.a79@nma.org.eg
[3] Zoology Department, Faculty of Science, Zagazig University, Zagazig 44519, Egypt; sabhael@zu.edu.eg
[4] Biochemistry Department, King Khalid University, Abha 61421, Saudi Arabia; sjamal@kku.edu.sa
[5] Chemistry Department, College of Science, Taif University, P.O. Box 11099, Taif 21944, Saudi Arabia; i.helmy@tu.edu.sa
[6] Department of Molecular Biosciences, The Wenner-Gren Institute, Stockholm University, S-10691 Stockholm, Sweden; shaden.khalifa@su.se
[7] International Research Center for Food Nutrition and Safety, Jiangsu University, Zhenjiang 212013, China
[8] Chemistry Department, Faculty of Science, Menoufia University, Shebin El-Kom, Menoufia 32511, Egypt
[9] Basic Science Center, Department of Biology, Misr University for Science and Technology, Giza 12511, Egypt; sawsan.alshamy@must.edu.eg
* Correspondence: dr.garawani@science.menofia.edu.eg (I.E.-G.); hesham.elseedi@su.se (H.R.E.-S.); Tel.: +2-010-64455948 (I.E.-G.); +46-700-434343 (H.R.E.-S.)

Citation: El-Garawani, I.; Hassab El-Nabi, S.; El Kattan, A.; Sallam, A.; Elballat, S.; Abou-Ghanima, S.; El Azab, I.H.; R. El-Seedi, H.; A. M. Khalifa, S.; El-Shamy, S. The Ameliorative Role of *Acacia senegal* Gum against the Oxidative Stress and Genotoxicity Induced by the Radiographic Contrast Medium (Ioxitalamate) in Albino Rats. *Antioxidants* **2021**, *10*, 221. https://doi.org/10.3390/antiox10020221

Academic Editor: Tim Hofer
Received: 6 January 2021
Accepted: 26 January 2021
Published: 2 February 2021

Abstract: Arabic gum (*Acacia senegal*, AG) is proven effective antioxidant and cytoprotective agent. The present study was designed to test this notion by investigating the possible role of AG against the radiographic contrast medium (Ioxitalamate, Telebrix-35®, TBX)-induced oxidative stress and genotoxicity. Albino rats were divided into four groups and supplied with either; distilled water, daily 10% (*w*/*v*) AG, an intravenous dose of TBX (1600 mg I/kg b.wt) and co-administration of TBX and AG. Rats were sacrificed and blood samples were collected to assess the genotoxicity employing the peripheral blood leucocytes fluorescent double staining; namely the acridine orange/ethidium bromide (AO/EB) staining and alkaline comet assay. Further, chromosomal analyses were done in bone marrow cells. Serum urea and creatinine levels, in addition to malondialdehyde (MDA), nitric oxide (NO), catalase (CAT) and glutathione (GSH) levels in kidney tissues were measured. Liquid chromatography-mass spectrophotometry (LC-MS-MS) was performed to identify the chemical composition of AG extract. Kidney functions, single/double-stranded DNA damage, chromosomal aberrations, mitotic index, MDA and NO levels were significantly ($p < 0.001$) increased in TBX-treated group compared to the control and AG-treated one. Meanwhile, CAT and GSH activities were significantly diminished and the AG supplementation significantly ($p < 0.001$) ameliorated these effects compared with the control and AG-treated groups. Five compounds have been identified using GNPS networking including 7,3′,4′-Trihydroxyisoflavone, Noscapine, Tetrahydropapaveroline, Costunolide, Hesperidin. In conclusion, results of the present study suggest that AG exerted a protective role against TBX-induced oxidative stress and genotoxicity which may be attributed to the active metabolites in the gum.

Keywords: Arabic gum; DNA damage; genotoxicity; Ioxitalamate; oxidative stress; LC-MS-MS

1. Introduction

Contrast media (CM) is a common group of chemicals commonly used in radiographic diagnosis and radiotherapy [1,2]. Iodine is a crucial element used in CM due to its high-contrast density [3]. It is a tri-iodinated derivative of benzoic acid [4]. Iodinated contrast agents have different osmolality [5]. The higher CM osmolality, the severe adverse effects [6]. For instance, hyperthyroidism or hypothyroidism may be correlated to the iodinated CM exposure. Moreover, nephropathy is certainly the most common adverse effect [3].

Ioxitalamate (Telebrix®), is an ionic iodinated CM widely used for radiography [7]. The iodinated radiographic CM-induced nephropathy is associated with renal apoptosis and oxidative stress [8–10]. Cytogenetic changes in bone marrow cells were reported after CM administrations in mice [1], manifested by the increase of micronucleated polychromatic erythrocytes [11].

Arabic gum (AG) is an edible, dried gummy exudate from the stems and branches of *Acacia senegal* tree. It is characterized by the richness of non-viscous soluble fibers. It is commonly used as an emulsifier and stabilizer in the food and pharmaceutical industries [12]. Among other rich natural products, Arabic gum has a history as a preventive supplementation in Arabic folk medicine and recently attracted more attention due to its protective impact against the effects of environmental and chemical exposure to hazards [13–19].

The high-molecular weight polysaccharides and their potassium, magnesium and calcium salts are the major constituents of AG. After AG hydrolysis, arabinogalactan, arabinogalactanprotein and glycoprotein were the main fractions obtained [20]. AG is considered a safe food additive as it did not induce genotoxicity or carcinogenicity when given intraperitoneally or orally [21]. Arabic gum plays an important antioxidant and protective role against experimental gentamicin and cisplatin-induced nephrotoxicity [22], doxorubicin cardiotoxicity [23] and acetaminophen hepatotoxicity [24]. It is also known to combat many diseases such as renal [12], cardiovascular [25], gastrointestinal [26] and respiratory diseases, due to its agonist effect on the oxidative stress and DNA damage [27]. Moreover, Badreldin et al. [28] confirmed the antioxidative and anti-inflammatory properties of AG. The antischistosomal and protective effects on some reproductive aspects in mice were also reported [29]. Clinically, in chronic renal failure patients, AG has been shown to be beneficial [30].

Thus, the present investigation aimed to evaluate the protective potentials of Arabic gum against the iodinated radiographic contrast medium (Ioxitalamate)-induced genotoxicity and oxidative stress in rats.

2. Materials and Methods

2.1. Chemicals

Radiographic contrast medium, Ioxitalamate ($C_{12}H_{11}I_3N_2O_5$); [CAS. 20978/00, Lyomit, Telebrix-35®, GASTRO, Guerbet, France] was enrolled in the current study. Agarose gel (Sigma-Aldrich, Darmstadt, Germany), DNA ladder (100 bp, O'gene ruler™, Thermo Fisher Scientific, Austin, TX, USA), normal melting point ultra-pure agarose (Sigma-Aldrich, Darmstadt, Germany) and ethidium bromide (Sigma-Aldrich, Darmstadt, Germany).

2.2. Arabic Gum

It is a soluble dietary fiber obtained naturally from the stems and branches of *Acacia senegal* trees (family: legume). It was purchased from El-Nobi Co. (Shebin El-Kom, Menoufia, Egypt) as crystals (commercial grade) and ground to a fine powder. An aqueous solution of Arabic Gum (10%, w/v) was prepared freshly every day and introduced to the rats in drinking water [31,32].

2.3. Experimental Animals

The study was carried out on apparently healthy 60 adult Wister male albino rats, weighing 180 ± 15 g (7–9 weeks old), obtained from the Holding Company for Biological

Products & Vaccines (VACSERA), Egypt. Animals were housed in cages in a controlled animal facility environment (room temperature about 25 °C, 12 h light and 12 h dark cycle), with free access to a standard commercial diet (El-Haramain® standard rodent diet, Egypt) and water was *ad libitum*. Animals were acclimatized for 14 days prior to the experiment. The study followed the Institutional Animal Ethical Committee (IAEC) guidelines at the Faculty of Science, Menoufia University (MUFS-F-GE-2-20).

2.4. Experimental Design

Animals were divided into four groups (15 rats/group). Group (1) animals served as a negative control group, Group (2) animals received 10% (*w*/*v*) of AG in drinking water daily, Group (3) animals were subjected to the codal vein, intravenous, single dose injection (1600 mg I/kg b.wt., equivalent to the higher range of the human dose) according to Li et al. [33] and Group (4) animals were injected with Telebrix (single dose, i.v.) and received 10% (*w*/*v*) of AG daily. Five rats per group, randomly selected, were sacrificed after 1 day (for liver and peripheral leucocytes analyses) and the other 5 rats/group were sacrificed after 14 days for liver sampling only.

In addition, further genotoxicity investigation was done in parallel using 5 rats/group which processed for bone marrow chromosomal preparations after 24 h of treatments.

2.5. General Health

All rats were monitored to assess mortality, skin irritation, food intake and activities during the period of the experiment.

2.6. Kidney Functions

Sera of treated and control rats were collected, by centrifugation at 900× *g* for 5 min, from submandibular peripheral venous blood samples (2 mL/rat) after clotting. Urea and creatinine levels were determined according to the method of Tabacco et al. [34] and Rartels and Böhmer [35] using Milton Roy spectrophotometer (Spectronic 1201, Houston, TX, USA).

2.7. Oxidative Status in Kidneys' Tissue

Kidney tissues homogenate (10%, *wt*/*v*) were prepared in phosphate buffer (0.1 M, pH 7.4). Supernatants were collected by centrifugation at 3600× *g* for 5 min and kept at −20 °C for further investigations.

Catalase activities were measured according to the method of Cohen et al. [36]. Lipid Peroxidation (MDA) was determined at 534 nm using the method of Mesbah et al. [37] The sulfhydryl group of GSH was measured colorimetrically at 412 nm according to Ellman [38]. Nitric oxide levels were determined following the procedure of Green et al. [39]. All investigations were performed using Milton Roy spectrophotometer (Spectronic 1201, Houston, TX, USA)

2.8. Total Genomic DNA Extraction and Apoptosis Detection in Kidney Tissue

DNA extraction and detection of apoptosis (DNA fragmentation assay) were done according to the salting out extraction method of Aljanabi and Martinez [40] and modifications by El-Nabi and Elhassaneen [41]. About 20 mg of tissue samples from treated and control kidneys were lysed in lysing buffer at 40 °C for 24 h. Then, proteins were precipitated by a solution of NaCl (4 M). Cold isopropanol was used for nucleic acids precipitation. The resuspended pellets of nucleic acids in Tris-EDTA buffer (10 mM Tris-HCl, 1 mM EDTA, pH 8) were incubated for 60 min. with a loading buffer supplemented with RNase. The samples were stained directly with ethidium bromide [42] and processed for electrophoresis on 1.8% agarose gel. Apoptotic bands of DNA were aligned, at 180 bp and its multiples 360, 540 and 720 bp, against 100 bp DNA ladder. The intensity of released DNA fragments was analyzed using Image J software at 256 grey levels.

2.9. Isolation of Peripheral Blood Leucocytes

Heparinized peripheral venous blood samples (2 mL/rat) were drawn from treated and control animals. Then leucocytes were isolated from rats' peripheral blood by incubation (1:3, *v/v*) with erythrocyte lysing buffer [NH_4C1 (0.015 M), $NaHCO_3$ (1 mM), EDTA (0.1 mM)]. They were centrifuged for 5 min at 900× *g*. These steps were repeated until a white pellet of leucocytes appeared [43]. Then, the pellets were resuspended in RPMI 1640 medium for further investigations.

2.10. Leucocytes Double Staining by Acridine Orange/Ethidium Bromide (AO/EB)

One microliter of florescent dye mixture (100 mg/mL of AO and 100 mg/mL of EB in distilled water) was incubated with 4 µL of leucocytes' suspension for 1 min. On a microscopic glass slide, the suspension was directly examined by a fluorescent microscope (Olympus CX23, Tokyo, Japan) at 400× magnification. A minimum of 400 cells were counted per a sample [44].

2.11. Alkaline Comet Assay in Peripheral Blood Leucocytes

In the present study, alkaline single cell gel electrophoresis (comet assay) was performed to assess the DNA damage in peripheral blood leucocytes of untreated and treated rats after 24 h [45]. Briefly, the mixed cells with low melting point agarose gel were layered between two layers of ultra-pure agarose with a normal melting point, on microscopic glass slides. In dark conditions, the slides were dipped in lysis buffer (2.5 M NaCl, 100 mM EDTA and 10 mM Tris, 1% Triton X-100, 10% DMSO, pH 10) for 1 h at 4 °C. The slides were kept in a cold and freshly prepared alkaline buffer (300 mM NaOH and 1 mM EDTA, pH > 13) for 10 min. Then, they were subjected to 25 V/300 mA electric current for 15 min. Neutralization of the slides was carried out using 0.4 M Tris-HCl buffer, pH 7.5 three times by rinsing. Next, they were labeled with ethidium bromide (20 µg/mL). Nuclei were visualized using Olympus, fluorescence microscope (BX41, Tokyo, Japan). For DNA damage assessment, about 50 randomly selected nuclei were inspected per field of total examined five fields per slide. The results were visually classified as normal nuclei with no tail, lightly damaged with a migrated tail less than the nucleus diameter and strongly damaged with a DNA migration more than the nucleus diameter.

2.12. Bone Marrow Chromosomal Preparations

Chromosomal preparations were carried out in bone marrow of treated and control rats [46]. Animals were injected 2 h before sacrificing with 0.2 mL (0.0012%) colchicine to arrest the cell division at metaphase. Briefly, bone marrow cells were collected from the femurs in isotonic solution (0.9% NaCl). Hypotonic solution (0.56% KCl) was added to the cells and they were incubated for 20 min at 37 °C. Then, cells were drop-wise fixed by 3–4 mL of cold and freshly prepared fixative (absolute methanol and glacial acetic acid, 3:1, *v/v*) for 15 min. This step was repeated three times to complete fixation. Then, the pellets were re-suspended in 100 µL of the fixative and dropped on a clean slide previously dipped in cold 70% ethyl alcohol then air-dried after flaming. Slides were stained with phosphate-buffered Giemsa solution. For each animal, 100 well spread metaphases were scored for chromosomal aberrations at 1000× magnification using a light microscope (Olympus CX23, Tokyo, Japan). Furthermore, the assessment of the mitotic index was carried out by evaluating five hundred cells per group at 400× magnifications using a light microscope (Olympus CX23, Tokyo, Japan). The mitotic index was calculated using the following equation:

Mitotic index (%) = (No. of metaphases + prophases)× 100/(No. of all counted nuclei).

The abnormal metaphases were evaluated also in the same slides and the calculations were carried out using the following equation:

Abnormal metaphases (%) = (No. of abnormal metaphases) ×100/(No. of normal metaphases + No. of abnormal metaphases).

2.13. Extraction of Acacia senegal Gum

The raw material was grounded and soaked in DCM-MeOH 1:1 on shaker overnight and evaporated using rotatory evaporator. The residue was redissolved in cold water overnight and was dried using freeze dryer. The crude extracts were ready for further analysis.

2.14. LC-MS-MS Analysis

For LC-MS analysis, the crude extract was dissolved in 50% acetonitrile (ACN), 0.1% formic acid (FA). The samples were injected by syringe through a PicoTip emitter at 0.3 μL/min connected to a Q-Tof Micro (Waters, Milford, MA, USA) with the voltage set at 1.4 kV. The analysis was carried out in positive ion mode and linear gradient from 10% (*v/v*) H_2O to 99% (*v/v*) ACN in 0.1% (*v/v*) FA at a flow rate of 0.3 μL/min over 75 min. The analysis was followed by LC/MS-MS where 0.1mg/mL in LC-MS solvent: 60% MeCN in 0.1% FA was dissolved with a linear gradient ranging from 10–60% (*v/v*) MeCN in 0.1% (*v/v*) FA at a flow rate of 0.3 mL/min over 75 min. The capillary temperature was set at 220 °C and the spray voltage at 4 kV [47]. GNPS molecular network analysis was performed to identify the metabolites [48,49].

2.15. Statistical Data Analysis

All experiments were done in triplicates ($n = 5$). Data were presented as Mean ± SE. Using IBM SPSS software version 21.1 (New York, NY, USA), data were statistically analyzed and $p < 0.001$ were considered significant.

3. Results

3.1. General Health

All treated and control rats showed no mortality during the period of the experiment. No irritation was observed at the TBX sites of injection.

3.2. Kidney Functions

In the present study, serum urea and creatinine were measured in the four groups. The administration of TBX induced a significant ($p < 0.001$) increase in serum urea and creatinine levels of treated rats with respect to the control. However, treatment with AG significantly ($p < 0.001$) ameliorated these effects. Serum urea levels were improved by ~11.6 & 23% in AG + TBX combined group relative to the TBX-treated group after 1 and 14 days of treatments respectively. Regarding serum creatinine, the improvements were ~53.5 & 65% in AG + TBX combined group relative to the TBX-treated group after 1 and 14 days of treatments respectively (Figure 1).

Following the treatments for 1 & 14 days, the oxidative markers were evaluated in the kidney tissue homogenates. Results revealed that TBX significantly ($p < 0.001$) elevated the lipid peroxidation (LPO), as measured by malondialdehyde (MDA) and nitric oxide (NO) concentrations (Figure 2A,B). Otherwise, a significant ($p < 0.001$) decrease in catalase (CAT) activities and glutathione (GSH) levels were observed as compared to control rats (Figure 2C,D). Whereas, rats in the AG+TBX combined group showed a significant amelioration exerted by AG against TBX effects after 1 and 14 days of treatments. The activities, after 1 and 14 days, of CAT and GSH were improved by about 41.6 & 56% and 20 & 100% respectively. However, LPO and NO levels were improved by approximately 18.4 & 30% and 21 & 26% after 1 and 14 days; respectively.

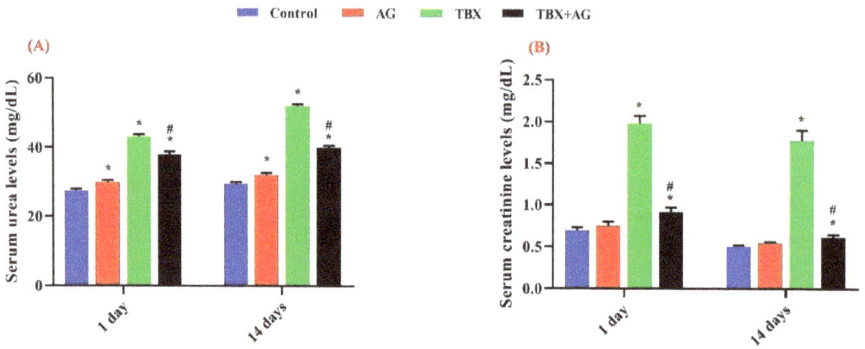

Figure 1. The protective effect of Arabic gum (*Acacia senegal*, AG) against the Telebrix (TBX)-induced alterations at the levels of serum urea (**A**) & creatinine (**B**) in treated rats after 1 and 14 days of the experiment. *: Significant ($p < 0.001$) relative to untreated groups. #: Significant ($p < 0.001$) relative to TBX-treated groups. Data were represented as mean ± SE of three independent experiments ($n = 5$). AG: 10% Arabic gum, TBX: Telebrix (1600 mg/kg b.wt.) and TBX+AG: co-treatment with 1600 mg/kg b.wt. and 10% Arabic gum.

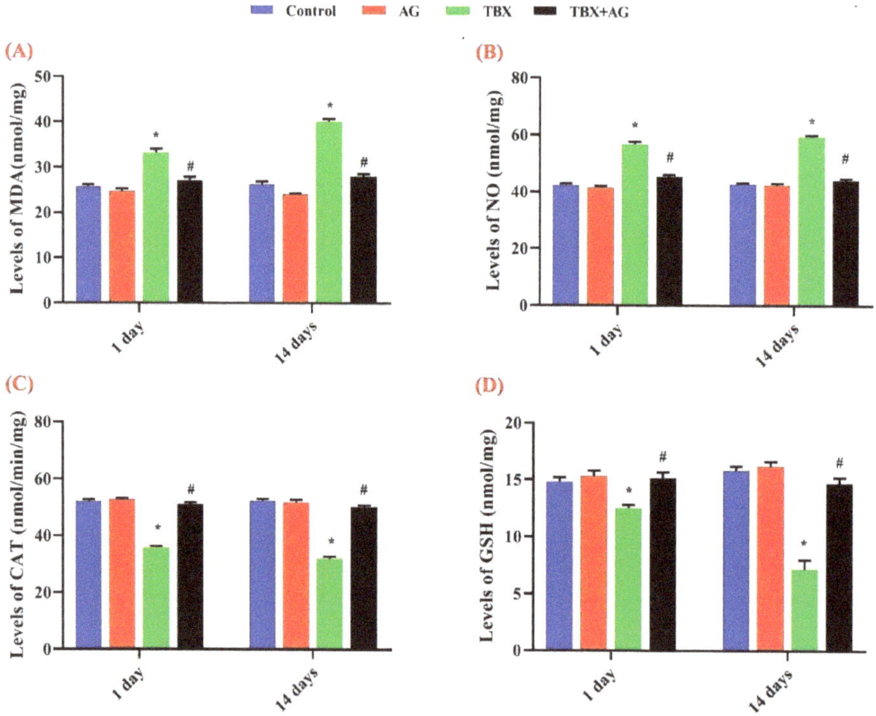

Figure 2. The protective effect of AG against TBX-induced alterations in oxidative stress markers after 1 and 14 days of treatments. (**A**) MDA: malondialdehyde, (**B**) NO: nitric oxide, (**C**) CAT: catalase and (**D**) GSH: glutathione. *: Significant ($p < 0.001$) relative to untreated groups. #: Significant ($p < 0.001$) relative to TBX-treated groups. Data were represented as mean ± SE of three independent experiments ($n = 5$). AG: 10% Arabic gum, TBX: Telebrix (1600 mg I/kg b.wt.) and TBX+AG: co-treatment with 1600 mg/kg b.wt. and 10% Arabic gum.

3.3. Genomic Double-Strand DNA Damage in Kidneys' Tissue

The double-strand breaks were evaluated in kidneys' tissue by agarose gel electrophoresis. TBX treatments induced internucleosomal cleavage as illustrated by the laddering pattern of DNA damage, a hallmark of apoptosis (Figure 3). A significant ($p < 0.001$) time-dependent increase in DNA fragmentation was recorded in the kidneys tissue of TBX-treated rats after 1 and 14 days when compared to control (Figure 4). Otherwise, in the case of AG combined treatments, a noticeable decrease in DNA damage (26 and 71.4%) was detected after 1 and 14 days of treatments respectively, when compared to TBX-treated rats.

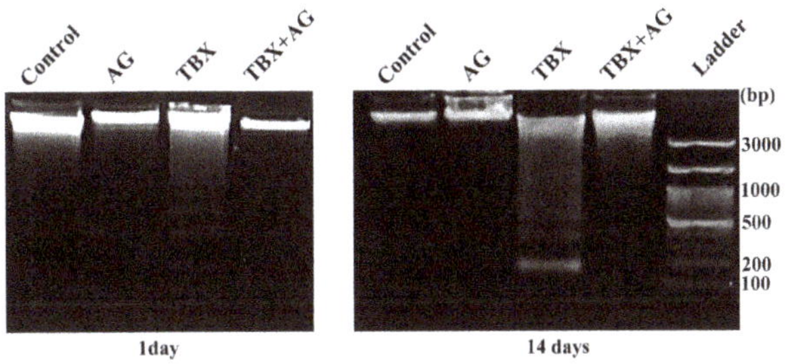

Figure 3. Representative photographs (1.8% ethidium bromide-stained agarose gel) show the protective effect of AG against TBX-induced DNA damage in kidney tissues after 1 and 14 days of treatment. AG: 10% Arabic gum, TBX: Telebrix (1600 mg I/kg b.wt.) and TBX+AG: co-treatment with 1600 mg/kg b.wt. and 10% Arabic gum.

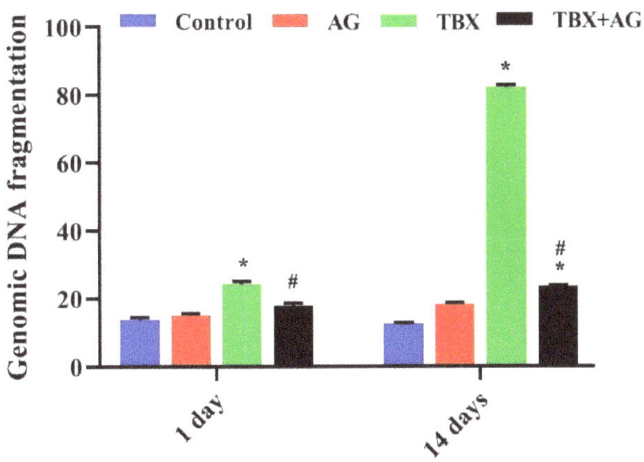

Figure 4. The protective effect of AG against TBX-induced DNA fragmentation employing 1.8% agarose gel electrophoresis of kidney tissues of treated rats after 1 and 14 days of treatments. *: Significant ($p < 0.001$) relative to untreated groups. #: Significant ($p < 0.001$) relative to TBX-treated groups. Data were represented as mean ± SE of three independent experiments ($n = 5$). AG: 10% Arabic gum, TBX: Telebrix (1600 mg I/kg b.wt.) and TBX+AG: co-treatment with 1600 mg/kg b.wt. and 10% Arabic gum.

3.4. Cytotoxicity on Leucocytes

Results of AO/EB double fluorescent labeling (Figure 5) revealed a significant ($p < 0.001$) time-dependent elevation of the number of dead cells in TBX-treated rats after 1 and 14 days when compared to the untreated rats. Moreover, the combined treatment with AG showed a significant protective effect on leucocytes as the percentage of damaged cells was reduced by approximately 43% and 60% after 1 and 14 days, respectively (Figure 6).

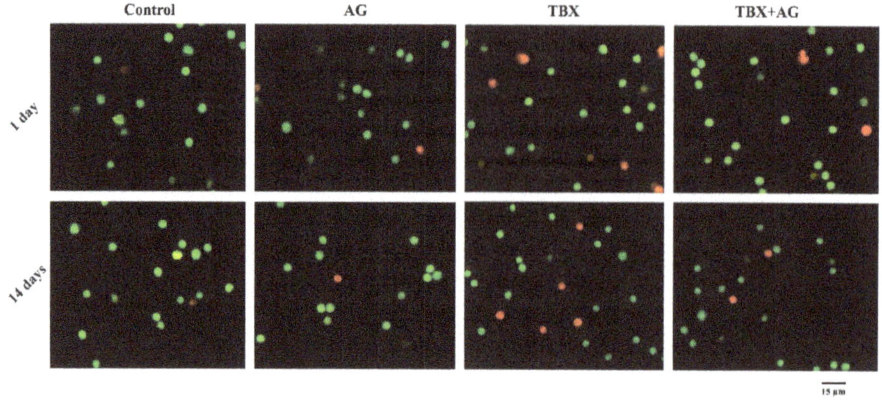

Figure 5. Representative photomicrographs show the protective effect of AG against TBX-induced cytotoxicity in peripheral blood leucocytes of treated rats after 1 and 14 days of the experiment. The isolated leucocytes were stained by AO/EB fluorescent dyes (Olympus CX23 microscope, Japan). AG: 10% Arabic gum, TBX: Telebrix (1600 mg/kg b.wt.) and TBX+AG: co-treatment with 1600 mg/kg b.wt. and 10% Arabic gum.

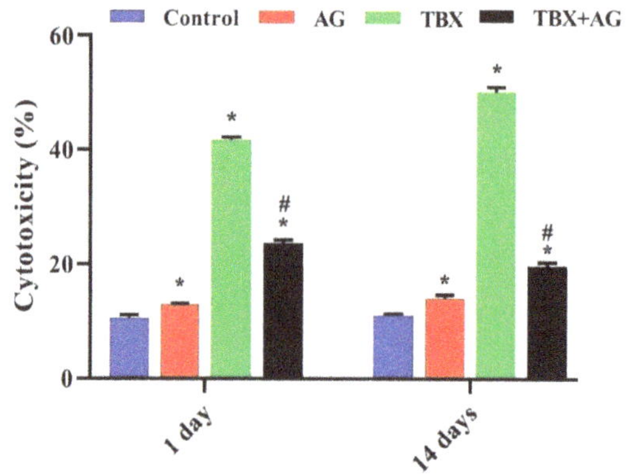

Figure 6. The protective effect of AG against TBX-induced alterations in the viability of peripheral blood leucocytes of treated rats after 1 and 14 days of the experiment. *: Significant ($p < 0.001$) relative to untreated groups. #: Significant ($p < 0.001$) relative to TBX-treated groups. Data were represented as mean ± SE of three independent experiments ($n = 5$). AG: 10% Arabic gum, TBX: Telebrix (1600 mg I/kg b.wt.) and TBX+AG: co-treatment with 1600 mg/kg b.wt. and 10% Arabic gum.

3.5. *Acute Genotoxicity in Leucocytes*

3.5.1. DNA Strand Breaks in Leucocytes

DNA strand breaks were detected using single-cell gel electrophoresis (alkaline comet assay). The extent of DNA damage was evaluated as a length of DNA tail migration towards the anode. In the present work, TBX treatments for 24 h induced a significant ($p < 0.001$) DNA strand breaks in rats leucocytes as compared to controls (Figure 7). The light damaged cells were 30.1 ± 0.67% and the strong damaged cells were scored as 54.8 ± 0.8% (Figure 8). However, the combined treatment with AG showed significant ($p < 0.001$) improvements in DNA strand breaks as compared with TBX-treated rats (Figure 7). The detected light DNA damage was 24.26 ± 0.49 with about 20 improvement relative to the TBX-treated group. Whereas, the strong damage was 4.6 ± 0.6% with approximately 91% improvement when compared with TBX-treated group (Figure 8).

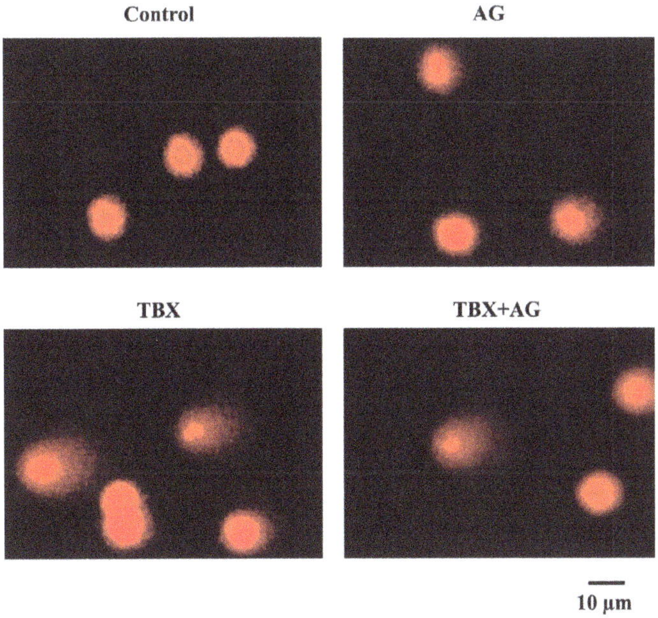

Figure 7. Representative photomicrographs of comet assay (ethidium bromide-stained) showing the protective effect of AG against TBX-induced DNA strand breaks in isolated peripheral blood leucocytes of treated rats after 24 h of treatments (Olympus CX23 microscope, Tokyo Japan). AG: 10% Arabic gum, TBX: Telebrix (1600 mg I/kg b.wt.) and TBX+AG: co-treatment with 1600 mg/kg b.wt. and 10% Arabic gum.

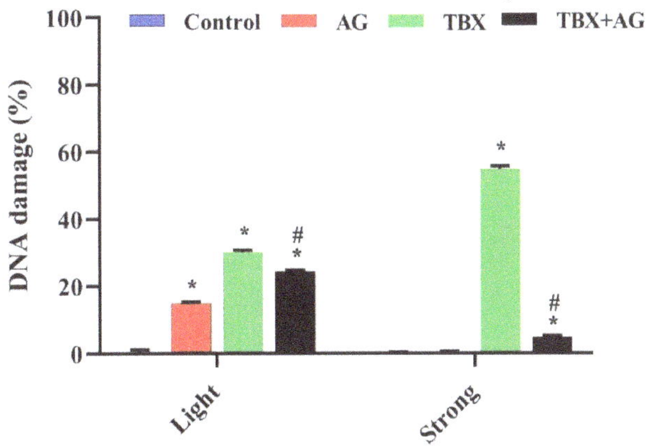

Figure 8. The protective effect of AG against TBX-induced DNA strand breaks in isolated peripheral blood leucocytes of treated rats after 24 h of treatments. *: Significant ($p < 0.001$) relative to untreated groups. #: Significant ($p < 0.001$) relative to TBX-treated groups. Data were represented as mean ± SE of three independent experiments ($n = 5$). AG: 10% Arabic gum, TBX: Telebrix (1600 mg I/kg b.wt.) and TBX+AG: co-treatment with 1600 mg/kg b.wt. and 10% Arabic gum.

3.5.2. Chromosomal Aberrations in Bone Marrow

In the current investigation, the structural chromosomal aberrations (deletion, fragmentation, breaking down or forming gap and ring) were scored in bone marrows among treated and untreated rats. Figure 9 showed a significant ($p < 0.001$) increase in the percentage of the evaluated aberrations among the TBX-treated rats where the total chromosomal aberrations without gaps was (TCA: 42 ± 0.7) with respect to the control (TCA: 0.6 ± 0.3). The highest recorded values were 63 ± 1.6% as detected as chromatid deletions and the lowest were gaps (2.4 ± 0.7%). However, the AG-treated animals revealed a significant ($p < 0.001$) decrease in the observed aberrations when compared to TBX-treated rats with approximately 69% improvement in TCA.

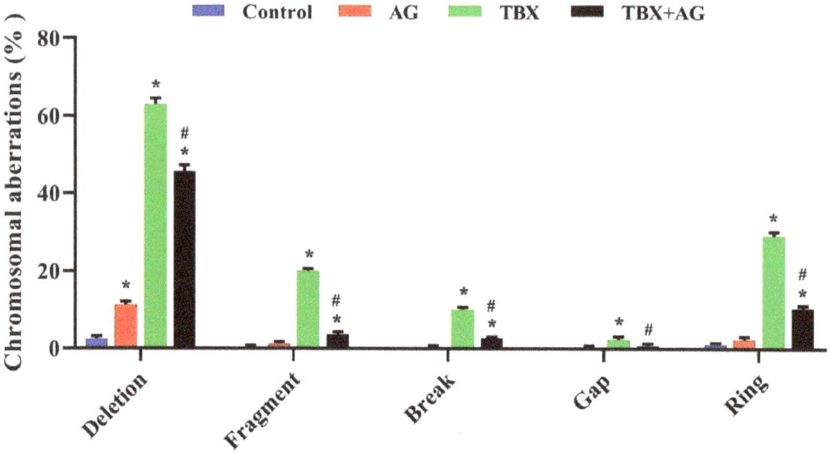

Figure 9. The protective effect of AG against TBX-induced chromosomal abnormalities in the bone marrow of treated rats after 24 h of treatments. *: Significant ($p < 0.001$) relative to untreated groups. #: Significant ($p < 0.001$) relative to TBX-treated groups. Data were represented as mean ± SE of three independent experiments ($n = 5$). AG: 10% Arabic gum, TBX: Telebrix (1600 mg I/kg b.wt.) and TBX+AG: co-treatment with 1600 mg/kg b.wt. and 10% Arabic gum.

3.5.3. Assessment of the Mitotic Index and Abnormal Metaphases

In order to assess the proliferation capacities, chromosomal preparations were carried out in bone marrows of treated and untreated rats then, the mitotic indices were calculated. Surprisingly, TBX treatments caused a significant ($p < 0.001$) increase in the dividing populations (pro- and metaphases). While, AG administration significantly ameliorated, about 50%, of this elevation (Figure 10A). Moreover, the abnormal metaphase percentages were assessed along with the mitotic index evaluation. Figure 10B illustrated the significant ($p < 0.001$) increase in abnormal metaphases among TBX treatments whereas the protective potentials of AG combined treatments significantly ($p < 0.001$) restored these elevated levels by about 48%.

3.6. *Chemical Investigation of Acacia senegal Extract*

The mass profile of *A. senegal* extract was analyzed to exhibit 195 nodes in the Global Natural Product Social Molecular Networking (GNPS), showed in Figure 11. All the metabolites detected in the raw mass file represented in the molecular network as nodes. The bigger the nodes the higher the intensity of their peaks in the raw mass file. The chemically related metabolites with similar common classes are clustered together. From 195 metabolites, only six were trustable when screened and compared with the previously isolated ones and found to have similar fragmentation patterns. One of the six identified metabolites was isolated previously from the same genus which was hesperidin. The five metabolites that were matched with other standards in the GNPS library, identified as 7,3′,4′-trihydroxyisoflavone, noscapine, tetrahydropapaveroline, costunolide and hesperidin. The precursor masses were as follows, $[M + H]^+$ (m/z; 271.698, 414.525, 289.652, 234.359 and 611.313) as shown in Table 1.

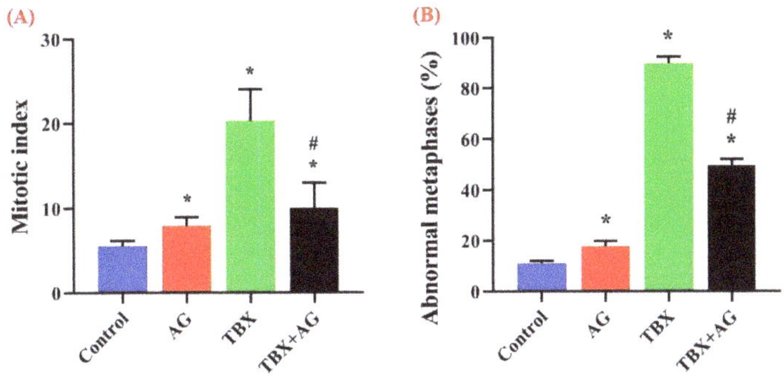

Figure 10. The protective effect of AG against TBX-induced mitotic index alterations (**A**) and the percentage of abnormal metaphases (**B**) in the bone marrow of treated rats after 24 h of treatments. *: Significant ($p < 0.001$) relative to untreated groups. #: Significant ($p < 0.001$) relative to TBX-treated groups. Data were represented as mean ± SE of three independent experiments ($n = 5$). AG: 10% Arabic gum, TBX: Telebrix (1600 mg I/kg b.wt.) and TBX+AG: co-treatment with 1600 mg/kg b.wt. and 10% Arabic gum.

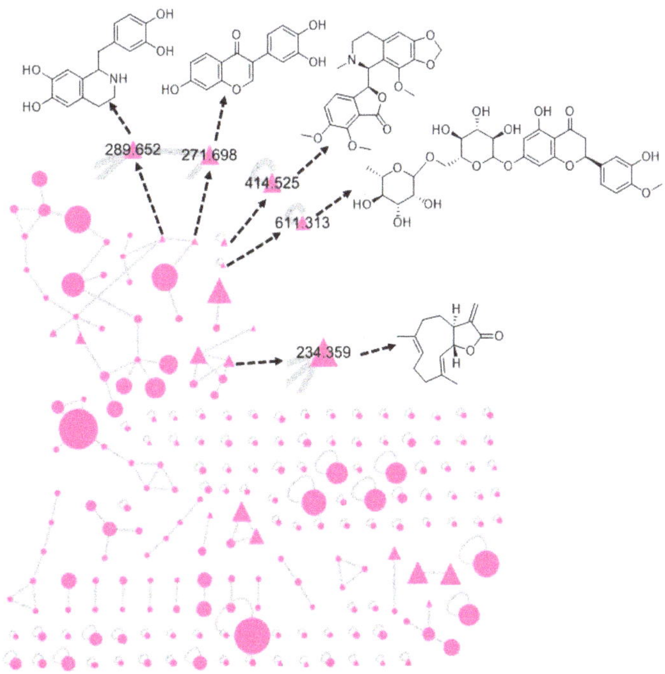

Figure 11. Molecular network of 195 parent ions produced from *A. senegal*. The circular fuchsia nodes refer to the whole parent masses that have unique detected peaks in the molecular network. The triangle fuchsia Nodes represent parent ions that have been identified in the GNPS molecular network.

Table 1. The precursor and fragments' masses of the identified metabolites compared with that of the standards from the molecular networking database.

	Precursor Mass	Raw Mass Fragments	Library Fragments	Reference
7,3′,4′-Trihydroxyisoflavone	271.698	252.94, 242.95, 224.94, 214.93, 160.93 and 136.91	271.06, 235.05, 225.05, 215.07 and 137.02	[50]
Noscapine	414.525	352.99, 219.94 and 205.03	353.10, 220.10 and 205.07	[51]
Tetrahydropapaveroline	289.652	270.92, 242.95 and 164.02	288.12, 271.09 and 164.07	[52]
Costunolide	234.359	215.02, 187.05 and 160.05	233.15, 215.14, 187.15 and 159.12	[53]
Hesperidin	611.313	593.07, 575.01, 556.97, 488.97, 464.92, 449.02, 430.99, 413.00, 345.03 and 303.01	593.00, 575.00, 557.00, 489.00, 465.00, 449.00, 431.00, 413.00, 345.00 and 303.00	[54]

4. Discussion

The present study was designed to assess the toxic effects of TBX (Ioxitalamate) on peripheral leucocytes and kidney tissues. It has thrown more light on the potential of AG treatments against the genotoxicity and oxidative stress induced by TBX in rats.

Ioxitalamate is an ionic iodinated contrast medium with a high osmolality of 1500–1800 mOsm/kg [55]. It has been shown to be associated with an incidence of nephrotoxicity [56] and directly damages the chemical bonds of DNA molecules [57]. However, radiation dose that presents in ioxitalamate contrast media may lead to the elevated number of DNA double-strand breaks [58,59]. Ionizing radiation has enough energy to release one or more electrons within biomolecules and subsequently change the chemistry of the DNA molecule leading to one or more single or double-strand DNA breaks [60]. Data of the present study showed that TBX induced DNA strand breaks in peripheral leucocytes and DNA double-strand internucleosomal cleavage in kidneys' tissues. This was in agreement with the previous studies that illustrated that contrast medium caused in vivo [61–63] and in vitro [61,64] DNA damage. The intravenous injection with ioxitalamate in rats was claimed to lead to DNA damage and apoptosis [65], in vivo chromosomal aberrations [66] and micronuclei [67]. Similar to another study [68], the damage caused by radiocontrast agents may be related to their hyperosmolar cytotoxicity leading to apoptosis. However, Chromosomal damage in human somatic cells can trigger mutations and thus possibly the development of tumor cells and they can exert clastogenic effects [67]. The elevated mitotic index and proliferation in TBX-treated rats were seen previously prior to cancer development [69,70].

The biochemical and oxidative stress recorded in the current study showed a significant increase in the levels of serum urea and creatinine [71], the concentrations of lipid peroxidation and nitric oxide, accompanied with the decrease in catalase activities and glutathione concentrations in kidney tissues after 1 and 14 days of TBX-intravenous injection. These results may be attributed to the effect of biologically active, free iodine ions in CM [3]. These ionic imbalances after TBX treatments can also affect the proteins structure and functions causing several cytotoxic mechanisms such as the damage to cell membranes and mitochondrial injury leading to apoptosis and necrosis [8–10]. Such effects have been reported in renal cells, involving cellular energy failure, impaired calcium homeostasis, compromised tubular cell polarity and apoptosis [72,73]. Similarly, lipid peroxidation was elevated in the liver, bladder and ovary tissues by contrast media [74]. The in vivo elevated NO levels [75] and diminished GSH-Px and CAT enzyme activities were also reported [76]. The two main mechanisms, probably acting in a synergic manner, are the renal medullary ischemia and direct toxicity to renal tubular cells leading to acute kidney injury by contrast media [77,78]. Consequently, hemodynamic shifts in contrast drugs are responsible, on the one hand, for a reduction in glomerular filtration rate and, on the other hand, for medullary hypoxia in a medullary region where O_2 supply is even low. Under normal conditions, nitric oxide (NO), prostaglandins and adenosine adjust tubular transport of sodium to adapt to this low O_2 supply [79]. A reduced blood supply due to vasoconstriction and sodium increased reabsorption due to the elevated sodium delivery to the distal tubule will alter this mechanism, thereby causing more severe hypoxia [2].

Arabic gum has been used in Arabic folk medicine to reduce both the frequency and the need for hemodialysis in chronic renal failure patients [22]. In the current study, the supplementation of AG was expected to ameliorate the TBX-induced DNA damage and oxidative stress in kidney tissues in addition to cytogenotoxic effects in blood and bone marrow [80]. Although the current work followed the literature in the choice of AG dosing [31,32], AG itself exerted little adverse effects when compared to untreated group. However, the significant ameliorating potency of AG against the TBX toxicity was proven, further investigations of dose dependency should be performed to specify the accurate dose of AG with no adverse effects.

The majority of altered genetic conditions in humans are attributed to chromosomal aberrations [81] and may lead to some diseases as well as cancer progression [66]. The higher records of chromosomal aberrations and mitotic index in TBX-treated rats were reversed also by AG treatments in the present study. The treatment with AG demonstrated a good anticarcinogenic potential against colon carcinoma in rodents, as well as antioxidant and cytoprotective effects [82]. The elevated levels of antioxidant enzymes were evidenced and the secondary decline of oxidative stress was discussed previously for AG [83]. Similar results were recognized also for SOD and GSH levels [28]. Further, AG prevents the release of pro-inflammatory cytokines in the plasma and kidney of rats suffering from adenine-induced renal failure. Its usefulness against chronic kidney disease was owned to its anti-inflammatory and anti-oxidative potentials [28]. However, GSH plays a vital role in the elimination of reactive species [84]. AG administration prevented pulmonary inflammation and restored the lung function [85,86], by inhibiting the free radical-mediated oxidative stress and DNA damage [87]. Amino acids such as histidine, methionine and tyrosine appear to be responsible for the antioxidant properties of AG against ROS [88]. The six metabolites were identified and included flavonoids known with their excellent antioxidant activities owned to the presence of ortho-dihydroxyl groups and their role in the structure-activity relationship [89]. The flavonoid 7,3′,4′-trihydroxyisoflavone (3′-hydroxydaidzein) was mentioned before to inhibit the lipid peroxidation in the rat liver with IC50 4.1 μM [54]. It exerted also the same1,1-diphenyl-2-picrylhydrazyl (DPPH) radical-scavenging activity as α-tocopherol and similar antimutagenic activity to 6-hydroxydaidzein [90]. In case of hesperidin which was previously isolated from the same genus, it was able to significantly decrease the free radical levels in DPPH assay with similar efficacy to the positive control (Trolox) [91]. Additionally, it offers a robust cellular antioxidant protection against the damage caused by paraquat and peroxide hydrogen [92]. The opium alkaloid papaverine was found to inhibit the lipid peroxidation when given with ethanol concurrently [93]. It also demonstrated a potential protective effect against oxidative stress and its antioxidant activity on rabbit testicular tissue [94]. After the treatment with the lipophilic alkaloid noscapine in the in vitro yeast assays, a considerable antioxidant activity and improvement of the cell tolerance against oxidative stress was observed [95]. Besides, it prevents lipid peroxidation in electrochemical techniques [96]. For the first time, the sesquiterpene lactone (costunolide) was evaluated to exhibit a protective effect against oxidative stress in a study conducted by Eliza and other colleagues in 2010 [97]. Collectively, all of the identified metabolites of *A. senegal* gum were previously suggested to participate in antioxidant and protective activities. The future investigations may be recommended to compare between AG and other well-known drugs which protect against radiographic contrast medium cytogenotoxicity such as atorvastatin [98] or serofendic acid [99].

5. Conclusions

The present study showed the possible protective effect of Arabic gum against the genotoxicity and oxidative stress induced by Telebrix in male albino rats. The findings may be attributed to the antioxidant properties of AG which restored the oxidative status in kidney tissues and exerted anti-genotoxic potential in DNA damage and mitotic index of bone marrow. These results can be attributed to the identified metabolites of Arabic gum.

Author Contributions: Conceptualization, S.H.E.-N., A.E.K. and I.E.-G.; Methodology, I.E.-G., A.S. and S.E.-S.; Software, I.H.E.A. and A.E.K.; Validation, I.E.-G., S.A.M.K. and S.E.; Formal Analysis, A.S.; Investigation, I.E.-G., S.E.-S. and A.S.; Resources, H.R.E.-S., A.S., I.H.E.A. and S.A.M.K.; Data Curation, S.A.M.K.; Writing-Original Draft Preparation, I.E.-G. and A.E.K.; Writing-Review & Editing, S.A.M.K., S.E., I.H.E.A. and H.R.E.-S.; Visualization, I.E.-G. and S.H.E.-N.; Supervision, I.E.-G. and S.H.E.-N.; project administration, I.E.-G.; funding acquisition, I.H.E.A., H.R.E.-S., A.S. and S.A.M.K.; Data curation, S.A.-G. All authors have read and agreed to the published version of the manuscript.

Funding: The funding support by Taif University Researchers Supporting Project number (TURSP-2020/27), Taif University, Taif, Saudi Arabia is highly appreciated. H.R.E.-S. is very grateful to the Swedish Research links Grant VR 2016–05885 and the Department of Molecular Biosciences, Wenner-Grens Institute, Stockholm University, Sweden, for the financial support.

Institutional Review Board Statement: The study followed the Institutional Animal Ethical Committee (IAEC) guidelines at the Faculty of Science, Menoufia University (MUFS-F-GE-2-20).

Informed Consent Statement: Not applicable.

Data Availability Statement: Not applicable.

Acknowledgments: I.H.E.A. thanks Taif University Researchers Supporting Project number (TURSP-2020/27), Taif University, Taif, Saudi Arabia.

Conflicts of Interest: The authors declare no conflict of interest.

References

1. Monica, B.; Leffa, D.D.; Mazzorana, D.; Andrade, V.M. Evaluation of the Mutagenic Effect of the Iodinated Contrast Medium Urografina® 292 Using the Micronucleus Test in Mouse Bone Marrow Cells. *An. Acad. Bras. Cienc.* **2013**, *85*, 737–744.
2. Andreucci, M. Side Effects of Radiographic Contrast Media. *BioMed Res. Int.* **2014**, *2014*. [CrossRef] [PubMed]
3. Andreucci, M.; Faga, T.; De Sarro, G.; Michael, A. The Toxicity of Iodinated Radiographic Contrast Agents in the Clinical Practice. *J. Nephrol. Adv.* **2015**, *1*, 6.
4. Berg, K.; Skarra, S.; Bruvold, M.; Brurok, H.; Karlsson, J.O.G.; Jynge, P. Iodinated Radiographic Contrast Media Possess Antioxidant Properties In Vitro. *Acta Radiol.* **2005**, *46*, 815–822. [CrossRef]
5. Katzberg, R.W. Urography into the 21st Century: New Contrast Media, Renal Handling, Imaging Characteristics, and Nephrotoxicity. *Radiology* **1997**, *204*, 297–312. [CrossRef] [PubMed]
6. Maddox, T.G. Adverse Reactions to Contrast Material. Recognition, Prevention and Treatment. *Am. Fam. Phys.* **2002**, *66*, 1229.
7. Kim, K.-H.; Kim, Y.-S.; Kuh, S.-U.; Park, H.-S.; Park, J.-Y.; Chin, D.-K.; Kim, K.-S.; Cho, Y.-E. Time-and Dose-Dependent Cytotoxicities of Ioxitalamate and Indigocarmine in Human Nucleus Pulposus Cells. *Spine J.* **2013**, *13*, 564–571. [CrossRef] [PubMed]
8. Sanaei-Ardekani, M.; Movahed, M.-R.; Movafagh, S.; Ghahramani, N. Contrast-Induced Nephropathy: A Review. *Cardiovasc. Revasc. Med.* **2005**, *6*, 82–88. [CrossRef]
9. Chang, C.-F.; Lin, C.-C. Current Concepts of Contrast-Induced Nephropathy: A Brief Review. *J. Chin. Med. Assoc.* **2013**, *76*, 673–681. [CrossRef]
10. Pisani, A.; Riccio, E.; Andreucci, M.; Faga, T.; Ashour, M.; Di Nuzzi, A.; Mancini, A.; Sabbatini, M. Role of Reactive Oxygen Species in Pathogenesis of Radiocontrast-Induced Nephropathy. *BioMed Res. Int.* **2013**, *2013*. [CrossRef]
11. Deimling, L.I.; Machado, F.L.S.; Welker, A.G.; Peres, L.M.; Santos-Mello, R. Micronucleus Induction in Mouse Polychromatic Erythrocytes by an X-Ray Contrast Agent Containing Iodine. *Mutat. Res. Genet. Toxicol. Environ. Mutagen.* **2009**. [CrossRef]
12. Ali, B.H.; Ziada, A.; Blunden, G. Biological Effects of Gum Arabic: A Review of Some Recent Research. *Food Chem. Toxicol.* **2009**, *47*, 1–8. [CrossRef] [PubMed]
13. Daoub, R.M.A.; Elmubarak, A.H.; Misran, M.; Hassan, E.A.; Osman, M.E. Characterization and Functional Properties of Some Natural Acacia Gums. *J. Saudi Soc. Agric. Sci.* **2018**. [CrossRef]
14. Dauqan, E.; Abdullah, A. Utilization of Gum Arabic for Industries and Human Health. *Am. J. Appl. Sci.* **2013**. [CrossRef]
15. Gashua, I.B.; Williams, P.A.; Baldwin, T.C. Molecular Characteristics, Association and Interfacial Properties of Gum Arabic Harvested from Both Acacia Senegal and Acacia Seyal. *Food Hydrocoll.* **2016**. [CrossRef]
16. Sanchez, C.; Nigen, M.; Mejia Tamayo, V.; Doco, T.; Williams, P.; Amine, C.; Renard, D. Acacia Gum: History of the Future. *Food Hydrocoll.* **2018**. [CrossRef]
17. Aoki, H.; Al-Assaf, S.; Katayama, T.; Phillips, G.O. Characterization and Properties of *Acacia senegal* (L.) Willd. var. Senegal with Enhanced Properties (Acacia (Sen) SUPER GUM™): Part 2—Mechanism of the Maturation Process. *Food Hydrocoll.* **2007**. [CrossRef]

18. Hamid, M.; Abdulrahim, Y.; Liu, D.; Qian, G.; Khan, A.; Huang, K. The Hepatoprotective Effect of Selenium-Enriched Yeast and Gum Arabic Combination on Carbon Tetrachloride-Induced Chronic Liver Injury in Rats. *J. Food Sci.* **2018**. [CrossRef]
19. Bliss, D.Z. Effect of a Gum Arabic Supplement on the Nitrogen Excretion and Serum Urea Nitrogen Concentration of Chronic Renal Failure Patients on a Low Protein Diet. 1992. AAI9227619. Available online: https://repository.upenn.edu/dissertations/AAI9227619 (accessed on 6 January 2021).
20. Desplanques, S.; Renou, F.; Grisel, M.; Malhiac, C. Impact of Chemical Composition of Xanthan and Acacia Gums on the Emulsification and Stability of Oil-in-Water Emulsions. *Food Hydrocoll.* **2012**, *27*, 401–410. [CrossRef]
21. Johnson, W. Final Report of the Safety Assessment of Acacia Catechu Gum, Acacia Concinna Fruit Extract, Acacia Dealbata Leaf Extract, Acacia Dealbata Leaf Wax, Acacia Decurrens Extract, Acacia Farnesiana Extract, Acacia Farnesiana Flower Wax, Acacia Farnesiana Gum, Acacia senegal extract, Acacia senegal gum, and Acacia senegal gum extract. *Int. J. Toxicol.* **2005**, *24*, 75–118.
22. Al-Majed, A.A.; Mostafa, A.M.; Al-Rikabi, A.C.; Al-Shabanah, O.A. Protective Effects of Oral Arabic Gum Administration on Gentamicin-Induced Nephrotoxicity in Rats. *Pharmacol. Res.* **2000**, *46*, 445–451. [CrossRef]
23. Abd-Allah, A.R.A.; Al-Majed, A.A.; Mostafa, A.M.; Al-Shabanah, O.A.; El Din, A.G.; Nagi, M.N. Protective Effect of Arabic Gum against Cardiotoxicity Induced by Doxorubicin in Mice: A Possible Mechanism of Protection. *J. Biochem. Mol. Toxicol.* **2002**, *16*, 254–259. [CrossRef]
24. Gamal El-Din, A.M.; Mostafa, A.M.; Al-Shabanah, O.A.; Al-Bekairi, A.M.; Nagi, M.N. Protective Effect of Arabic Gum against Acetaminophen-Induced Hepatotoxicity in Mice. *Pharmacol. Res.* **2003**, *48*, 631–635. [CrossRef]
25. Glover, D.A.; Ushida, K.; Phillips, A.O.; Riley, S.G. Acacia (Sen) SUPERGUM^{TM} (Gum Arabic): An Evaluation of Potential Health Benefits in Human Subjects. *Food Hydrocoll.* **2009**, *23*, 2410–2415. [CrossRef]
26. Wapnir, R.A.; Sherry, B.; Codipilly, C.N.; Goodwin, L.O.; Vancurova, I. Modulation of Rat Intestinal Nuclear Factor NF-KB by Gum Arabic. *Dig. Dis. Sci.* **2008**, *53*, 80–87. [CrossRef] [PubMed]
27. Nemmar, A.; Al-Salam, S.; Beegam, S.; Yuvaraju, P.; Ali, B.H. Waterpipe Smoke Exposure Triggers Lung Injury and Functional Decline in Mice: Protective Effect of Gum Arabic. *Oxid. Med. Cell. Longev.* **2019**, *2019*. [CrossRef]
28. Ali, B.H.; Al-Husseni, I.; Beegam, S.; Al-Shukaili, A.; Nemmar, A.; Schierling, S.; Queisser, N.; Schupp, N. Effect of Gum Arabic on Oxidative Stress and Inflammation in Adenine–Induced Chronic Renal Failure in Rats. *PLoS ONE* **2013**, *8*, 55242. [CrossRef]
29. El-Garawani, I.M.; El-Nabi, S.E.H.; Mohamed, A.H.; El-Esawy, H.M. Molecular Amelioration of Acacia Arabica Gum on Some Male Reproductive Aspects in Schistosoma Mansoni Infected Mice. *Res. J. Pharm. Biol. Chem. Sci.* **2016**, *7*, 505–512.
30. Bliss, D.Z.; Stein, T.P.; Schleifer, C.R.; Settle, R.G. Supplementation with Gum Arabic Fiber Increases Fecal Nitrogen Excretion and Lowers Serum Urea Nitrogen Concentration in Chronic Renal Failure Patients Consuming a Low-Protein Diet. *Am. J. Clin. Nutr.* **1996**, *63*, 392–398. [CrossRef]
31. Amanullah, M. Protective Effect of Long Term Administration of Gum Arabic on Oxidative Stress in Hepatic Tissue of Diabetic Rats. *Biomed. J. Sci. Tech. Res.* **2018**. [CrossRef]
32. Nasir, O.; Umbach, A.T.; Rexhepaj, R.; Ackermann, T.F.; Bhandaru, M.; Ebrahim, A.; Artunc, F.; Kempe, D.S.; Puchchakayala, G.; Siraskar, B.; et al. Effects of Gum Arabic (*Acacia senegal*) on Renal Function in Diabetic Mice. *Kidney Blood Press. Res.* **2012**, *35*, 365–372. [CrossRef] [PubMed]
33. Li, L.P.; Thacker, J.; Lu, J.; Franklin, T.; Zhou, Y.; Papadopoulou, M.V.; Solomon, R.; Prasad, P.V. Efficacy of Preventive Interventions for Iodinated Contrast-Induced Acute Kidney Injury Evaluated by Intrarenal Oxygenation as an Early Marker. *Investig. Radiol.* **2014**. [CrossRef] [PubMed]
34. Tabacco, A.; Meiattini, F.; Moda, E.; Tarli, P. Simplified Enzymic/Colorimetric Serum Urea Nitrogen Determination. *Clin. Chem.* **1979**, *25*, 336–337. [CrossRef] [PubMed]
35. Rartels, H.; Böhmer, M. Eine Mikromethode 7air Kreatininbestimmung. *Clin. Chim. Acta* **1971**, *32*, 81–85. [CrossRef]
36. Cohen, G.; Dembiec, D.; Marcus, J. Measurement of Catalase Activity in Tissue Extracts. *Anal. Biochem.* **1970**, *34*, 30–38. [CrossRef]
37. Mesbah, L.; Soraya, B.; Narimane, S.; Jean, P.F. Protective Effect of Flavonides against the Toxicity of Vinblastine Cyclophosphamide and Paracetamol by Inhibition of Lipid–Peroxydation and Increase of Liver Glutathione. *Haematology* **2004**, *7*, 59–67.
38. Ellman, G.L. Tissue Sulfhydryl Groups. *Arch. Biochem. Biophys.* **1959**. [CrossRef]
39. Green, L.C.; Wagner, D.A.; Glogowski, J.; Skipper, P.L.; Wishnok, J.S.; Tannenbaum, S.R. Analysis of Nitrate, Nitrite, and [15N] Nitrate in Biological Fluids. *Anal. Biochem.* **1982**, *126*, 131–138. [CrossRef]
40. Aljanabi, S.M.; Martinez, I. Universal and Rapid Salt-Extraction of High Quality Genomic DNA for PCR-Based Techniques. *Nucleic Acids Res.* **1997**, *25*, 4692–4693. [CrossRef]
41. El-Nabi, S.E.H.; Elhassaneen, Y.A. Detection of DNA Damage, Molecular Apoptosis and Production of Home-Made Ladder by Using Simple Techniques. *Biotechnology* **2008**, *7*, 514–522.
42. El-Garawani, I.M.; Hassab El Nabi, S.E. Increased Sensitivity of Apoptosis Detection Using Direct Staining Method and Integration of Acridine Orange as an Alternative Safer Fluorescent Dye in Agarose Gel Electrophoresis and Micronucleus Test. *Can. J. Pure Appl. Sci.* **2016**, *10*, 3865–3871.
43. El-Garawani, I.M. Ameliorative Effect of Cymbopogon Citratus Extract on Cisplatin-Induced Genotoxicity in Human Leukocytes. *J. Biosci. Appl. Res.* **2015**, *1*, 304–310. [CrossRef]
44. Liu, K.; Liu, P.-C.; Liu, R.; Wu, X. Dual AO/EB Staining to Detect Apoptosis in Osteosarcoma Cells Compared with Flow Cytometry. *Med. Sci. Monit. Basic Res.* **2015**, *21*, 15–20. [CrossRef] [PubMed]

45. Singh, N.P.; McCoy, M.T.; Tice, R.R.; Schneider, E.L. A Simple Technique for Quantitation of Low Levels of DNA Damage in Individual Cells. *Exp. Cell Res.* **1988**, *175*, 184–191. [CrossRef]
46. Evans, H.J. Cytological Methods for Detecting Chemical Mutagens. In *Chemical Mutagens*; Springer: Berlin, Germany, 1976; pp. 1–29. [CrossRef]
47. El-Seedi, H.R.; Burman, R.; Mansour, A.; Turki, Z.; Boulos, L.; Gullbo, J.; Göransson, U. The Traditional Medical Uses and Cytotoxic Activities of Sixty-One Egyptian Plants: Discovery of an Active Cardiac Glycoside from Urginea Maritima. *J. Ethnopharmacol.* **2013**, *145*, 746–757. [CrossRef]
48. El-Garawani, I.M.; El-Sabbagh, S.M.; Abbas, N.H.; Ahmed, H.S.; Eissa, O.A.; Abo-Atya, D.M.; Khalifa, S.A.M.; El-Seedi, H.R. A Newly Isolated Strain of Halomonas sp. (HA1) Exerts Anticancer Potential via Induction of Apoptosis and G2/M Arrest in Hepatocellular Carcinoma (HepG2) Cell Line. *Sci. Rep.* **2020**. [CrossRef]
49. Song, M.; Xu, X.; Hang, T.; Wen, A.; Yang, L. Development of an LC–MS/MS Method for the Simultaneous Quantification of Aripiprazole and Dehydroaripiprazole in Human Plasma. *Anal. Biochem.* **2009**, *385*, 270–277.
50. Zhong, B.; Robinson, N.A.; Warner, R.D.; Barrow, C.J.; Dunshea, F.R.; Suleria, H.A.R. Lc-Esi-Qtof-Ms/Ms Characterization of Seaweed Phenolics and Their Antioxidant Potential. *Mar. Drugs* **2020**, *18*, 331. [CrossRef]
51. Li, Y.; Smolke, C.D. Engineering Biosynthesis of the Anticancer Alkaloid Noscapine in Yeast. *Nat. Commun.* **2016**, *7*, 1–14. [CrossRef]
52. Hawkins, K.M.; Smolke, C.D. Production of Benzylisoquinoline Alkaloids in Saccharomyces Cerevisiae. *Nat. Chem. Biol.* **2008**, *4*, 564–573. [CrossRef]
53. Hu, F.; Feng, S.; Wu, Y.; Bi, Y.; Wang, C.; Li, W. Quantitative Analysis of Costunolide and Dehydrocostuslactone in Rat Plasma by Ultraperformance Liquid Chromatography–Electrospray Ionization–Mass Spectrometry. *Biomed. Chromatogr.* **2011**, *25*, 547–554. [CrossRef] [PubMed]
54. Lee, J.-T.; Pao, L.-H.; Hsieh, C.-D.; Huang, P.-W.; Hu, O.Y.-P. Development and Validation of an LC-MS/MS Method for Simultaneous Quantification of Hesperidin and Hesperetin in Rat Plasma for Pharmacokinetic Studies. *Anal. Methods* **2017**, *9*, 3329–3337. [CrossRef]
55. Dusaj, R.; Reiner, J.S. Iodinated Contrast Media—A Safety Review. *Interv. Cardiol. Rev.* **2009**, *4*, 22. [CrossRef]
56. Juchem, B.C.; Dall'Agnol, C.M. Immediate Adverse Reactions to Intravenous Iodinated Contrast Media in Computed Tomography. *Rev. Lat. Am. Enfermagem* **2007**, *15*, 78–83. [CrossRef]
57. Fisher-Wellman, K.; Bloomer, R.J. Acute Exercise and Oxidative Stress: A 30 Year History. *Dyn. Med.* **2009**, *8*, 1. [CrossRef]
58. Rothkamm, K.; Balroop, S.; Shekhdar, J.; Fernie, P.; Goh, V. Leukocyte DNA Damage after Multi–Detector Row CT: A Quantitative Biomarker of Low-Level Radiation Exposure. *Radiology* **2007**, *242*, 244–251. [CrossRef]
59. Roch-Lefèvre, S.; Mandina, T.; Voisin, P.; Gaëtan, G.; Mesa, J.E.G.; Valente, M.; Bonnesoeur, P.; García, O.; Voisin, P.; Roy, L. Quantification of γ-H2AX Foci in Human Lymphocytes: A Method for Biological Dosimetry after Ionizing Radiation Exposure. *Radiat. Res.* **2010**, *174*, 185–194. [CrossRef]
60. Reisz, J.A.; Bansal, N.; Qian, J.; Zhao, W.; Furdui, C.M. Effects of Ionizing Radiation on Biological Molecules—Mechanisms of Damage and Emerging Methods of Detection. *Antioxid. Redox Signal.* **2014**, *21*, 260–292. [CrossRef]
61. Grudzenski, S.; Kuefner, M.A.; Heckmann, M.B.; Uder, M.; Löbrich, M. Contrast Medium–Enhanced Radiation Damage Caused by CT Examinations. *Radiology* **2009**, *253*, 706–714. [CrossRef]
62. Pathe, C.; Eble, K.; Schmitz-Beuting, D.; Keil, B.; Kaestner, B.; Voelker, M.; Kleb, B.; Klose, K.J.; Heverhagen, J.T. The Presence of Iodinated Contrast Agents Amplifies DNA Radiation Damage in Computed Tomography. *Contrast Media Mol. Imaging* **2011**, *6*, 507–513. [CrossRef]
63. Wang, L.; Li, Q.; Wang, X.-M.; Hao, G.-Y.; Hu, S.; Hu, C.-H. Enhanced Radiation Damage Caused by Iodinated Contrast Agents during CT Examination. *Eur. J. Radiol.* **2017**, *92*, 72–77. [CrossRef] [PubMed]
64. Deinzer, C.K.W.; Danova, D.; Kleb, B.; Klose, K.J.; Heverhagen, J.T. Influence of Different Iodinated Contrast Media on the Induction of DNA Double-strand Breaks after in Vitro X-ray Irradiation. *Contrast Media Mol. Imaging* **2014**, *9*, 259–267. [CrossRef] [PubMed]
65. Azimi, S.; Mozdarani, H.; Mahmoudzadeh, A. Induction of DNA Damage, Apoptosis and Micronuclei in Peripheral Blood Lymphocytes Following Injection of Contrast Media in Patients with Abdominal CT Scan. *Int. J. Radiat. Res.* **2017**. [CrossRef]
66. Mozdarani, H.; Fadaei, S. Similar Cytogenetic Effects of Sodium-Meglumine Diatrizoate and Sodium-Meglumine Ioxithalamate in Lymphocytes of Patients Undergoing Brain CT Scan. *Toxicol. Lett.* **1998**, *98*, 25–30. [CrossRef]
67. Matsubara, S.; Suzuki, S.; Suzuki, H.; Kuwabara, Y.; Okano, T. Effects of Contrast Medium on Radiation-Induced Chromosome Aberrations. *Radiology* **1982**, *144*, 295–301. [CrossRef] [PubMed]
68. Hizoh, I.; Sträter, J.; Schick, C.S.; Kübler, W.; Haller, C. Radiocontrast-Induced DNA Fragmentation of Renal Tubular Cells in Vitro: Role of Hypertonicity. *Nephrol. Dial. Transplant. Off. Publ. Eur. Dial. Transpl. Assoc. Ren. Assoc.* **1998**, *13*, 911–918. [CrossRef]
69. Gabelmann, A.; Haberstroh, J.; Weyrich, G. Ionic and Non-Ionic Contrast Agent-Mediated Endothelial Injury: Quantitative Analysis of Cell Proliferation during Endothelial Repair. *Acta Radiol.* **2001**, *42*, 422–425.
70. Auerbach, O.; Garfinkel, L. Histologic Changes in the Urinary Bladder in Relation to Cigarette Smoking and Use of Artificial Sweeteners. *Cancer* **1989**, *64*, 983–987. [CrossRef]
71. Suliman, S.M.; Hamdouk, M.I.; Elfaki, M.B. *GA Fiber as a Supplement to Low Protein Diet in Chronic Renal Failure Patients*; Sudan Association of Physicians: Friendship Hall, Khartoum, Sudan, 2000; pp. 21–23.

72. Higgins, C.B. Mechanism of Cardiovascular Effects of Contrast Media: Evidence for Transient Myocardial Calcium Ion Imbalance. *J. Am. Coll. Cardiol.* **1985**, *6*, 854–855. [CrossRef]
73. Haller, C.; Hizoh, I. The Cytotoxicity of Iodinated Radiocontrast Agents on Renal Cells in Vitro. *Investig. Radiol.* **2004**, *39*, 149–154. [CrossRef]
74. Ergüder, B.I.; Çetin, M.; Namuslu, M.; Kılıçoğlu, S.; Devrim, E.; Çetin, R.; Durak, İ. High Osmolar Contrast Medium Causes Mild Oxidation in Liver, Bladder, and Ovary Tissues from Rats: Vitamin C Has Protective Role. *Med. Chem. Res.* **2010**, *19*, 515–523. [CrossRef]
75. Çetin, M.; Devrim, E.; Serin Kilicoglu, S.; Ergüder, I.B.; Namuslu, M.; Çetin, R.; Durak, I. Ionic High-Osmolar Contrast Medium Causes Oxidant Stress in Kidney Tissue: Partial Protective Role of Ascorbic Acid. *Ren. Fail.* **2008**, *30*, 567–572. [CrossRef] [PubMed]
76. Colbay, M.; Yuksel, S.; Uslan, I.; Acarturk, G.; Karaman, O.; Bas, O.; Mollaoglu, H.; Yagmurca, M.; Ozen, O.A. Novel Approach for the Prevention of Contrast Nephropathy. *Exp. Toxicol. Pathol.* **2010**, *62*, 81–89. [CrossRef] [PubMed]
77. Barrett, B.J. Contrast Nephrotoxicity. *J. Am. Soc. Nephrol.* **1994**, *5*, 125–137. [PubMed]
78. Bbarrett, B.J.; Parfrey, P.S. Contrast Nephropathy. *Ther. Nephrol. Hypertens.* **2008**, *11*, 41–46. [CrossRef]
79. Heyman, S.N.; Rosen, S.; Rosenberger, C. Renal Parenchymal Hypoxia, Hypoxia Adaptation, and the Pathogenesis of Radiocontrast Nephropathy. *Clin. J. Am. Soc. Nephrol.* **2008**, *3*, 288–296.
80. Ali, B.H.; Al Balushi, K.; Al-Husseini, I.; Mandel, P.; Nemmar, A.; Schupp, N.; Ribeiro, D.A. Gum Acacia Mitigates Genetic Damage in Adenine-Induced Chronic Renal Failure in Rats. *Eur. J. Clin. Investig.* **2015**, *45*, 1221–1227. [CrossRef]
81. Santaguida, S.; Amon, A. Short-and Long-Term Effects of Chromosome Mis-Segregation and Aneuploidy. *Nat. Rev. Mol. Cell Biol.* **2015**, *16*, 473–485. [CrossRef]
82. Al-Yahya, A.A.; Al-Majed, A.A.; Gado, A.M.; Daba, M.H.; Al-Shabanah, O.A.; El-Azab, A.S.; Abd-Allah, A.R.A. Acacia Senegal Gum Exudate Offers Protection against Cyclophosphamide-Induced Urinary Bladder Cytotoxicity. *Oxid. Med. Cell. Longev.* **2009**, *2*, 207–213. [CrossRef]
83. Kaddam, L.; Fadl-Elmula, I.; Eisawi, O.A.; Abdelrazig, H.A.; Salih, M.A.; Lang, F.; Saeed, A.M. Gum Arabic as Novel Anti-Oxidant Agent in Sickle Cell Anemia, Phase II Trial. *BMC Hematol.* **2017**, *17*, 4. [CrossRef]
84. Jemal, A.; Bray, F.; Center, M.M.; Ferlay, J.; Ward, E.; Forman, D. Global Cancer Statistics. CA. *Cancer J. Clin.* **2011**, *61*, 69–90. [CrossRef] [PubMed]
85. Nemmar, A.; Al-Salam, S.; Yuvaraju, P.; Beegam, S.; Ali, B.H. Exercise Training Mitigates Water Pipe Smoke Exposure-Induced Pulmonary Impairment via Inhibiting NF-KB and Activating Nrf2 Signalling Pathways. *Oxid. Med. Cell. Longev.* **2018**, *2018*. [CrossRef] [PubMed]
86. Al Za'abi, M.; Al Salam, S.; Al Suleimani, Y.; Manoj, P.; Nemmar, A.; Ali, B.H. Gum Acacia Improves Renal Function and Ameliorates Systemic Inflammation, Oxidative and Nitrosative Stress in Streptozotocin-Induced Diabetes in Rats with Adenine-Induced Chronic Kidney Disease. *Cell. Physiol. Biochem.* **2018**, *45*, 2293–2304. [CrossRef] [PubMed]
87. Gado, A.M.; Aldahmash, B.A. Antioxidant Effect of Arabic Gum against Mercuric Chloride-Induced Nephrotoxicity. *Drug Des. Dev. Ther.* **2013**, *7*, 1245. [CrossRef] [PubMed]
88. Rodrigues, E.; Mariutti, L.R.B.; Faria, A.F.; Mercadante, A.Z. Microcapsules Containing Antioxidant Molecules as Scavengers of Reactive Oxygen and Nitrogen Species. *Food Chem.* **2012**, *134*, 704–711. [CrossRef]
89. Chang, T.-S. Isolation, Bioactivity, and Production of Ortho-Hydroxydaidzein and Ortho-Hydroxygenistein. *Int. J. Mol. Sci.* **2014**, *15*, 5699–5716. [CrossRef]
90. Chen, Y.-C.; Sugiyama, Y.; Abe, N.; Kuruto-Niwa, R.; Nozawa, R.; Hirota, A. DPPH Radical-Scavenging Compounds from Dou-Chi, a Soybean Fermented Food. *Biosci. Biotechnol. Biochem.* **2005**, *69*, 999–1006. [CrossRef]
91. Salman, A.A.; El-Aleem, I.M.A.; El-Rahman, A.A.A.; El-Husseiny, T.S.; El-Hadary, A.E. Assessment of Antioxidant Traits and Protective Action of Egyptian Acacia Pods Extracts against Paracetamol-induced Liver Toxicity in Rats. *J. Food Biochem.* **2020**, *44*, e13392. [CrossRef]
92. Wilmsen, P.K.; Spada, D.S.; Salvador, M. Antioxidant Activity of the Flavonoid Hesperidin in Chemical and Biological Systems. *J. Agric. Food Chem.* **2005**, *53*, 4757–4761.
93. Chandra, R.; Aneja, R.; Rewal, C.; Konduri, R.; Dass, S.K.; Agarwal, S. An Opium Alkaloid-Papaverine Ameliorates Ethanol-Induced Hepatotoxicity: Diminution of Oxidative Stress. *Indian J. Clin. Biochem.* **2000**, *15*, 155–160. [CrossRef]
94. Benko, F.; Lenický, M.; Ďuračka, M.; Lukáč, N.; Tvrdá, E. In Vitro Antioxidant Activity of Papaverine on Rabbit Testicular Tissue. *Anim. Reprod. Sci.* **2020**, *220*, 106362. [CrossRef]
95. Kawadkar, M.; Mandloi, A.S.; Saxena, V.; Tamadaddi, C.; Sahi, C.; Dhote, V.V. Noscapine Alleviates Cerebral Damage in Ischemia-Reperfusion Injury in Rats. *Naunyn. Schmiedebergs. Arch. Pharmacol.* **2020**, 1–15. [CrossRef] [PubMed]
96. Luna, M.A.; Gutierrez, J.A.; Solis, A.K.C.; Molina, P.G.; Correa, N.M. Vehiculization of Noscapine in Large Unilamellar Vesicles. Study of Its Protective Role against Lipid Peroxidation by Electrochemical Techniques. *J. Electroanal. Chem.* **2019**, *833*, 26–32. [CrossRef]
97. Eliza, J.; Daisy, P.; Ignacimuthu, S. Antioxidant Activity of Costunolide and Eremanthin Isolated from Costus Speciosus (Koen Ex. Retz) Sm. *Chem. Biol. Interact.* **2010**, *188*, 467–472. [CrossRef]

98. Liu, L.Y.; Liu, Y.; Wu, M.Y.; Sun, Y.Y.; Ma, F.Z. Efficacy of Atorvastatin on the Prevention of Contrast-Induced Acute Kidney Injury: A Meta-Analysis. *Drug Des. Devel. Ther.* **2018**. [CrossRef]
99. Kitamura, O.; Uemura, K.; Kitamura, H.; Sugimoto, H.; Akaike, A.; Ono, T. Serofendic Acid Protects from Iodinated Contrast Medium and High Glucose Probably against Superoxide Production in LLC-PK1 Cells. *Clin. Exp. Nephrol.* **2009**. [CrossRef]

antioxidants

MDPI

Article

Weaning Mice and Adult Mice Exhibit Differential Carbon Tetrachloride-Induced Acute Hepatotoxicity

Tae Bin Jeong [1,†], Doyoung Kwon [2,†], Seung Won Son [1], Sou Hyun Kim [1], Yun-Hee Lee [3], Min-Soo Seo [4], Kil Soo Kim [4,5] and Young-Suk Jung [1,*]

1 College of Pharmacy, Pusan National University, Busan 46241, Korea; jtaebin0504@naver.com (T.B.J.); wellsu123@naver.com (S.W.S.); hyunie0731@naver.com (S.H.K.)
2 Department of Cellular and Molecular Pharmacology, University of California San Francisco, San Francisco, CA 94158, USA; Doyoung.Kwon@ucsf.edu
3 College of Pharmacy and Research Institute of Pharmaceutical Sciences, Seoul National University, Seoul 08826, Korea; yunhee.lee@snu.ac.kr
4 Laboratory Animal Center, Daegu-Gyeongbuk Medical Innovation Foundation, Daegu 41061, Korea; msseo@dgmif.re.kr (M.-S.S.); kskim728@knu.ac.kr (K.S.K.)
5 College of Veterinary Medicine, Kyungpook National University, Daegu 41566, Korea
* Correspondence: youngjung@pusan.ac.kr; Tel.: +82-51-510-2816
† These authors contributed equally to this work.

Received: 12 January 2020; Accepted: 16 February 2020; Published: 1 March 2020

Abstract: Age is a risk factor for drug-induced liver injury (DILI). However, there is a limited understanding of pediatric DILI. Here, 2-week-old weaning and 8-week-old adult male ICR mice were intraperitoneally injected with CCl_4 (0.1 mmol/kg equal to 15.4 mg/kg) to comparatively evaluate the time-dependent liver damage and cellular events. CCl_4 significantly enhanced the serum alanine aminotransferase/aspartate aminotransferase levels and hepatic centrilobular necrosis in the weaning mice, whereas it induced mild liver injury in the adult mice. CCl_4-treated weaning mice exhibited higher hepatic levels of pro-apoptotic proteins (Bax, cleaved caspase-3, -7, and -9), activated MAPKs (p-JNK and p-Erk), and endoplasmic reticulum stress indicators (ATF6 and CHOP) and lower hepatic anti-apoptotic Bcl-2 levels than the adult mice. The weaning mice exhibited enhanced basal hepatic glutathione (GSH) levels due to high glutamate cysteine ligase (GCL) and low anti-cysteine dioxygenase (CDO) enzyme levels. However, CCl_4 markedly reduced the hepatic GSH levels only in the weaning mice. Furthermore, higher hepatic levels of oxidative stress-induced malondialdehyde, 4-hydroxynonenal, nitrotyrosine-protein adducts, and oxidized proteins were observed in CCl_4-treated weaning mice than in CCl_4-treated adult mice. The enhanced levels of hepatic cytochrome P450 (CYP) 2E1 and CYP3A, and decreased hepatic GSH S-transferase (GST)-π and GSH reductase (GR) levels in the weaning mice may contribute to their enhanced susceptibility to liver damage.

Keywords: drug-induced liver injury; carbon tetrachloride; hepatotoxicity; cytochrome P450; glutathione

1. Introduction

The liver is the primary organ for metabolism of xenobiotics, including toxins, environmental pollutants, chemicals, and drugs [1,2]. Xenobiotics can be detoxified and converted to hydrophilic forms for excretion by phase I and phase II drug-metabolizing enzymes in the liver. However, metabolic reactions can also generate reactive products to cause liver damage [2].

Drug-induced liver injury (DILI) is the leading cause of acute liver failure and drug withdrawal from the market [3,4]. The risk factors for DILI are age, sex, and genetic polymorphisms [5]. Among the risk factors, age-related changes in hepatic drug metabolism and antioxidant defense system are reported to be associated with the incidence of DILI in different age groups [6–9].

Although the incidence of DILI among children is reported to be less than 10%, the mortality rate is in the range of 4–31% [4]. The hepatic responses of children to drugs can be markedly different from those of adults mainly due to developmental immaturity [10]. Children are reported to be less sensitive to DILI than the sensitivity exhibited by adults [4]. However, controversial evidence suggests that children are at a greater risk for valproic acid- and propylthiouracil-induced liver injuries than the risk observed in adults subjected to a similar injury [11,12]. There are limited studies on pediatric DILI due to the difficulties in data collection and an insufficient number of research participants [4]. In animal studies, the effects of aging on DILI are examined mostly by comparing the drug-induced hepatotoxic phenotypes of young rodents and old rodents [8,13,14]. However, the susceptibilities of infant animals and juvenile animals to hepatotoxic chemicals/drugs are not fully understood.

CCl_4, a well-known experimental hepatotoxin, is reported to cause necrosis, apoptosis, fibrosis, cirrhosis, and cancer in the liver [15,16]. CCl_4 is mainly metabolized by hepatic cytochrome P450s and is converted into toxic metabolites, such as trichloromethyl radical ($CCl_3\bullet$), trichloromethylperoxy radical ($CCl_3OO\bullet$), and phosgene (Figure 1), which results in lipid peroxidation that causes hepatocellular damage [15,16]. The reactive metabolites are detoxified non-enzymatically by the cellular antioxidant, glutathione (GSH) or by GSH S-transferase (GST)-catalyzed conjugation reaction [15,16]. However, the excessive generation of free radicals depletes the GSH levels and causes oxidative stress [15,16].

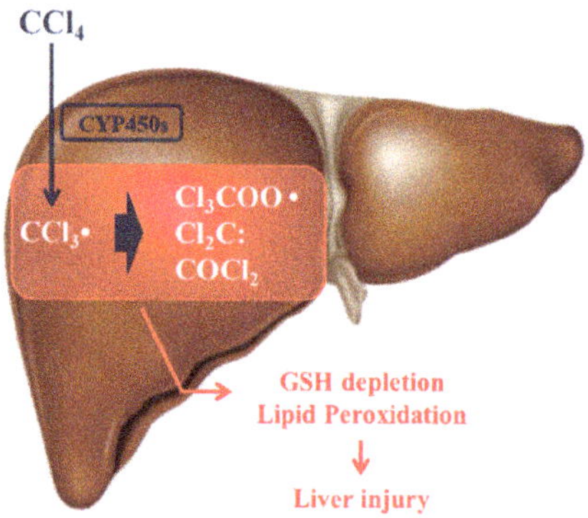

Figure 1. Mechanism of carbon tetrachloride (CCl_4) hepatotoxicity. CCl_4 is metabolized by cytochrome P450s (CYP450s) to $CCl_3\bullet$, which is further converted into $Cl_3COO\bullet$, Cl_2C, and $COCl_2$. These toxic metabolites are detoxified by glutathione (GSH). However, excessive exposure to CCl_4 causes GSH depletion and oxidative stress via reactions, such as membrane lipid peroxidation that can cause liver damage.

Some rodent studies have investigated the age-related CCl_4 hepatotoxicity. However, the findings of various rodent studies are contradictory, which may be due to the species- and strain-dependent differences in the hepatic responses [17–21]. Moreover, there are limited studies evaluating the sensitivity of infant/juvenile animals to CCl_4 hepatotoxicity. In this study, weaning mice and adult mice were acutely intoxicated by a single dose of CCl_4 to comparatively evaluate the time-dependent severity of liver injury. The weaning mice exhibited more severe CCl_4-induced liver damages than the one exhibited by the adult mice, that was mainly due to the high levels of hepatic cytochrome P450.

2. Materials and Methods

2.1. Animal and Treatments

Male Korl: ICR (referred as ICR) mice were purchased from Koatech (Gyeonggi-do, Korea). The animals were maintained in compliance with the guidelines established and approved by the institutional animal care and use committee at the Pusan National University (PNU-2018-1860). The mice were allowed to acclimatize to the animal facility conditions (room temperature, 22 °C ± 2 °C; humidity, 55% ± 5%; 12 h light-dark cycle) for one week prior to the experiment. Thirty weaning (2-week-old) and thirty young adult (8-week-old) mice were divided into ten groups (six mice per group). After 12 h fasting, the mice were intraperitoneally injected with a single dose of CCl_4 (0.1 mmol/kg equal to 15.4 mg/kg). The mice were sacrificed at 0 h, 3 h, 6 h, 12 h, and 24 h post-CCl_4 treatment (six mice per time point). The serum and liver samples were collected for measuring the biochemical parameters. The liver samples stored at −80 °C were used for determining lipid peroxidation/protein oxidation, immunoblotting analysis, and quantifying levels of cysteine, GSH, and taurine. Additionally, some liver samples were fixed in neutral-buffered formalin immediately after the surgery for histopathological examination.

2.2. CCl_4-Induced Hepatotoxicity

The serum sample was collected using the BD Microtainer Serum Collection Tube (BD Life Sciences, Franklin Lakes, NJ, USA). The serum alanine aminotransferase (ALT) and aspartate aminotransferase (AST) levels were measured using the Multiskan GO reader (Thermo Scientific, Waltham, MA, USA), following the method of Reitman and Frankel [22]. Serum total protein and albumin levels were examined with an Automated Chemistry Analyzer (Prestige 24I; Tokyo Boeki Medical System, Tokyo, Japan). The left lateral lobe of liver fixed in neutral-buffered formalin were embedded in paraffin and sectioned. The sections were mounted on the glass slides and stained with hematoxylin and eosin (H&E). The histopathological examination was performed under the Olympus CX41RF microscope (Olympus Co., Tokyo, Japan).

2.3. Immunoblotting Analysis

The right posterior lobe of liver was homogenized with chilled ProEX™ CETi Lysis Buffer (TransLab Biosciences, Daejeon, Korea). The protein concentration in the tissue lysates was determined by the bicinchoninic acid (BCA) assay (BioVision Research Products, Mountain View, CA, USA). Equal amounts of proteins were subjected to sodium dodecyl sulfate-polyacrylamide gel electrophoresis (SDS-PAGE). The resolved proteins were transferred onto a nitrocellulose membrane (Bio-Rad Laboratories). The membrane was blocked with 5% skim milk in Tris-buffered saline (pH 7.5) containing 0.2% Tween-20 (TBST) for 1 h at room temperature. Next, the membrane was incubated with the following antibodies for 12 h: anti-cysteine dioxygenase (CDO), anti-glutamate cysteine ligase catalytic subunit (GCLC), anti-GSH peroxidase (GPx), anti-GSH reductase (GR), anti-nitrotyrosine (Santa Cruz Biotechnology, Santa Cruz, CA, USA), anti-CYP2E1, anti-CYP3A (Detroit R&D, Detroit, MI, USA), anti-c-Jun N terminal kinase (JNK), anti-phospho-JNK, anti-extracellular regulated protein kinase (Erk), anti-phospho-Erk, anti-p38, anti-phospho-p38, anti-caspase-3, anti-cleaved caspase-3, anti-caspase-7, anti-cleaved caspase-7, anti-caspase-9, anti-cleaved caspase-9, anti-caspase-12, anti-cleaved caspase-12, anti-poly (ADP-ribose) polymerase (PARP) (Cell Signaling Technology, Danvers, MA, USA), anti-GSH S-transferase-α (GST-α), anti-GST-μ, and anti-GST-π (Detroit R&D). The membrane was washed with TBST and incubated with the appropriate horseradish peroxidase (HRP)-conjugated secondary antibodies. The protein bands were visualized using an enhanced chemiluminescent HRP substrate kit (Western Bright, Advansta, Menlo Park, CA, USA).

2.4. *Hepatic Lipid Peroxidation Induced by CCl_4*

Hepatic lipid peroxidation was determined by measuring malondialdehyde (MDA) in the thiobarbituric acid reactive substances (TBARS) assay as previously described [23]. The liver homogenate was incubated with 0.2% thiobarbituric acid in 2 M sodium acetate buffer containing 5% butylated hydroxytoluene at 95 °C for 45 min. The reaction mixture was centrifuged. The supernatant was injected into the high-performance liquid chromatography (HPLC) system equipped with a 5-μm Symmetry C18 reversed-phase column (4.6 mm × 150 mm; Eka Chemicals, Bohus, Sweden). The levels of TBARS were monitored by a fluorescence detector (excitation at 515 nm and emission at 553 nm, FLD-3100; Thermo Scientific).

2.5. *Measurement of Hepatic Protein Oxidation*

Protein oxidation assay was performed using OxyBlot ™ (EMD Millipore, Billerica, MA, USA), following the manufacturer's instructions. The hepatic carbonyl-containing proteins were cross-reacted with 2,4-dinitrophenylhydrazone (DNPH) to produce DNP adducts. After derivatization, equal amounts of proteins were subjected to SDS-PAGE. The oxidized proteins were analyzed by immunoblotting as described in Section 2.3.

2.6. *Determination of Levels of Hepatic cysteine, GSH, and Taurine*

Hepatic taurine was derivatized with o-phthalaldehyde/2-mercaptoethanol and quantified using an HPLC system equipped with a fluorescence detector (excitation 338 nm and emission 425 nm; FLD-3100, Thermo Scientific) [24] after separation using a Hector T-C18 column (3 μm × 4.6 mm × 100 mm; Chiral Technology Korea, Daejeon, Korea). The levels of cysteine and GSH were analyzed by the SBD-F derivatization method [25] using an HPLC equipped with a Hector M-C18 column (3 μm × 4.6 mm × 150 mm: Chiral Technology Korea) and a fluorescence detector (excitation 385 nm and emission 515 nm; FLD-3100, Thermo Scientific).

2.7. *Statistical Analysis*

All data have been expressed as mean ± standard deviation (SD). The data were analyzed by two-tailed unpaired Student's *t*-test. The difference was considered to be statistically significant if the *p*-value was found to be less than 0.05.

3. Results

3.1. *CCl_4-Induced Liver Injuries in Weaning Mice and Adult Mice*

The weaning mice and adult mice exhibited differential susceptibility to CCl_4 hepatotoxicity. The serum levels of ALT and AST, which indicate liver injury, in the weaning mice were found to be significantly higher than those in the adult mice post-CCl_4 administration (Figure 2A,B). Total protein and albumin levels in serum, reflecting the biosynthetic function of the liver, did not show a significant difference in age-dependent changes as well as time-dependent changes for 24 h after CCl_4 treatment (Figure 2C,D). The analysis of H&E-stained liver tissues indicated that the weaning mouse liver exhibited severe centrilobular necrosis at 24 h post-CCl_4 injection, whereas the adult mouse liver exhibited mild liver injury (Figure 2E). This suggested that the weaning mice were more susceptible to CCl_4-induced hepatotoxicity than the susceptibility observed in adult mice.

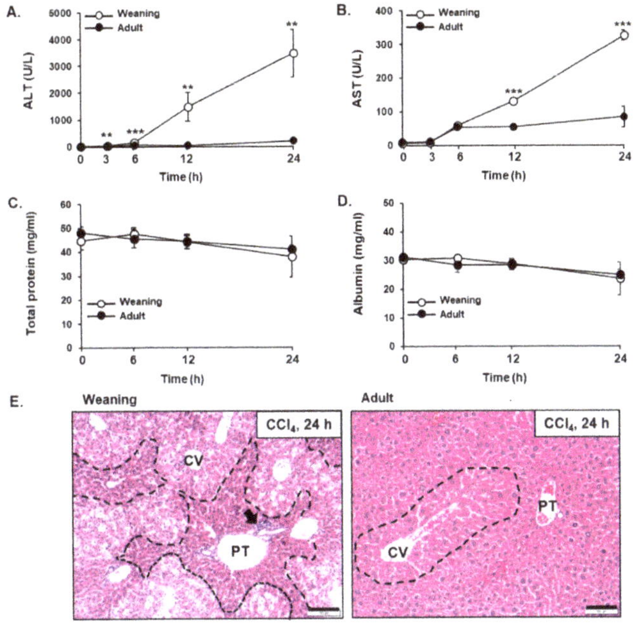

Figure 2. Carbon tetrachloride (CCl_4)-induced liver injury in the weaning mice and adult mice. The serum levels of (**A**) alanine aminotransferase (ALT), (**B**) aspartate aminotransferase (AST), (**C**) total protein, and (**D**) albumin. Male mice were injected with CCl_4 (0.1 mmol/kg equal to 15.4 mg/kg, intraperitoneal (IP)) and each parameter was measured at indicated time point after CCl_4 treatment. The results have been presented as mean ± standard deviation (SD). **, *** Significantly different from the corresponding weaning mice (Student's *t*-test, $p < 0.01$ and $p < 0.001$, respectively). (**E**) The histopathological analysis of hematoxylin and eosin (H&E)-stained liver tissues at 24 h post-CCl_4 treatment. Black dashed line, degeneration area; Black arrow, inflammatory cells; CV, central vein; PT, portal triad. Scale bar, 100 μm.

3.2. CCl_4-Induced Hepatic Apoptosis and Endoplasmic Reticulum (ER) Stress in Weaning Mice and Adult Mice

CCl_4 is reported to induce hepatic cell death by apoptosis [15,26]. To comparatively evaluate the CCl_4-induced hepatic apoptosis between weaning mice and adult mice, the expression levels of proteins involved in the apoptotic signaling pathway were determined at 24 h post-CCl_4 injection. The weaning mice exhibited higher pro-apoptotic Bax hepatic levels and lower anti-apoptotic Bcl-2 hepatic levels than the ones exhibited by the adult mice (Figure 3A). Compared to the adult mice, the weaning mice exhibited higher hepatic levels of caspase-9 (initiator caspase) and caspase-7 (executioner caspase) and their active cleaved forms (Figure 3A). Although the hepatic levels of caspase-3 were not markedly different between the two groups, the weaning mice exhibited higher hepatic levels of cleaved caspase 3 (active form) than the levels exhibited by the adult mice (Figure 3A). The hepatic levels of PARP, which is cleaved by caspases, in the weaning mice were found to be higher than those in the adult mice (Figure 3A). These results indicate that CCl_4 induces higher hepatic apoptosis in the weaning mice than the one induced in the adult mice. Endoplasmic reticulum (ER) stress is a cellular condition of disturbed ER homeostasis, which is characterized by the accumulation of unfolded proteins and misfolded proteins in the ER [27,28]. During ER stress, unfolded protein responses (UPRs) are activated to restore the ER functions. However, prolonged ER stress promotes apoptosis to remove the damaged cells [27,28]. Activating transcription factor 6 (ATF6) is a UPR sensor protein that is induced and activated in response to ER stress [27,28]. Treatment with CCl_4 in a time-dependent manner increased the hepatic levels of C/EBP homologous protein (CHOP), an indicator of ER stress, and ATF6 in the weaning mice after CCl_4 treatment. However, CCl_4 treatment did not markedly affect the hepatic levels of CHOP and ATF6 in the adult mice (Figure 3B). Cellular stresses activate mitogen-activated protein kinases (MAPKs), which mediate the intracellular signaling involved in cell proliferation, differentiation, survival, and apoptosis [29,30]. Three MAPKs, namely JNK, Erk, and p38, are reported to be activated in response to toxic stimuli, such as oxidative stress and inflammation [29,30]. Treatment with CCl_4 increased the hepatic levels of activated/phosphorylated forms of JNK, Erk, and p38 in both weaning mice and adult mice (Figure 3B). However, the weaning mice exhibited higher hepatic p-JNK and p-Erk levels than the ones exhibited by the adult mice. There was no significant difference in the hepatic p-p38 levels between the two groups. ER stress can activate JNK, which subsequently triggers apoptosis [27,28]. Thus, the enhanced activation of JNK in weaning mice may explain the ER stress-induced apoptotic cell death in the livers.

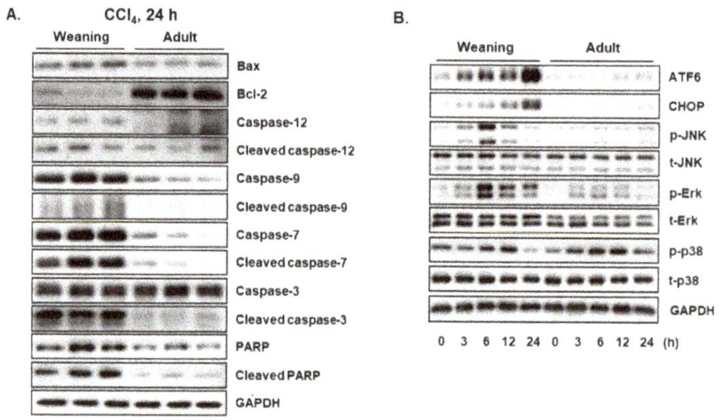

Figure 3. Carbon tetrachloride (CCl_4)-induced hepatic apoptosis and endoplasmic reticulum (ER) stress in the weaning mice and adult mice. (**A**) Proteins involved in the apoptotic signaling pathway were measured at 24 h post-CCl_4 injection. (**B**) Time-dependent changes in the expression levels of ER stress markers and MAPKs. Male mice were injected with CCl_4 (0.1 mmol/kg equal to 15.4 mg/kg, intraperitoneal (IP)). The liver tissues were obtained at 0 h, 3 h, 6 h, 12 h, and 24 h post-CCL_4 injection. Hepatic proteins were detected by immunoblotting and GAPDH was used as the loading control.

3.3. CCl_4-Induced Oxidative Stress in the Liver of Weaning Mice and Adult Mice

Oxidative damages of cellular molecules are reported to be the main mechanism underlying CCl_4-mediated hepatotoxicity [15,16]. The reactive radicals generated from CCl_4 not only oxidize proteins, lipids, and DNA but also cause GSH depletion, which exacerbates oxidative stress [15,16]. Treatment with CCl_4 enhanced the hepatic levels of MDA, a lipid peroxidation end product of polyunsaturated fatty acids (PUFAs), in the weaning mice but not in the adult mice (Figure 4A). The hepatic GSH level in the weaning mice was higher than that in the adult mice prior to CCl_4 injection (Figure 4B). However, CCl_4 decreased the hepatic GSH levels in the weaning mice in a time-dependent manner. The hepatic GSH levels in the adult mice were slightly decreased upon CCl_4 administration (Figure 4B). The hepatic levels of oxidized proteins, 4-hydroxynonenal (HNE), and nitrotyrosine-protein adducts in the weaning mice were higher than those in the adult mice (Figure 4C). HNE, another lipid peroxidation product from PUFAs, and protein tyrosine residues are nitrated by peroxynitrite ($ONOO^-$), which is produced by the reaction between superoxide anion and nitric oxide. Thus, the severity of oxidative stress induced by CCl_4 appears to be closely associated with the differential susceptibility to hepatotoxicity between the two groups.

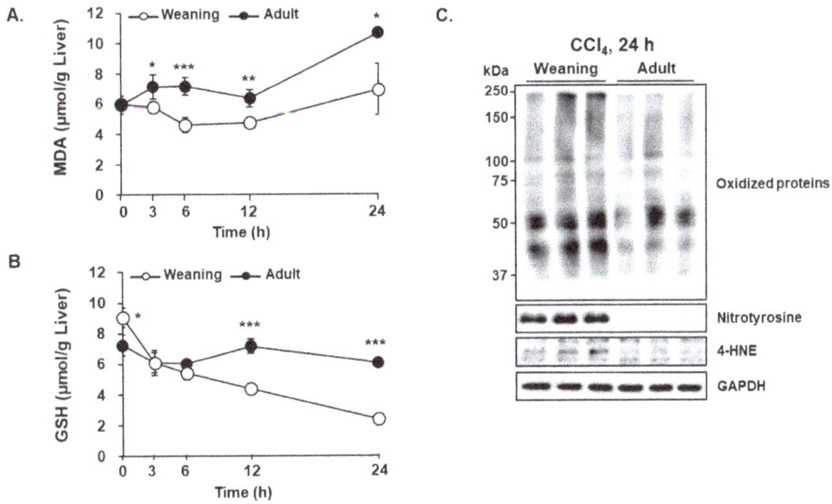

Figure 4. Carbon tetrachloride (CCl_4)-induced oxidative stress in the liver of weaning mice and adult mice. Time-dependent changes in the (**A**) hepatic malondialdehyde (MDA) and (**B**) hepatic glutathione (GSH) levels. (**C**) The levels of oxidized proteins, nitrotyrosine-protein adducts, 4-hydroxynonenal (4-HNE) at 24 h post-CCl_4 treatment. Male mice were injected with CCl_4 (0.1 mmol/kg equal to 15.4 mg/kg, intraperitoneal (IP)). The liver tissues were obtained at 0 h, 3 h, 6 h, 12 h, and 24 h post-CCl_4 treatment. The results have been presented as mean ± standard deviation (SD). *, **, *** Significantly different from the corresponding weaning mice (Student's *t*-test, $p < 0.05$, $p < 0.01$, and $p < 0.001$, respectively).

3.4. Hepatic CYP Levels in Weaning Mice and Adult Mice

The major CYP isozymes for CCl_4 metabolism are CYP2E1 and CYP3A. The weaning mice exhibited higher hepatic CYP2E1 and CYP3A levels than the levels exhibited by the adult mice (Figure 5). This suggested that the enhanced susceptibility of weaning mice to CCl_4 may be due to the enhanced metabolic activation of CCl_4 via the CYPs.

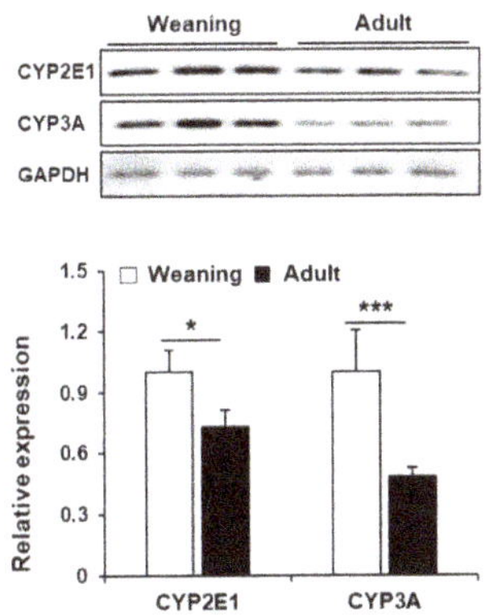

Figure 5. Hepatic CYP2E1 and CYP3A levels in the weaning mice and adult mice without carbon tetrachloride (CCl_4) treatment. Hepatic CYPs were detected by immunoblotting and GAPDH was used as the loading control. The relative protein amounts are presented as mean ± standard deviation (SD). *, *** Significantly different from the corresponding weaning mice (Student's *t*-test, $p < 0.05$ and $p < 0.001$, respectively).

3.5. Hepatic Enzymes Involved in GSH Metabolism in Weaning Mice and Adult Mice

GSTs catalyze the detoxification reaction by conjugating GSH with the electrophiles [31]. The hepatic levels of GST-α and GST-μ classes were similar between the weaning mice and adult mice. However, the hepatic levels of GST-π class in the adult mice were about 9-fold higher than those in the weaning mice. This suggested that the liver of adult mice may exhibit higher GST-π-mediated detoxification, that can also be a reason for the lower basal level of GSH in these mice as GSH is consumed by GSTs. The levels of GPx, which reduces lipid peroxides and H_2O_2 using GSH [32], were significantly different between the two groups. However, the hepatic levels of GR, which converts oxidized GSH (GSSG) to reduced GSH [32], in the weaning mice were lower than those in the adult mice. This can also explain the enhanced vulnerability of weaning mice to oxidative stress (Figure 6).

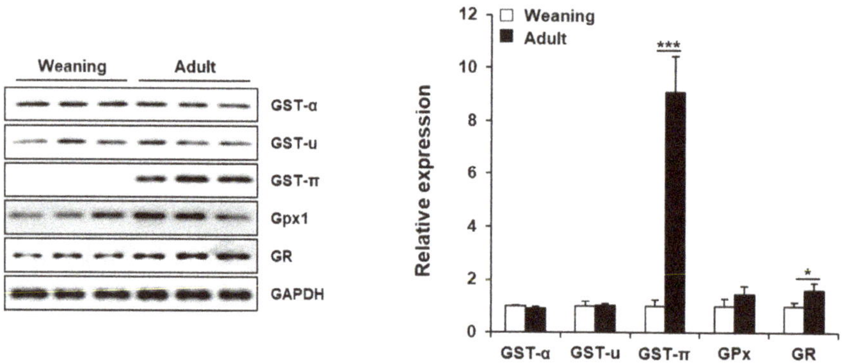

Figure 6. Hepatic proteins involved in glutathione (GSH) metabolism in the weaning mice and adult mice without carbon tetrachloride (CCl_4) treatment. Hepatic proteins were detected by immunoblotting and GAPDH was used as the loading control. The relative protein amounts have been presented as mean ± standard deviation (SD). *, *** Significantly different from the corresponding weaning mice (Student's *t*-test, $p < 0.05$ and $p < 0.001$, respectively).

3.6. Hepatic Synthesis of GSH and Taurine from Cysteine in Weaning Mice and Adult Mice

As described previously, the weaning mice exhibited higher hepatic GSH levels that the adult mice even without CCl_4 treatment (Figures 4B and 7A). To examine the GSH synthetic pathway, the hepatic levels of cysteine (precursor of GSH) and GCLC (rate-limiting enzyme for GSH synthesis) were comparatively evaluated between the weaning mice and adult mice. There was no significant difference in the hepatic cysteine level between the two groups. The hepatic GCLC levels in the weaning mice were higher than those in the adult mice (Figure 7B). Interestingly, the hepatic levels of taurine (another metabolic product from cysteine) and CDO (the first enzyme utilizing cysteine for taurine synthesis) in the weaning mice were significantly lower than those in the adult mice (Figure 7A,B). Therefore, the higher rate of GSH synthesis and the lower consumption of cysteine for taurine production may contribute to the enhanced hepatic GSH level in the weaning mice.

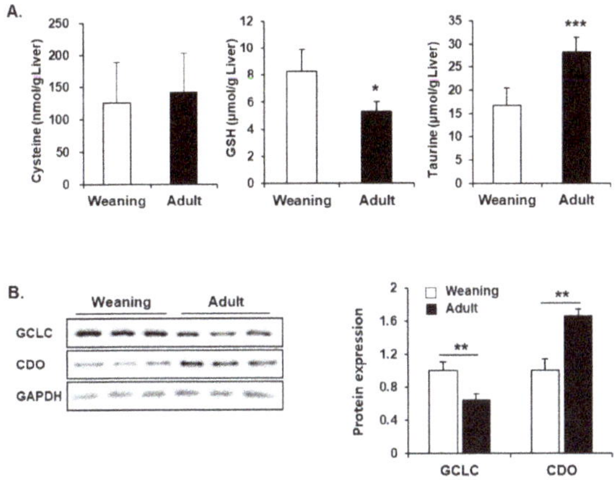

Figure 7. Hepatic levels of glutathione (GSH), taurine, and the respective synthetic enzymes in the weaning mice and adult mice without carbon tetrachloride (CCl_4) treatment. (**A**) Hepatic levels of cysteine, GSH, and taurine. (**B**) Relative glutamate cysteine ligase catalytic subunit (GCLC) and cysteine dioxygenase (CDO) protein levels. The results have been presented as mean ± standard deviation (SD). *, **, *** Significantly different from the corresponding weaning mice (Student's *t*-test, $p < 0.05$, $p < 0.01$, and $p < 0.001$, respectively).

4. Discussion

Aging is reported to be a risk factor for DILI. Compared to the adults, children are less sensitive to DILI because of the lower CYP-mediated metabolic rates and higher GSH synthetic rate in the liver [4,33]. However, the database collected from 1997 to 2012 by FDA's Adverse Event Reporting System show that the clinical incidences of liver injury in children (0 year–9 years, 187 cases; 10 years–19 years, 236 cases) induced by methotrexate is higher than those in young adults (20 years–29 years, 120 cases) [34]. Moreover, lower blood GSH level in <1-year-old individuals than that in older age (2 to 40 years) groups was found in a research of Turkish hospital [35]. These suggest that infants and children are also at a risk for DILI. CCl_4-induced hepatotoxicity in <20-day-old mice is more severe than that in the adult mice [17,19]. The underlying reason for the differential hepatic sensitivity has been not elucidated. In this study, we comparatively evaluated CCl_4-induced hepatotoxicity between 14-day-old mice, that represents infant age before weaning, and 8-week-old mice that represents adulthood. The results indicated that the differential hepatic levels of CYPs, GST-π, and GR proteins can potentially contribute to the differential susceptibility to CCl_4 between the weaning mice and adult mice.

CYPs, which are heme-containing monooxygenases, are the major hepatic phase I drug-metabolizing enzymes. CYPs have an important role in xenobiotic biotransformation and excretion. However, CYPs can also have a deleterious role in the metabolic activation of diverse chemicals/drugs, such as carbon tetrachloride, alcohol, acetaminophen, and halothane [2,4]. Among the 57 isozymes of human CYPs, CYP2E1 and CYP3A are mainly responsible for mediating the hepatotoxicity of CCl_4 [15,36,37]. CYP-mediated reductive cleavage of a carbon-chlorine bond of CCl_4 results in the generation of noxious trichloromethyl radical (CCl_3•), which is further converted into trichloromethylperoxy radical (CCl_3OO•) upon reaction with oxygen [15]. During the catalytic cycle of CYPs, reactive oxygen species (ROS), such as superoxide anion (O_2^-•) and hydrogen peroxide (H_2O_2) are generated that can also induce oxidative stress [38]. In this study, CCl_4-treated weaning mice exhibited enhanced hepatic necrosis (Figure 3) and apoptosis (Figure 3), which was due to the induction of oxidative stress in the mice livers as evidenced by enhanced lipid peroxidation and protein oxidation, as well as decreased GSH levels (Figure 4). However, CCl_4-treated adult mice exhibited mild liver injury. Higher susceptibility of weaning mice to CCl_4 may be attributed mainly to higher protein levels of CYP2E1 and CYP3A (Figure 5). The extensive centrilobular necrosis observed in the CCl_4-injected weaning mouse livers also supports the role of CYPs in mediating CCl_4-induced hepatotoxicity. The centrilobular region exhibits higher expression levels of CYPs than the levels exhibited by the periportal region [39,40].

Several clinical and animal studies have demonstrated the age-related changes in the hepatic levels of CYPs. An animal study demonstrated that hepatic activities of CYP2B and CYP3A in 6 week-old rats were significantly higher than those in the 21-month-old rats [41]. Declines of hepatic CYP2E1 and/or CYP3A correlated with the alleviations of acetaminophen- and isoniazid-induced hepatotoxicities with aging have been reported in rodents [13,14,42] as similar as our results (Figure 5). In contrast to our results, human liver CYP2E1 and CYP3A4 gradually increased after birth, while CYP3A7 decreased [43,44], suggesting that human CYP system in infant functions more slowly than in adult. However, only limited research in humans has been reported and thus it is not clear whether there is species difference in CYP development.

Apoptosis is a programmed and self-destructive cell death characterized by cellular condensation and cellular shrinkage. In contrast, necrosis is characterized by immediate and uncontrolled cell death accompanied by cell lysis. Apoptosis is reported to be an early process of CCl_4-induced hepatic cell death, which is followed by necrosis [26,45]. In studies using mouse models, acute administration of CCl_4 increased the levels of activated caspase-3, -9, and -12 with the release of cytochrome c in the liver [26,45]. Caspase-12-deficient mice exhibited markedly mitigated CCl_4-induced apoptosis and liver injury when compared to the wild type mice. This indicated that apoptosis is also an important mechanism of cell death in CCl_4-treated liver [26]. These studies demonstrated that CCl_4-induced ER stress, which is indicated by the upregulated expression of GRP78, CHOP, IRE1, and XBP1, is the mechanism underlying apoptosis [26,45]. As CYPs are mainly localized at the hepatocyte ER, reactive CCl_4 metabolites and ROS generated in this organelle can perturb ER calcium homeostasis and protein folding function, which results in ER stress [26]. In this study, CCl_4 treatment increased the levels of CHOP in a time-dependent manner, that is upregulated by ATF6, in the weaning mice (Figure 3). This indicated that CCL_4 induced ER stress in the liver. CHOP, a pro-apoptotic transcription factor, down-regulates levels of Bcl-2 and JNK. JNK is activated by ER stress, which results in the phosphorylation of Bcl-2 and the release of Bax [27,28]. The free Bax can form pores on the ER and mitochondrial membranes resulting in the release of ER calcium and mitochondrial cytochrome c, which mediates the activation of caspases [27,28]. Thus, the increased levels of ATF6, CHOP, p-JNK, and activated caspase suggest the association of ER stress and apoptosis in the livers of CCl_4-treated weaning mice. However, the adult mice appear to be relatively resistant to CCl_4-induced ER stress and apoptosis as evidenced by the unaltered levels of ER stress markers and high levels of anti-apoptotic Bcl-2 level in the liver.

GSH (γ-L-glutamyl-L-cysteinylglycine) is an essential non-protein thiol that has crucial roles in the detoxification of xenobiotics and the inhibition of oxidative stress by neutralizing free radicals and ROS in the liver [32]. In this study, the weaning mice exhibited high hepatotoxicity in spite of high hepatic GSH level after CCl_4 administration. This indicated that hepatic CYPs, rather than GSH, can be the primary factor for liver damage. The low levels of GST-π and GR proteins in the weaning mouse livers cannot be ruled out as the potential reasons for greater sensitivity as these enzymes are important for GSH conjugation and GSH regeneration from oxidized GSH (GSSG), respectively. GSTs are phase II detoxification enzymes that catalyze the thioether conjugation reaction of GSH with electrophiles [31]. The hepatic GST levels are reported to gradually increase from 2 weeks to 23 months of age in rats [41,46]. Xu et al. demonstrated that rat hepatic GST-π protein and mRNA expression continuously increased from 14 days up to 2 years after birth [47], which concurred with the results of our study. GST-π is reported to inhibit JNK phosphorylation by interacting with the C-terminus of JNK [48,49]. Moreover, oxidative stress dissociates the JNK-GST-π complex due to the oligomerization of GST-π, which results in the loss of JNK inhibition and subsequently apoptosis [48,50]. The enhanced expression of hepatic GST-π in the adult mice can partly contribute to the decreased activation of JNK. Thus, the adult mice exhibited higher resistance to apoptosis than the resistance exhibited by the weaning mice.

Age-related decline of GSH levels is reported in mammalian cells, which increases the risk of oxidative tissue damages in the elderly [51]. The main reason for the declined GSH levels is reported to be the reduction of de novo synthesis of GSH in aged cells [51]. GCL, the rate-limiting enzyme for GSH biosynthesis, condenses cysteine and glutamic acid to produce γ-glutamylcysteine, which reacts with glycine in the presence of GSH synthase to form GSH [32]. Older animals are reported to exhibit lower hepatic GCL activity due to the down-regulation of GCL expression [52–54]. CDO is another major enzyme consuming cysteine for taurine synthesis [55]. This enzyme adds dioxygen to cysteine thiol to form cysteine sulfinate, which is further converted to taurine [55]. The metabolic fate of cysteine depends on the activities of CDO and GCL. Thus, the cellular GSH concentration can also be affected by CDO in the liver. Previously, the levels of hepatic taurine were reported to decrease with aging due to the decreased CDO activity in middle-aged (9 months-10 months) and old (26 months-28 months) rats. However, the levels of hepatic taurine were not determined in the younger animals before they reached adulthood [56,57]. Our study, for the first time, demonstrates that infant mice have lower hepatic CDO level, that decreases the taurine concentration, than hepatic CDO levels observed in the adult mice (Figure 7). The low levels of CDO and high levels of GCL proteins in the weaning mouse livers imply that cysteine may be more efficiently utilized for the synthesis of GSH rather than the synthesis of taurine, that appears to contribute to the enhanced GSH levels in the weaning mice.

5. Conclusions

The results of this study have been summarized in Figure 8, that illustrates that weaning mice were found to be more vulnerable to CCl_4-induced liver injury than the extent of damage observed in the adult mice. The enhanced CYP2E1 and CYP3A protein levels in the weaning mice may contribute to severe necrosis and apoptosis through cellular dysfunctions induced by oxidative and ER stresses in the liver. Interestingly, the high GSH level and GSH synthesis rate in the liver of weaning mice cannot inhibit the CCl_4-mediated hepatotoxicity, that also indicates that CYPs are the major contributors to liver injury. Therefore, hepatic CYP activities in infants/children should be carefully considered for prescribing medication to prevent DILI. These findings demonstrated the possible reasons for the differential susceptibility of infant animals and adult animals using the CCl_4 hepatotoxicity model. The limitation of this study is that it does account for the species-dependent differences in the drug metabolism and antioxidant defense systems. Thus, further studies will be needed for a detailed understanding of pediatric DILI.

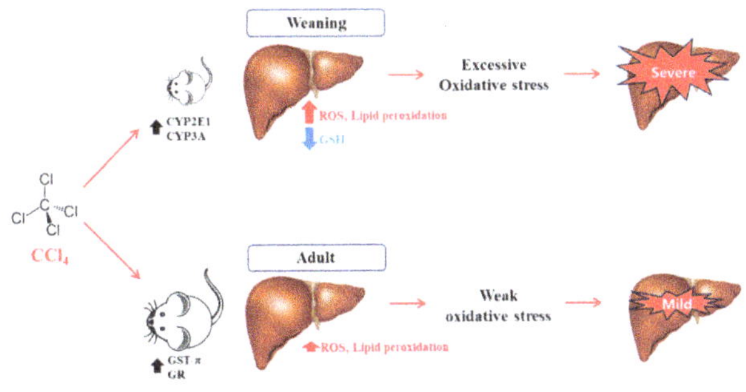

Figure 8. Summary and hypothetical mechanisms of age-related susceptibility to carbon tetrachloride (CCl_4) hepatotoxicity in weaning mice and adult mice. Higher metabolic rate of CCl_4 by hepatic CYP2E1 and CYP3A in weaning mice can cause excessive oxidative stress via reactive oxygen species (ROS) generation and glutathione (GSH) depletion, which can lead to significant liver injury. However, adult mice exhibit less CYP levels and high levels of GST-π and GR in their livers. This can contribute to resistance to CCl_4 hepatotoxicity.

Author Contributions: T.B.J., D.K., S.H.K., K.S.K., and Y.-S.J. designed the study. T.B.J., D.K., S.W.S., S.H.K., and M.-S.S. carried out experiment. T.B.J., D.K., S.W.S., S.H.K., Y.-H.L., M.-S.S., K.S.K., and Y.-S.J. analyzed experimental results. T.B.J., D.K., S.H.K., Y.-H.L., and Y.-S.J. prepared draft. D.K. and Y.-S.J. finalized the manuscript. All authors have read and agreed to the published version of the manuscript.

Funding: This work was supported by the National Research Foundation of Korea (NRF) funded by the Korean government (MSIT) (NRF-2018M3A7B4071233 and NRF-2019R1I1A3A01058584).

Conflicts of Interest: The authors declare that there is no conflict of interest.

References

1. Croom, E. Metabolism of xenobiotics of human environments. *Prog. Mol. Biol. Transl. Sci.* **2012**, *112*, 31–88. [CrossRef] [PubMed]
2. Gu, X.; Manautou, J.E. Molecular mechanisms underlying chemical liver injury. *Expert Rev. Mol. Med.* **2012**, *14*, e4. [CrossRef] [PubMed]
3. Corsini, A.; Bortolini, M. Drug-induced liver injury: The role of drug metabolism and transport. *J. Clin. Pharmacol.* **2013**, *53*, 463–474. [CrossRef] [PubMed]
4. Shi, Q.; Yang, X.; Greenhaw, J.J.; Salminen, A.T.; Russotti, G.M.; Salminen, W.F. Drug-Induced Liver Injury in Children: Clinical Observations, Animal Models, and Regulatory Status. *Int. J. Toxicol.* **2017**, *36*, 365–379. [CrossRef] [PubMed]
5. Chalasani, N.; Bjornsson, E. Risk factors for idiosyncratic drug-induced liver injury. *Gastroenterology* **2010**, *138*, 2246–2259. [CrossRef]
6. Sotaniemi, E.A.; Arranto, A.J.; Pelkonen, O.; Pasanen, M. Age and cytochrome P450-linked drug metabolism in humans: An analysis of 226 subjects with equal histopathologic conditions. *Clin. Pharmacol. Ther.* **1997**, *61*, 331–339. [CrossRef]
7. Parkinson, A.; Mudra, D.R.; Johnson, C.; Dwyer, A.; Carroll, K.M. The effects of gender, age, ethnicity, and liver cirrhosis on cytochrome P450 enzyme activity in human liver microsomes and inducibility in cultured human hepatocytes. *Toxicol. Appl. Pharmacol.* **2004**, *199*, 193–209. [CrossRef]
8. Rikans, L.E.; Moore, D.R. Acetaminophen hepatotoxicity in aging rats. *Drug Chem. Toxicol.* **1988**, *11*, 237–247. [CrossRef]
9. Palomero, J. Effects of aging on the susceptibility to the toxic effects of cyclosporin A in rats. Changes in liver glutathione and antioxidant enzymes. *Free Radic. Biol. Med.* **2001**, *30*, 836–845. [CrossRef]
10. Faa, G.; Ekstrom, J.; Castagnola, M.; Gibo, Y.; Ottonello, G.; Fanos, V. A developmental approach to drug-induced liver injury in newborns and children. *Curr. Med. Chem.* **2012**, *19*, 4581–4594. [CrossRef]
11. Bryant, A.E., III; Dreifuss, F.E. Valproic acid hepatic fatalities. III. U.S. experience since 1986. *Neurology* **1996**, *46*, 465–469. [CrossRef] [PubMed]
12. Rivkees, S.A.; Szarfman, A. Dissimilar hepatotoxicity profiles of propylthiouracil and methimazole in children. *J. Clin. Endocrinol. Metab.* **2010**, *95*, 3260–3267. [CrossRef] [PubMed]
13. Mach, J.; Huizer-Pajkos, A.; Cogger, V.C.; McKenzie, C.; Le Couteur, D.G.; Jones, B.E.; de Cabo, R.; Hilmer, S.N. The effect of aging on acetaminophen pharmacokinetics, toxicity and Nrf2 in Fischer 344 rats. *J. Gerontol. Biol. Sci. Med. Sci.* **2014**, *69*, 387–397. [CrossRef] [PubMed]
14. Mach, J.; Huizer-Pajkos, A.; Mitchell, S.J.; McKenzie, C.; Phillips, L.; Kane, A.; Jones, B.; de Cabo, R.; Cogger, V.; Le Couteur, D.G.; et al. The effect of ageing on isoniazid pharmacokinetics and hepatotoxicity in Fischer 344 rats. *Fundam. Clin. Pharmacol.* **2016**, *30*, 23–34. [CrossRef] [PubMed]
15. Brautbar, N.; Williams, J., II. Industrial solvents and liver toxicity: Risk assessment, risk factors and mechanisms. *Int. J. Hyg. Environ. Health* **2002**, *205*, 479–491. [CrossRef] [PubMed]
16. Weber, L.W.; Boll, M.; Stampfl, A. Hepatotoxicity and mechanism of action of haloalkanes: Carbon tetrachloride as a toxicological model. *Crit. Rev. Toxicol.* **2003**, *33*, 105–136. [CrossRef]
17. Jahn, F.; Reuter, A.; Karge, E.; Danz, M.; Klinger, W. Age dependent different influence of carbon tetrachloride on biotransformation of xenobiotics, glutathione content, lipid peroxidation and histopathology of rat liver. *Exp. Toxicol. Pathol.* **1993**, *45*, 101–107. [CrossRef]
18. Rikans, L.E.; Hornbrook, K.R.; Cai, Y. Carbon tetrachloride hepatotoxicity as a function of age in female Fischer 344 rats. *Mech. Ageing Dev.* **1994**, *76*, 89–99. [CrossRef]

19. Dalu, A.; Warbritton, A.; Bucci, T.J.; Mehendale, H.M. Age-related susceptibility to chlordecone-potentiated carbon tetrachloride hepatotoxicity and lethality is due to hepatic quiescence. *Pediatr. Res.* **1995**, *38*, 140–148. [CrossRef]
20. Collins, B.H.; Holzknecht, Z.E.; Lynn, K.A.; Sempowski, G.D.; Smith, C.C.; Liu, S.; Parker, W.; Rockey, D.C. Association of age-dependent liver injury and fibrosis with immune cell populations. *Liver Int.* **2013**, *33*, 1175–1186. [CrossRef]
21. Hong, I.H.; Lewis, K.; Iakova, P.; Jin, J.; Sullivan, E.; Jawanmardi, N.; Timchenko, L.; Timchenko, N. Age-associated change of C/EBP family proteins causes severe liver injury and acceleration of liver proliferation after CCl4 treatments. *J. Biol. Chem.* **2014**, *289*, 1106–1118. [CrossRef] [PubMed]
22. Reitman, S.; Frankel, S. A colorimetric method for the determination of serum glutamic oxalacetic and glutamic pyruvic transaminases. *Am. J. Clin. Pathol.* **1957**, *28*, 56–63. [CrossRef] [PubMed]
23. Volpi, N.; Tarugi, P. Improvement in the high-performance liquid chromatography malondialdehyde level determination in normal human plasma. *J. Chromatogr. B Biomed. Sci. Appl.* **1998**, *713*, 433–437. [CrossRef]
24. Lee, Y.H.; Kim, S.H.; Lee, S.; Kim, K.M.; Jung, J.C.; Son, T.G.; Ki, S.H.; Seo, W.D.; Kwak, J.H.; Hong, J.T.; et al. Antioxidant Effect of Barley Sprout Extract via Enhancement of Nuclear Factor-Erythroid 2 Related Factor 2 Activity and Glutathione Synthesis. *Nutrients* **2017**, *9*, 1252. [CrossRef] [PubMed]
25. Kim, S.H.; Kwon, D.Y.; Kwak, J.H.; Lee, S.; Lee, Y.H.; Yun, J.; Son, T.G.; Jung, Y.S. Tunicamycin-Induced ER Stress is Accompanied with Oxidative Stress via Abrogation of Sulfur Amino Acids Metabolism in the Liver. *Int. J. Mol. Sci.* **2018**, *19*, 4114. [CrossRef]
26. Liu, H.; Wang, Z.; Nowicki, M.J. Caspase-12 mediates carbon tetrachloride-induced hepatocyte apoptosis in mice. *World J. Gastroenterol.* **2014**, *20*, 18189–18198. [CrossRef]
27. Vannuvel, K.; Renard, P.; Raes, M.; Arnould, T. Functional and morphological impact of ER stress on mitochondria. *J. Cell Physiol.* **2013**, *228*, 1802–1818. [CrossRef]
28. Iurlaro, R.; Munoz-Pinedo, C. Cell death induced by endoplasmic reticulum stress. *FEBS J.* **2016**, *283*, 2640–2652. [CrossRef]
29. Suh, Y. Cell signaling in aging and apoptosis. *Mech. Ageing Dev.* **2002**, *123*, 881–890. [CrossRef]
30. Kim, E.K.; Choi, E.J. Pathological roles of MAPK signaling pathways in human diseases. *Biochim. Biophys Acta* **2010**, *1802*, 396–405. [CrossRef]
31. Hayes, J.D.; Flanagan, J.U.; Jowsey, I.R. Glutathione transferases. *Annu. Rev. Pharmacol. Toxicol.* **2005**, *45*, 51–88. [CrossRef] [PubMed]
32. Forman, H.J.; Zhang, H.; Rinna, A. Glutathione: Overview of its protective roles, measurement, and biosynthesis. *Mol. Aspects Med.* **2009**, *30*, 1–12. [CrossRef] [PubMed]
33. Arana, A.; Morton, N.S.; Hansen, T.G. Treatment with paracetamol in infants. *Acta Anaesthesiol. Scand.* **2001**, *45*, 20–29. [CrossRef] [PubMed]
34. O'Donnell, J.T.; Marks, D.H.; Danese, P.; O'Donnell, J.J., III. Drug-induced liver disease: Primer for the primary care physician. *Dis. Mon.* **2014**, *60*, 55–104. [CrossRef]
35. Erden-Inal, M.; Sunal, E.; Kanbak, G. Age-related changes in the glutathione redox system. *Cell Biochem. Funct.* **2002**, *20*, 61–66. [CrossRef]
36. Wong, F.W.; Chan, W.Y.; Lee, S.S. Resistance to carbon tetrachloride-induced hepatotoxicity in mice which lack CYP2E1 expression. *Toxicol. Appl. Pharmacol.* **1998**, *153*, 109–118. [CrossRef]
37. Zangar, R.C.; Benson, J.M.; Burnett, V.L.; Springer, D.L. Cytochrome P450 2E1 is the primary enzyme responsible for low-dose carbon tetrachloride metabolism in human liver microsomes. *Chem. Biol. Interact.* **2000**, *125*, 233–243. [CrossRef]
38. Veith, A.; Moorthy, B. Role of Cytochrome P450s in the Generation and Metabolism of Reactive Oxygen Species. *Curr. Opin. Toxicol.* **2018**, *7*, 44–51. [CrossRef]
39. Oinonen, T.; Lindros, K.O. Zonation of hepatic cytochrome P-450 expression and regulation. *Biochem. J.* **1998**, *329*, 17–35. [CrossRef]
40. Kietzmann, T. Metabolic zonation of the liver: The oxygen gradient revisited. *Redox Biol.* **2017**, *11*, 622–630. [CrossRef]
41. Vyskocilova, E.; Szotakova, B.; Skalova, L.; Bartikova, H.; Hlavacova, J.; Bousova, I. Age-related changes in hepatic activity and expression of detoxification enzymes in male rats. *Biomed. Res. Int.* **2013**, *2013*, 408573. [CrossRef] [PubMed]

42. Lu, Y.; Zhang, C.; Chen, Y.H.; Wang, H.; Zhang, Z.H.; Chen, X.; Xu, D.X. Immature mice are more susceptible than adult mice to acetaminophen-induced acute liver injury. *Sci. Rep.* **2017**, *7*, 42736. [CrossRef]
43. Vieira, I.; Sonnier, M.; Cresteil, T. Developmental expression of CYP2E1 in the human liver. Hypermethylation control of gene expression during the neonatal period. *Eur. J. Biochem.* **1996**, *238*, 476–483. [CrossRef] [PubMed]
44. Johnson, T.N. The development of drug metabolising enzymes and their influence on the susceptibility to adverse drug reactions in children. *Toxicology* **2003**, *192*, 37–48. [CrossRef]
45. Iracheta-Vellve, A.; Petrasek, J.; Gyongyosi, B.; Satishchandran, A.; Lowe, P.; Kodys, K.; Catalano, D.; Calenda, C.D.; Kurt-Jones, E.A.; Fitzgerald, K.A.; et al. Endoplasmic Reticulum Stress-induced Hepatocellular Death Pathways Mediate Liver Injury and Fibrosis via Stimulator of Interferon Genes. *J. Biol. Chem.* **2016**, *291*, 26794–26805. [CrossRef] [PubMed]
46. Jang, I.; Chae, K.; Cho, J. Effects of age and strain on small intestinal and hepatic antioxidant defense enzymes in Wistar and Fisher 344 rats. *Mech. Ageing Dev.* **2001**, *122*, 561–570. [CrossRef]
47. Xu, S.; Hou, D.; Liu, J.; Ji, L. Age-associated changes in GSH S-transferase gene/proteins in livers of rats. *Redox Rep.* **2018**, *23*, 213–218. [CrossRef]
48. Adler, V.; Yin, Z.; Fuchs, S.Y.; Benezra, M.; Rosario, L.; Tew, K.D.; Pincus, M.R.; Sardana, M.; Henderson, C.J.; Wolf, C.R.; et al. Regulation of JNK signaling by GSTp. *EMBO J.* **1999**, *18*, 1321–1334. [CrossRef]
49. Tew, K.D.; Townsend, D.M. Regulatory functions of glutathione S-transferase P1-1 unrelated to detoxification. *Drug Metab. Rev.* **2011**, *43*, 179–193. [CrossRef]
50. Bernardini, S.; Bernassola, F.; Cortese, C.; Ballerini, S.; Melino, G.; Motti, C.; Bellincampi, L.; Iori, R.; Federici, G. Modulation of GST P1-1 activity by polymerization during apoptosis. *J. Cell Biochem.* **2000**, *77*, 645–653. [CrossRef]
51. Ferguson, G.; Bridge, W. Glutamate cysteine ligase and the age-related decline in cellular glutathione: The therapeutic potential of gamma-glutamylcysteine. *Arch. Biochem. Biophys.* **2016**, *593*, 12–23. [CrossRef] [PubMed]
52. Liu, R.; Choi, J. Age-associated decline in gamma-glutamylcysteine synthetase gene expression in rats. *Free Radic Biol. Med.* **2000**, *28*, 566–574. [CrossRef]
53. Toroser, D.; Sohal, R.S. Age-associated perturbations in glutathione synthesis in mouse liver. *Biochem. J.* **2007**, *405*, 583–589. [CrossRef] [PubMed]
54. Jiang, P.; Sheng, Y.; Ji, L. The age-related change of glutathione antioxidant system in mice liver. *Toxicol. Mech. Methods* **2013**, *23*, 396–401. [CrossRef] [PubMed]
55. Schuller-Levis, G.B.; Park, E. Taurine: New implications for an old amino acid. *FEMS Microbiol. Lett.* **2003**, *226*, 195–202. [CrossRef]
56. Eppler, B.; Dawson, R., Jr. Cysteine sulfinate decarboxylase and cysteine dioxygenase activities do not correlate with strain-specific changes in hepatic and cerebellar taurine content in aged rats. *Mech. Ageing Dev.* **1999**, *110*, 57–72. [CrossRef]
57. Eppler, B.; Dawson, R., Jr. Dietary taurine manipulations in aged male Fischer 344 rat tissue: Taurine concentration, taurine biosynthesis, and oxidative markers. *Biochem. Pharmacol.* **2001**, *62*, 29–39. [CrossRef]

antioxidants

Review

Interaction of Neuromelanin with Xenobiotics and Consequences for Neurodegeneration; Promising Experimental Models

Andrea Capucciati [1], Fabio A. Zucca [2], Enrico Monzani [1], Luigi Zecca [2], Luigi Casella [1] and Tim Hofer [3,*]

[1] Department of Chemistry, University of Pavia, 27100 Pavia, Italy; andrea.capucciati@gmail.com (A.C.); enrico.monzani@unipv.it (E.M.); luigi.casella@unipv.it (L.C.)
[2] Institute of Biomedical Technologies, National Research Council of Italy, Segrate, 20054 Milan, Italy; fabio.zucca@itb.cnr.it (F.A.Z.); luigi.zecca@itb.cnr.it (L.Z.)
[3] Department of Environmental Health, Norwegian Institute of Public Health, P.O. Box 222 Skøyen, N-0213 Oslo, Norway
* Correspondence: tim.hofer@fhi.no; Tel.: +47-21076671

Abstract: Neuromelanin (NM) accumulates in catecholamine long-lived brain neurons that are lost in neurodegenerative diseases. NM is a complex substance made of melanic, peptide and lipid components. NM formation is a natural protective process since toxic endogenous metabolites are removed during its formation and as it binds excess metals and xenobiotics. However, disturbances of NM synthesis and function could be toxic. Here, we review recent knowledge on NM formation, toxic mechanisms involving NM, go over NM binding substances and suggest experimental models that can help identifying xenobiotic modulators of NM formation or function. Given the high likelihood of a central NM role in age-related human neurodegenerative diseases such as Parkinson's and Alzheimer's, resembling such diseases using animal models that do not form NM to a high degree, e.g., mice or rats, may not be optimal. Rather, use of animal models (i.e., sheep and goats) that better resemble human brain aging in terms of NM formation, as well as using human NM forming stem cellbased in vitro (e.g., mid-brain organoids) models can be more suitable. Toxicants could also be identified during chemical synthesis of NM in the test tube.

Keywords: adverse outcome pathway (AOP); iron; locus coeruleus; MPTP; quinone; substantia nigra

Citation: Capucciati, A.; Zucca, F.A.; Monzani, E.; Zecca, L.; Casella, L.; Hofer, T. Interaction of Neuromelanin with Xenobiotics and Consequences for Neurodegeneration; Promising Experimental Models. *Antioxidants* **2021**, *10*, 824. https://doi.org/10.3390/antiox10060824

Academic Editor: Stanley Omaye

Received: 26 April 2021
Accepted: 18 May 2021
Published: 21 May 2021

1. Introduction—Neuromelanin in Neurodegenerative Diseases

The incidence and prevalence of neurological disorders such as Parkinson's disease (PD) and Alzheimer's disease (AD) increase with age and in most cases ($\geq$85% for PD and $\geq$90% for AD), they are sporadic (without a known genetic cause), indicating that environmental factors such as chemical exposures play strong roles. Neuromelanin (NM) is a dark pigment characterizing neurons of two particular brain areas known as the substantia nigra (SN) and the locus coeruleus (LC) [1–3]. NM is practically ubiquitous in brain neurons although it is diffused and in smaller amounts in different neurons of other brain areas [3,4]. Interestingly, the two regions richest in NM: SN (mid-brain) and LC (in the pons of the brainstem) are both damaged in PD and AD [5]. SN is strongly affected in PD and LC is early affected in both PD and AD [6].

The NM pigment has a protective function for neurons, at least in normal conditions, because it acts as a scavenger of toxic molecules and metal ions [7,8]. A peculiar feature of NM is its ability to tightly bind metal ions, in particular iron [8] but also copper and zinc, especially in NM of LC [9], as described below.

Potentially toxic organic molecules include the unwanted accumulation of endogenous cytosolic dopamine (DA), or norepinephrine, in dopaminergic or noradrenergic neurons of SN and LC, respectively, i.e., the neurotransmitters particularly rich in neurons of these brain areas [10]. Such accumulation may result from dysfunctioning of the catecholamine processing enzymes downstream or leaking from the catecholamine storage vesicles. In addition, their metabolites (e.g., quinones) can be toxic as they can react and cause protein adducts. Furthermore, exogenous chemicals (xenobiotics) can cause toxicity through NM interactions. The situation may worsen dramatically with the development of neuropathologies, especially PD, where NM is released in the extraneuronal space by degenerating neurons discharging the toxic compounds and metal ions accumulated over many years of life. In addition, NM itself activates microglia inducing further release of toxic (e.g., reactive oxygen species, ROS) and pro-inflammatory molecules that cause chronic neuroinflammation and neurodegeneration [11].

Below, we review the latest knowledge on NM formation, the neuroinflammatory pathway involving extracellular NM leading to neuronal cell death, review chemically-induced animal PD models, list identified chemicals that bind to NM and finally discuss models to study NM related toxicity.

2. Neuromelanin Composition, Structure and Biosynthesis

In humans, the amount of NM pigment steadily increases in aging and starts to accumulate in early life [12]. The pigment is compartmentalized within specialized cytosolic organelles surrounded by a double membrane and such organelles also contain lipid bodies mainly consisting of dolichols and dolichoic acids [13,14]. This structural organization indicates that formation of NM follows a designed pathway, although the details of NM biosynthesis are not fully understood. Even though NM is often associated with peripheral melanins, it should be pointed out that both the composition, structural organization and biosynthesis of the two types of pigments are completely different, making the neuronal pigment unique, although a lot of confusion still exists in the literature. Peripheral melanins are products of the monophenolase activity of tyrosinases or catecholase activity of both tyrosinases and catechol oxidases (both copper enzymes) in animals, fungi and insects [15–17], and consist of packed eumelanic oligomers organized in π-stacked layers with typical aromatic π-interactions [18,19], without addition of other components. The layers auto assemble upon oxidative oligomerization of the phenolic substrates promoted by the copper enzymes, and in vitro rapidly separate out as insoluble products from the solution as soon as the enzymatic reaction is started. The biosynthetic pathway of brain NM is much more complex and leads to pigments of complex composition consisting of melanic, protein and lipidic components [4,7,14], see Scheme 1. In particular, the melanic portion of NM contains polymerized catecholamine (e.g. DA) residues (eumelanin) and polymerized cysteinyldopamine (CysDA) residues (pheomelanin), in an approximate 3:1 ratio, characterized by dihydroxyindole and benzothiazine units, respectively [20]. In NM probably a pheomelanic core is contained, while eumelanin is localized on the surface, forming spherical electron-dense aggregates [3,21].

We can trace the beginning of the pathway leading to mature NM organelles to the formation of dopaminoquinone by oxidation of cytosolic DA leaking from storage vesicles or accumulating due to malfunctioning of downstream regulatory mechanisms [8,10,22,23]. The direct link between DA oxidation and mitochondrial and lysosomal dysfunction in human SN neurons leading to PD pathogenesis has been demonstrated [24]. DA oxidation is then one of the effects of oxidative stress and it is likely promoted by redox metal ions, whereas it can be excluded that the quinone, and its subsequent evolution products such as aminochrome and further oligomers, result from enzymatic oxidation, since tyrosinase is not expressed in the catecholamine neurons of SN [25]. Other DA metabolites could be toxic as well, if their formation is not regulated; one important example is given by 3,4-dihydroxyphenyl acetaldehyde (DOPAL), which is the product of DA oxidation by monoamine oxidase, together with hydrogen peroxide [26]. Dopaminoquinone, as well as related quinones, are very reactive towards nucleophiles, and in particular, toward the cysteine, histidine and lysine residues of proteins [22]; what is peculiar in the pathway to NM is that the target residues of the quinones are those of fibrillated protein seeds present in the cytosol and results in protein/peptide loss of structure and aggregation. The current view is that DOPAL is the most reactive DA metabolite with α-synuclein, readily forming a Schiff base with the protein lysine residues. In fact, α-synuclein does not contain cysteine and the reaction of DOPAL with the lysines is much faster than that of dopaminoquinone. The surface residues on these structured β-sheets react faster, as we recently have shown by comparing the reactivity of dopaminoquinone with monomeric or fibrillated β-lactoglobulin [27]. These complex iron-melanin-β-sheet proteins cannot be degraded by proteasome but are engulfed by macroautophagy to form an autophagosome. The latter fuses with lysosomes and other vesicles carrying proteins and dolichol lipids to generate an autolysosome. Within this autolysosome the complex iron-melanin-β-sheet protein reacts with dolichols to produce the final NM inside the so named NM-containing organelle [8]. In fact, the trace of the protein seeds on which the melanic oligomers are linked is the only structural feature of NM recognizable by X-ray powder analysis, which shows the characteristic 4.7 Å separation of the backbones [3] typical for cross-β structure of amyloid fibrils [28]. On the contrary, synthetic melanins obtained by oxidative oligomerization of DA [29] and natural peripheral melanins [30], which lack protein components, exhibit the signature of π-stacked layers with 3.5 Å separation. This difference is very important also in the context of the present review, because the binding properties of NM towards exogenous molecules, xenobiotics and pharmaceuticals will be strongly affected by the structural organization of the pigment. For obvious reasons, given the limited availability of NM from natural origin, all in vitro studies aiming at determining the affinity of NM for exogenous compounds typically use synthetic melanins derived from DA oxidative oligomerization in the presence of cysteine and, therefore, we have to take into account that the synthetic pigments are only a rough approximation of natural NM. The effects of exogenous compounds (xenobiotics) may not be limited to the interaction with pre-formed NM, and some of them could interfere with any of the steps leading to mature NM granules, as indicated in Scheme 1.

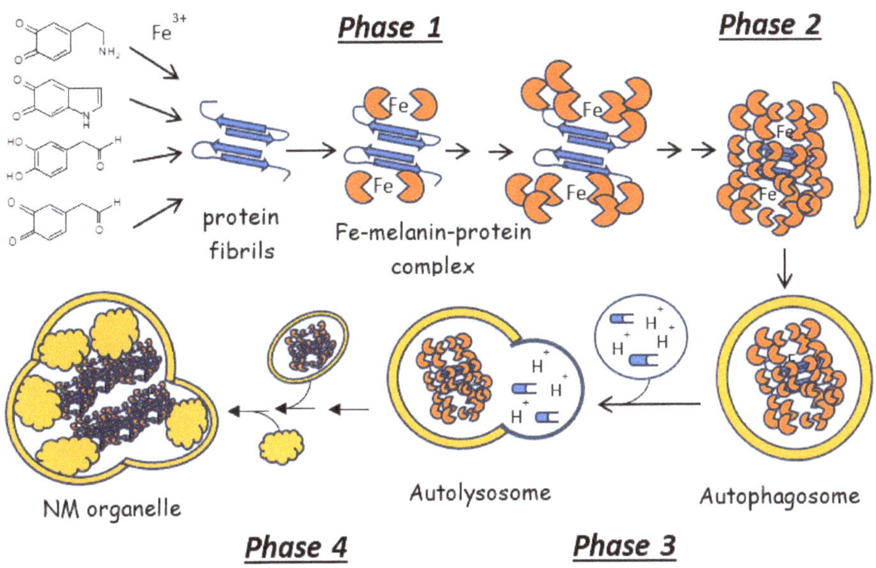

Scheme 1. The hypothesized biosynthetic pathway of NM is schematically shown to proceed through four main phases. In the initial Phase 1, the oxidation products of DA, or the oxidized DA metabolites [22], generated by catalytic oxidation mediated by Fe^{3+}, or Cu^{2+}, react with fibrillary seeds of aggregated peptides/proteins (blue) at the exposed side chains of Cys, His and Lys forming melanin (orange)-protein conjugates maintaining the fibrillary core as the only structural element. In Phase 2, the undegradable conjugates are engulfed by the phagophore and isolated from the cytosol through a double lipid membrane (yellow). The autophagic vacuoles fuse with lysosomes (blue circle) in Phase 3, causing interaction between

the melanin-protein conjugates with other proteins, enzymes and lipids carried by lysosomes, and are transformed into autolysosomes. The resulting autolysosomes fuses with other vesicles containing lipids that attach to melanin-protein conjugates to form NM. Finally, in Phase 4 these organelles fuse with other autophagic vacuoles containing NM precursors or with mature NM organelles. A more detailed description of the entire process can be found in Zucca et al., 2018 [14].

3. Neuroinflammation Leads to Degeneration of Dopaminergic Neurons of the Nigro-Striatal Pathway

A typical neuropathological feature of PD is the presence of extracellular NM in the degenerating SN tissue [31], due to the conspicuous and selective death of pigmented dopaminergic neurons over the disease [32]. Notably, in SN of PD patients and those with Parkinsonian syndromes numerous activated microglia cells have been observed in close proximity of degenerating neurons and free NM debris [33,34] which were also observed inside reactive microglia [31]. Therefore microglia, due to high phagocytic capability, are responsible for the disappearance of NM in SN following to neuronal death.

These observations clearly indicated the key role of NM in microglia activation in the disease progression, which was deeply investigated over the last two decades. In vitro experiments have shown that human NM added to microglia is able to activate the proinflammatory transcription factor nuclear factor κB while inducing chemotactic effects: microglia move toward NM granules to actively phagocytose them, and release neurotoxic mediators such as tumor-necrosis factor α, interleukin 6 and nitric oxide [35], as well as reactive species like superoxide and hydrogen peroxide [11]. This effect was nicely replicated by using synthetic analogues of human NM, which were able to induce microglia activation as natural NM does in pathological processes of PD [27]. This phagocytic process was proved to be very rapid and efficient for young microglia in contrast to aged microglia that showed a rapid NM phagocytosis in spite of a reduced ability to break down NM [11]. The slower degradation of NM in addition to its high insolubility could mean that extracellular NM can remain longer in the tissue, sustaining microglial activation as observed in the PD brain [31,34], acting as agent of chronic inflammation and slowly releasing the metals and toxic compounds encased in its structure.

In PD pathogenesis, neuroinflammation is an important contributor to disease progression leading to SN dopaminergic neurons loss, and NM likely contributes to this noxious process [36]. In vitro experiments have shown that in neuron/microglia co-cultures, microglia activated by NM induced neuronal death by releasing pro-inflammatory molecules and reactive species and this effect was further reproduced in vivo, upon injection of NM into rat SN [11,37]. Since a moderate astrocytosis was also observed in the SN in proximity of the NM injection site [11], the effect of NM on human astroglia has been investigated showing that NM exposure caused inhibition of the tumor necrosis factor α induced expression of the chemokine interferon γ inducible protein-10 (CXCL-10) [38]. This likely suggests an impairment of astrocytes function, although further investigation is needed since less is known about the role of astrocytes than microglia in neuroinflammatory and neurodegenerative mechanisms of the disease. Cell cultures studies clearly demonstrated that the neurodegeneration induced by NM-activated microglia likely involves macrophage antigen complex-1 and phagocytic oxidase of microglia [11], even though other factors could play a key role.

Therefore, the role of NM in neuroinflammatory/neurodegenerative processes of PD can start after an initial neuronal damage (due to environmental or genetic factor), then released NM induces microglial activation with production of neurotoxic molecules which damage other neurons with further release of NM, establishing a chronic condition of neuroinflammation and neurodegeneration. In addition, toxic compounds and redox active metals immobilized into NM structure can be released, which could further exacerbate microglial activation and neuronal death.

Recently, intraneuronal NM has been shown to increase neuronal vulnerability due to high content of major histocompatibility complex class I (MHC-I) in NM-containing organelles of neurons in human SN and LC, highly pigmented regions that selectively degenerate in PD [39]. These neurons express MHC-I, while other neurons not targeted by PD have low or absent MHC-I expression. Interestingly, $CD8^+$ cells were sometimes observed in SN and LC of PD brains in proximity to neurons containing NM and expressing MHC-I, suggesting an involvement in neuronal death. To clarify these mechanisms, expression of MHC-I was induced in cultured neurons by exposure to factors commonly released by microglia activated by NM or α-synuclein, known microglia activators [36] and both present extracellularly in PD brains, or in condition of high oxidative stress derived by high cytosolic DA. In these in vitro experiments, MHC-I was proved to bind antigenic peptides presenting them on neuronal membrane, so that cytotoxic $CD8^+$ lymphocytes could target neuron inducing neuronal death [39].

In AD, the first cells to degenerate are the noradrenergic neurons containing NM of LC as shown by neuropathological studies and Magnetic Resonance Imaging of NM [5,40,41].

4. Underlying Causes of PD—Searching for NM Modulators

It is well established that environmental factors, such as pesticides, insecticides and neurotoxins, in addition to genetic factors and toxic metal ions, are involved in PD pathogenesis [42,43]. For example, it appears that environmental exposure to lead, manganese, mercury, cobalt or cadmium increases the incidence of PD [44], and NM apparently blocks the toxicity of these metals in the brain. NM binds and accumulates high amounts of these metals [3,45], and synthetic NM halted lipid peroxidation in vitro, acting as an antioxidant [46]. Several previously-used pesticides (e.g., chlorpyrifos, maneb, paraquat and rotenone; now banned in the EU but can be allowed elsewhere) are strongly linked to PD. In particular, such toxins are thought to cause mitochondrial dysfunction through the formation of ROS. Endogenous redox metal ions beyond the physiological range of concentration and altered intercellular distribution would of course have a direct role in ROS production, by reacting e.g., with hydrogen peroxide in Fenton-type reactions [47,48], but also through some indirect mechanisms such as damaging proteins of e.g., Complex I [49], or detoxifying enzymes, thus impairing their activity [50] and cell function. This redox activity can be strongly promoted by the increase of cytosolic DA concentration in neurons of SN with formation of reactive quinones, unless DA is sequestered into vesicles or converted into NM [51]. The key point is trying to understand how NM is affected in the pathogenic processes induced by toxic species, both in the degradation of NM, which eliminates an important protective substance, and in the biosynthesis of NM, through the possible interference in some steps of its build-up.

In other instances, a chemical species may actively participate in the formation of the melanic derivatization of the protein fibrils, in competition with DA, through quinone groups of similar or even higher reactivity (Phase 1 in Scheme 1). This is the case, for instance, of 6-hydroxydopamine (6-OHDA) [52], an important toxin that may be formed endogenously by hydroxylation of DA by ROS produced through Fenton chemistry [53].

Certain PD-inducing xenobiotics exert neurotoxicity (e.g., oxidative stress and cell death) in NM rich brain compartments such as SN, and molecular studies suggest that they interact and/or disturb NM's protective function. It has been suggested that NM first rapidly binds xenobiotics that thereafter slowly diffuse away causing toxicity. However, it is unknown if this is the case.

5. If NM Is Involved in Neurodegenerative Diseases, What Are Suitable Animal Models?

Interestingly, some xenobiotics are only neurotoxic in humans and animals having NM, suggesting that choosing a relevant model that resembles the human high NM accumulation in aging is important when studying human neurological disease development. However, a recent paper states that >90% of all studies performed in years 2000–2019 had used rodents (low or devoid in NM) for neurotoxin-induced PD research [54] which is quite surprizing. A similar situation likely presents also for AD models. Importantly, rodents do not naturally develop neurodegenerative diseases or even signs thereof as they age. To resemble human neurodegenerative diseases in some way (pathologically or behaviourally, but often not the full disease spectrum or chronology), often double or even triple trans-genetic rodent constructs are required with unnatural protein overexpression. Often, young adult or mid-aged rodents are studied, seldomly old rodents. Moreover, mice (important because transgenic mice are frequently used to model human AD and PD) entirely lack NM and also differ considerably from humans in many other aspects, e.g., mice have differences in brain regions of DA system (striatum compared to putamen-caudatus in human) with respect to human brain, and mice also lack entire brain compartments that humans have.

More suitable model species can likely be found in monkeys, dogs, sheep and goats that are more similar to humans in size and genetics, and for which there are often more suitable (than for rodents) cognitive tests available. The NM-like pigment is present in SN of primates [55], cats [55], dogs [55–57], sheep [55], goats [58] and horses [59,60] in low amount (in 5% of neurons) in middle-aged rats [56], but devoid in mice, Indian fruit bats [55], pigs [55,61] and guinea-pigs (are rodents) [55]. However, monkey facilities are very limited and toxicity studies in monkeys and dogs undergo several ethical restrictions.

6. Chemically-Induced PD-Resembling Animal Models

Dopaminergic neurons in the SN can be damaged in several ways both physically and chemically, resulting in PD-related motor abnormalities and cognitive disturbances. Only a few toxin-induced (chemical) models presently exist that to various extent resemble human PD disease progression and pathology. In human PD, age is a major factor and pathological SN effects include neuronal iron accumulation, oxidative stress, cellular degeneration and α-synuclein protein aggregation inside Lewy Bodies (LB) in remaining SN neurons. The role of LBs in causing death of DA neurons in SN is unclear, as well as what the eventual mechanistic link between NM and LBs is (possibly iron accumulation as an atypical iron response element (IRE) is found in the 5′-untranslated region of α-synuclein messenger RNA transcripts and α-synuclein was found to possess ferrireductase activity). In addition, also non-pigmented neurons build up LBs and die as they age [62].

Table 1 lists the four most commonly used PD-inducers in animals, and their effects in brief. Common PD-inducing chemical toxicants are often taken up through catecholaminergic plasma membrane transporters. A common phenomenon is radical and/or quinone formation resulting in ROS production (also inside the mitochondrial respiratory chain), causing oxidative stress and cell death to the exposed brain region (sometimes rapidly, just hours after administration). However, most PD inducers are effective toxicants also in young animals, wherefore it is questionable how well these resemble the human situation.

Table 1. Common PD-modeling chemicals and their mode of action (in general).

Substance	Induces Parkinsonism in	Age Dependent Effect	Region (SN) Specificity	NM Affinity	BBB Passage	Mitochondrial Electron-Chain Inhibitor	Forms Quinones	ROS Generator	Induces α-Synuclein/ Lewy Bodies
MPTP/MPP+		 ✓	✓	✓	✓	✓	✗	✓	✓ ✗ species-dependent
Paraquat		✗	✗	✓	✓	✗	✗	✓	✓
Rotenone		✗	✗	✗	✓	✓	✗	✓	✓
6-OHDA		✗	✗	✗	✗ 	✓	✓	✓	✗

✓ = yes, ✗ = no.

6.1. MPTP/MPP$^+$

In the late 1970s, the toxic, highly lipophilic, substance 1-methyl-4-phenyl-1,2,3,6-tetrahydropyridine (MPTP) was found to induce parkinsonism in humans. It was later shown that toxicity was dependent upon conversion into the pyridinium cation (MPP$^+$), followed by uptake by dopaminergic neurons through DA transporters and subsequent loading into NM granules, as MPP$^+$ binds with high affinity to NM [63]. MPTP/MPP$^+$ causes more damage to neurons in SN than to neurons in other regions. In humans and monkeys, MPTP caused extracellular NM accumulation (resulting from cell death) with activated microglia in the SN, indicative of long-lasting effects [34,64]. MPP$^+$ may be accidentally produced during manufacture of the synthetic opioid MPPP, and MPP$^+$ induces Parkinson-like syndromes in humans, primates [64] and aged (7–8 years old), but not young (1–3 years old), sheep [65] by damaging SN neurons rich in NM organelles [64,65]. In addition, male minipigs developed PD symptoms after MPTP administration, which lowered striatal DA concentrations and resulted in loss of SN cells [66], and cats were severely parkinsonian after MPTP administration but recovered with time [67]. Direct brain (into SN) MPP$^+$ injections in rats also caused nigral lesions [68]. Rats, however, have shown to be relatively resistant to MPTP-induced neurotoxicity [69,70]. Furthermore, in mice, low or even devoid of NM in the SN, rapid severe behavioral effects (e.g., tremor) were observed with damage to dopaminergic neurons after systemic MPTP administration. Aged (12–14 mo old) mice were more susceptible than young (2–3 mo old) adults [71,72] who also recovered better [72]. However, since MPTP was potent also in young adult mice [71,72], this indicates that MPTP also targets dopaminergic neurons lacking NM, at least at higher doses. MPTP/MPP$^+$ induces LBs in humans, but often not in animals (also often not in monkeys), although one study found that more LBs are formed in older MPTP-injected monkeys [73].

In human brain organoids, MPTP causes neuron-specific cell death to midbrain dopaminergic neurons [74].

One suggested mechanism involves oxidative stress (MPP$^+$ interacts with the mitochondrial respiratory chain, acting as an electron transport chain Complex I inhibitor, producing ROS) where oxidation of DA leads to the formation of a potentially toxic dopaminoquinone that can redox-cycle forming toxic ROS or react and form DA protein adducts that ultimately induces cell death [8,75]. The adverse outcome pathway (AOP) "Inhibition of the mitochondrial complex I of nigrostriatal neurons leads to parkinsonian motor deficits" (https://aopwiki.org/aops/3, accessed on 15 April 2021) is mainly based on MPTP findings, but also lists several mechanistic uncertainties.

6.2. Paraquat

The herbicide paraquat has a striking structural similarity to MPP$^+$, both being lipophilic amines. Paraquat undergoes extensive redox-cycling, producing ROS. It has been reported to induce α-synuclein aggregation inside dopaminergic SN neurons and cause nigral cell loss. The herbicide paraquat accumulated in NM as the structurally related MPP$^+$ [76]. Paraquat displayed affinity to NM, although considerably lower affinity for isolated natural NM than MPP$^+$ and antimalarial drugs [77]. Paraquat is sometimes co-administrated together with maneb, a ROS producing manganese-containing polymeric complex used as a fungicide, to exert synergistic toxicity.

6.3. Rotenone

Rotenone is a natural compound extracted from plants. The systemic uptake is low and it is broken down by the liver. After intravenous injection, rotenone (highly lipophilic) reaches all organs including brain. Rotenone impairs mitochondrial oxidative phosphorylation, producing ROS. The distribution inside brain is heterogenous and rotenone damages different cell types, not necessarily those of the nigrostriatal pathway. Specific injection into, or close to SN, increases the selective toxicity. It has been reported to induce α-synuclein aggregation/LB formation inside dopaminergic SN neurons. Rotenone causes PD development in rats and mice. Rotenone induced PD in mice was followed by ROS and α-synuclein aggregate formation [78].

6.4. 6-OHDA

6-OHDA is a catecholaminergic neurotoxin which does not pass the BBB readily and needs to be injected into the brain, into, or close to, the SN. If systemic, it damages the peripheral nervous system. 6-OHDA is readily oxidized into a toxic para-quinone which induces ROS production. 6-OHDA has been reported to be an inhibitor of mitochondrial electron chain Complex I and IV. It causes rapid cell death if reaching the SN in several animal species including mice, rats and monkeys. It is indifferent to age and there is no evidence of NM binding. The model can be useful for PD therapies.

There are also other less used PD-models. In rodents, permethrin (inhibitor of mitochondrial complex I) induced PD symptoms (striatal α-synuclein aggregation) already in young rats [79]. Another model involves administration of the DA oxidation product aminochrome, which after brain injection induced PD symptoms in rats [80].

7. Known NM-Binding Metals and Chemicals

Due to its negative charge having catecholate groups, NM attracts (and binds) positively charged metal ions and substances with cationic properties such as basic organic amines (e.g., MPTP that metabolizes/oxidizes into MPP^+).

7.1. Interaction of NM in Human Brain with Metals

NM has high affinity for physiological heavy metals like iron, copper and zinc [3,45,81–83]. In particular, iron and copper have been suggested to be involved in NM synthesis, due to their redox activity. Iron is abundant in SN and the content of iron in NM isolated from SN of normal subjects is as high as 11.0 μg/mg of NM [3,9]. Both iron and copper appear to be involved in NM synthesis in LC and significant amount of these metals are present in NM isolated from this area (iron 1.8 μg/mg and copper 0.6 μg/mg of pigment). Iron is present as high spin iron (III) centers bound to oxygen atoms in octahedral configuration with melanic component of NM and is partially coordinated by catecholate groups of the dihydroxyindole residues, as shown by X-ray Absorption Spectroscopy and Electron Paramagnetic Resonance Spectroscopy [84–86]. The major fraction of bound iron is also in the iron (III) state but associated in oxy-hydroxy clusters as demonstrated by Mössbauer spectroscopy [83]. The iron fraction bound to catechol groups seems to be present in low affinity sites and can be removed by treatment with chelators [84,87,88], while iron bound in oxy-hydroxy clusters is likely in higher affinity sites. In any case, iron is effectively bound to NM in an inactive form that prevents Fenton chemistry and ascorbate oxidation [88]. NM is only partially saturated with iron in SN and has residual chelating ability for iron [86,87,89]. NM appears to be the main iron compound in dopaminergic neurons of the SN, since ferritin expression is very low in these neurons [9,83]. In conditions of iron overload, like those in PD [90–92], the binding sites of NM could be saturated and an increased content of redox-active iron has been found in NM of SN of PD patients [93,94]. Zinc is also found in NM of SN although at lower concentrations than iron [3,45,83].

Other metals that have been found into NM include aluminum, lead, manganese, mercury, chromium, molybdenum and cadmium [3,45,83,95]. Except for manganese, these metals derive mainly from environmental exposure. Higher levels of aluminum were found in NM of PD patients compared to controls [96]. Enormous accumulation of lead (~1400-fold with respect to the tissue concentration) has been observed in the NM of SN [3] and mercury (~2550-fold with respect to the tissue concentration) in NM isolated from cerebellum [3,45,83]. This high affinity of NM for cadmium, mercury and lead is due to the presence of high amount of sulfur in NM pigment [97], that is 2.69% in benzothiazine and 0.46% in cysteine/cystine moieties, and it is known that the heavy metals preferably bind to sulfur [98]. Therefore, NM clearly shows the specific ability to differentially accumulate metals in brain regions. This accumulation of metals in NM could be also the consequence of high influx into specific neurons, due to occupational and environmental exposure. The occupational exposure to lead has been associated with a higher risk of PD [99]. These observations suggest that NM play a neuroprotective role because of its ability to immobilize toxic metals forming stable complexes.

7.2. *Confirmed and Presumed NM-Binding Organic Xenobiotics*

Several studies have shown that multiple chemicals and drugs bind to melanins (various types) in vitro and are retained in pigmented cells. However, few studies have measured and confirmed xenobiotics on NM in human brains. In addition to the NM-binding parkinsonian neurotoxins MPTP and paraquat (covered above), examples of other NM-binding xenobiotics are:

7.2.1. β-Carbolines

β-Carbolines are a type of indole alkaloids present in some plant and processed foods that distribute into the brain where they can (depending on structure) act protectively as antioxidants [100]. Some β-carbolines, however, show structural resemblance to MPTP/MPP^+ [101]. β-carboline (norharmane) and harmane bound to NM in brains of frogs but not in mice [101]. Detectable in most human brains analyzed, harman and norharman as well as the norharman derivatives 2-methyl-norharmanium ion (2-MeNH) and 2,9-dimethyl-norharmanium ion (2,9-Me_2NH) were higher in SN than in parietal association cortex [102]. After stereotaxical injections of various β-carbolines and MPP^+ separately into the SN of adult male rats, some β-carbolines were nearly as effective as MPP^+ in lowering striatal DA levels and causing lesions [68].

7.2.2. Quinolines

Isoquinolines are naturally present in high amounts in food and distribute into brain where they generally act protectively. However, some tetrahydroisoquinoline derivatives, e.g., 1,2,3,4-tetrahydroisoquinoline (TIQ), 1-benzyl-TIQ and 1-methyl-5,6-dihydroxy-TIQ (salsolinol) structurally resemble MPTP and some of them could be responsible for inducing PD [103]. The antimalarial quinolines chloroquine, hydroxychloroquine, and quinacrine show strong affinity for isolated natural NM from human SN [77], and chloroquine pretreatment protects from MPTP toxicity in monkeys [104].

7.2.3. Chlorpromazine

Following incubations of human brain slices with the antipsychotic drug ^{35}S-chlorpromazine in vitro, it accumulated on NM in human SN and LC [105].

7.2.4. Imipramine

Binding of the antidepressant drug imipramine to human NM has been described [106].

7.2.5. ^{18}F-AV-1451

Even though not a typical xenobiotic but a positron emission tomography tau tangle ligand, ^{18}F-AV-1451 was found to strongly bind to NM in human midbrain SN [107–109]. However, to what structure ^{18}F-AV-1451 specifically binds is presently unclear.

7.2.6. Haloperidol

The antipsychotic drug haloperidol can cause extrapyramidal symptoms (EPS) such as tardive dyskinesia in humans. Haloperidol accumulated in melanin containing tissues in mice [110] and bound to synthetic DA melanin in vitro [110]. After short (30 min) [^{3}H]haloperidol incubation with either human brain SN or superior cerebellar peduncle homogenates in vitro, centrifugation revealed that about twice as much haloperidol bound to pigmented bands (containing NM) from SN than non-pigmented bands from superior cerebellar peduncle (similar results were obtained for [^{3}H]imipramine and [^{3}H]chlorpromazine) [106]. Haloperidol can undergo biotransformation into 4-(4-

chlorophenyl)-1-[4-(4-fluorophenyl)-4-oxobutyl]pyridinium ion (HPP^+, resembling MPP^+) which bound to synthetic melanin in a reversible manner [111]. HPP^+ damaged rat embryonic primary cultured dopaminergic neurons [111], but which does not demonstrate NM involvement of toxicity (young cultured cells are considered not to have NM)). HPP^+ was moderately taken up into brains of rats [112].

7.2.7. Nicotine

Several chemicals interact/bind to NM (often synthetic) in vitro. Examples include nicotine [52,113] which also was neuroprotective in rats and monkeys when administrated before, not after, 6-OHDA [114].

7.2.8. L-BMAA

One natural environmental neurotoxin strongly suspected of causing a number of neurodegenerative diseases including AD, amyotrophic lateral sclerosis and PD is β-methylamino-L-alanine (L-BMAA) [115]. L-BMAA is thought to be produced by all known groups of cyanobacteria (often called blue-green algae) [116] that are ubiquitously present in various aquatic environments. Structurally, L-BMAA resembles a number of natural amino acids (particularly L-serine) and incorporates into proteins [117–119] causing protein misfolding and aggregation [117]. L-BMAA can also disturb amino-acid mediated (e.g., glutamate) receptor signaling. L-BMAA was a developmental neurotoxicant in rodents at low doses after single administration [120]. Interestingly, a recent long-term L-BMAA exposure study in adult monkeys [121] resulted in AD-like pathology including the neurofibrillary tangles (NFT; aggregates of hyper-phosphorylated tau proteins) and amyloid-beta (Aβ) plaques that were also found in Guam patients (after World War II), who also had L-BMAA in their brains and hair. However, all repeated low-dose L-BMAA studies using adult mice are negative [122] (no neurotoxic effect observed), which could be due to the fact that L-BMAA acts on brain NM that is not present in mice [123]. Recent studies indicate that L-BMAA concentrates in NM-rich neurons [124,125]. ^{3}H-BMAA was retained in NM-containing neurons of frogs, and ^{3}H-BMAA bound to Sepia melanin and interacted with synthesis of melanin in vitro [124]. When bicarbonate ions are present (bicarbonate is a major physiological buffer in our bodies), carbamate adducts form with L-BMAA that can chelate divalent metals such as iron, manganese and zinc [126], which could implicate that L-BMAA-carbamate may interfere with NM's metal-chelating ability [75]. However, being hydrophilic, L-BMAA can likely not reach already incapsulated NM residing inside double membranes.

7.2.9. Lipophilic Xenobiotics (Various)

The lipid bodies of NM organelles in humans supposedly contain various dissolved lipophilic xenobiotics, but few studies exist. It has been suggested that non-covalently NM bound xenobiotics may slowly diffuse (e.g., along fatty membranes) both in and out of NM organelles. Bases ($pK_a > 7$) have high volume of distribution due to base (e.g., R_3N^+) interactions with acidic membrane phospholipids ($R\text{-}PO_4^-$) and may bind NM. Furthermore, positively charged organometals (e.g., mercury- or tin-based) can be suspected of binding NM. Bioaccumulating substances with long elimination half-lives in brain are of particular interest [127,128], and may interact with NM.

8. Promising Experimental Models to Identify NM Binders/Modulators

8.1. In Chemico NM Synthesis

Synthetic model NM are typically prepared by oxidative polymerization of DA and/or CysDA [46], generally using iron ions as promoters of DA oxidation [129]. In other cases, the melanins are produced by enzymatic oxidation [97]. However, as it was pointed out in Section 2, these DA polymers can be considered as models of peripheral melanins but lack essential components characterizing brain NMs, i.e., covalently bound proteins and lipids [4,7,8]. A better approach to model NM takes into account that the biosynthetic pathway of NM involves the non-enzymatic reaction of oxidized DA with fibrillar protein seeds, and generates a melanin-protein conjugate with an unstructured melanic component [27,130]. The lipid portion of NM has not been included in these synthetic NM models because it will severely depress the solubility of the conjugates making their characterization much more difficult. The main structural element of DA/cysteine melanins is the extended π-stacking of aromatic sheets characterizing the eumelanic component of the polymer, whereas in the conjugates containing melanin bound to fibrillar proteins, as well in natural NMs, this arrangement is completely missing and the melanic component is unstructured [8,27,130]. The structural difference between synthetic DA/cysteine melanins and melanin-protein conjugates is important for studies aiming at simulating the interaction between NM and xenobiotics because both the affinity and mode of interaction will be different. This problem should be taken into account when the NM-xenobiotic interaction is modeled using DA/cysteine melanins [52]. The peculiar biosynthetic pathway of NM is important also because it is not enzymatically controlled and in certain situations the rate of NM formation may get overwhelmed and xenobiotics could enter and disrupt any step of NM synthesis [22].

In chemico and in vitro studies can possibly identify potential NM-disturbing toxicants. Xenobiotics can be added during chemical synthesis of NM analogues obtained by oxidation of DA by iron (III) in the presence of cysteine and aggregated proteins, in the test tube, to assess whether NM synthesis is halted and/or its structure/function is altered (e.g., its metal binding capacity and redox properties). Xenobiotic-NM binding studies are also possible, either using synthetic or natural (isolated from human brains) NM.

8.2. NM Binding In Vitro

Magnetic beads [131] and affinity chromatography [132,133] can be tools for studying the interactions of small molecules and metal ions with various types of melanin.

8.3. Stem Cell Derived Human 3D Mid-Brain Models

Recent in vitro differentiation protocols of pluripotent stem cells (PSCs) allows generation of human midbrain-like organoids (hMLOs) that have NM-like granules similar to NM organelles found in human brain SN [134–136]. Midbrain organoids have been claimed suitable for neurotoxin-based PD disease modeling [74]. The hMLOs secrete DA and neurons within the hMLOs form functional synapses. In 2D cultures, a NM-containing SN does not form as occurs in human SN and in hMLOs. Similar models may exist for NM studies in LC (LC is not present in the mid-brain).

Neurons in the organoids form a heterogeneous network surrounded by supportive cells, i.e., oligodendrocytes, astrocytes and ependymal cells like inside the human brain. NM-like organelles were clearly visible in the hMLOs after 2 months and the NM amount gradually increased up to 4 months of cultivation [74,134]. It is presently unclear how long the hMLOs can be kept in culture. In human brains, many post-mitotic neurons survive the entire life span. Organoids shall not be allowed to grow large since they become anoxic in the middle.

8.4. Neurodegeneration and Cognitive Effects in Cats, Goats and Sheep

More suitable non-rodent model species can possibly be found in monkeys, dogs, cats, sheep and goats that are more similar to humans in size, genetics and physiology, for which there are often cognitive tests available. MPTP administration lowered striatal DA levels and made cats parkinsonian when studied using behavioral/motor function [67]. However, performing toxicological testing in house pets such as cats and dogs, as well as monkeys, is ethically challenging, and monkey facilities are very few. Cats apparently also recover better than humans after MPTP exposure. Contrary to rodents that do not develop signs of AD and PD as they age, sheep (like humans) develop both Aβ-plaques and tau tangles naturally as they age [137,138], and aged sheep (like humans and primates) also develop PD-symptoms after MPTP exposure [65]. In sheep, MPTP administration selectively induces necrosis in both SN and LC [65], regions high in NM. Thus, sheep could be a suitable species in which to study age-related effects by toxicants on NM functioning. Several cognitive tests, e.g., facial recognition [139,140] for testing of learning/memory, are available for sheep. Compared to rodents, sheep are closer to humans in terms of genetics, physiology and size [141], have a more similar brain structure and organization [142], and have a relative long-life span. Studying NM effects in middle aged (7–10 years) sheep may be ideal [65]. Goats can be another alternative since goats may have more NM than sheep [58], and also develop signs of brain aging [143]. Goats are more independent than flocking sheep. However, the availability of suitable behavioral tests for goats is less certain.

Funding: This research was funded by the Italian Ministry of Education, University, and Research (MIUR)—Research Projects of National Interest (PRIN) 2015 prot. 2015T778JW (A.C., E.M., L.C.).

Data Availability Statement: Data is contained within the article.

Conflicts of Interest: The authors declare no conflict of interest.

References

1. Sulzer, D.; Mosharov, E.; Talloczy, Z.; Zucca, F.A.; Simon, J.D.; Zecca, L. Neuronal pigmented autophagic vacuoles: Lipofuscin, neuromelanin, and ceroid as macroautophagic responses during aging and disease. *J. Neurochem.* **2008**, *106*, 24–36. [CrossRef] [PubMed]
2. Vila, M. Neuromelanin, aging, and neuronal vulnerability in Parkinson's disease. *Mov. Disord.* **2019**, *34*, 1440–1451. [CrossRef] [PubMed]
3. Zecca, L.; Bellei, C.; Costi, P.; Albertini, A.; Monzani, E.; Casella, L.; Gallorini, M.; Bergamaschi, L.; Moscatelli, A.; Turro, N.J.; et al. New melanic pigments in the human brain that accumulate in aging and block environmental toxic metals. *Proc. Natl. Acad. Sci. USA* **2008**, *105*, 17567–17572. [CrossRef]
4. Engelen, M.; Vanna, R.; Bellei, C.; Zucca, F.A.; Wakamatsu, K.; Monzani, E.; Ito, S.; Casella, L.; Zecca, L. Neuromelanins of human brain have soluble and insoluble components with dolichols attached to the melanic structure. *PLoS ONE* **2012**, *7*, e48490. [CrossRef] [PubMed]
5. Zarow, C.; Lyness, S.A.; Mortimer, J.A.; Chui, H.C. Neuronal loss is greater in the locus coeruleus than nucleus basalis and substantia nigra in Alzheimer and Parkinson diseases. *Arch. Neurol.* **2003**, *60*, 337–341. [CrossRef] [PubMed]
6. Fu, H.; Hardy, J.; Duff, K.E. Selective vulnerability in neurodegenerative diseases. *Nat. Neurosci.* **2018**, *21*, 1350–1358. [CrossRef] [PubMed]
7. Zucca, F.A.; Basso, E.; Cupaioli, F.A.; Ferrari, E.; Sulzer, D.; Casella, L.; Zecca, L. Neuromelanin of the human substantia nigra: An update. *Neurotox. Res.* **2014**, *25*, 13–23. [CrossRef]
8. Zucca, F.A.; Segura-Aguilar, J.; Ferrari, E.; Munoz, P.; Paris, I.; Sulzer, D.; Sarna, T.; Casella, L.; Zecca, L. Interactions of iron, dopamine and neuromelanin pathways in brain aging and Parkinson's disease. *Prog. Neurobiol.* **2017**, *155*, 96–119. [CrossRef]
9. Zecca, L.; Stroppolo, A.; Gatti, A.; Tampellini, D.; Toscani, M.; Gallorini, M.; Giaveri, G.; Arosio, P.; Santambrogio, P.; Fariello, R.G.; et al. The role of iron and copper molecules in the neuronal vulnerability of locus coeruleus and substantia nigra during aging. *Proc. Natl. Acad. Sci. USA* **2004**, *101*, 9843–9848. [CrossRef]
10. Sulzer, D.; Bogulavsky, J.; Larsen, K.E.; Behr, G.; Karatekin, E.; Kleinman, M.H.; Turro, N.; Krantz, D.; Edwards, R.H.; Greene, L.A.; et al. Neuromelanin biosynthesis is driven by excess cytosolic catecholamines not accumulated by synaptic vesicles. *Proc. Natl. Acad. Sci. USA* **2000**, *97*, 11869–11874. [CrossRef]
11. Zhang, W.; Phillips, K.; Wielgus, A.R.; Liu, J.; Albertini, A.; Zucca, F.A.; Faust, R.; Qian, S.Y.; Miller, D.S.; Chignell, C.F.; et al. Neuromelanin activates microglia and induces degeneration of dopaminergic neurons: Implications for progression of Parkinson's disease. *Neurotox. Res.* **2011**, *19*, 63–72. [CrossRef]
12. Zecca, L.; Fariello, R.; Riederer, P.; Sulzer, D.; Gatti, A.; Tampellini, D. The absolute concentration of nigral neuromelanin, assayed by a new sensitive method, increases throughout the life and is dramatically decreased in Parkinson's disease. *FEBS Lett.* **2002**, *510*, 216–220. [CrossRef]
13. Fedorow, H.; Tribl, F.; Halliday, G.; Gerlach, M.; Riederer, P.; Double, K.L. Neuromelanin in human dopamine neurons: Comparison with peripheral melanins and relevance to Parkinson's disease. *Prog. Neurobiol.* **2005**, *75*, 109–124. [CrossRef]
14. Zucca, F.A.; Vanna, R.; Cupaioli, F.A.; Bellei, C.; De Palma, A.; Di Silvestre, D.; Mauri, P.; Grassi, S.; Prinetti, A.; Casella, L.; et al. Neuromelanin organelles are specialized autolysosomes that accumulate undegraded proteins and lipids in aging human brain and are likely involved in Parkinson's disease. *NPJ Park. Dis.* **2018**, *4*, 17. [CrossRef]
15. Kanteev, M.; Goldfeder, M.; Fishman, A. Structure-function correlations in tyrosinases. *Protein Sci.* **2015**, *24*, 1360–1369. [CrossRef]
16. Lai, X.; Wichers, H.J.; Soler-Lopez, M.; Dijkstra, B.W. Structure of Human Tyrosinase Related Protein 1 Reveals a Binuclear Zinc Active Site Important for Melanogenesis. *Angew. Chem. Int. Ed. Engl.* **2017**, *56*, 9812–9815. [CrossRef]
17. Sugumaran, M.; Barek, H. Critical Analysis of the Melanogenic Pathway in Insects and Higher Animals. *Int. J. Mol. Sci.* **2016**, *17*, 1753. [CrossRef]

18. Arzillo, M.; Mangiapia, G.; Pezzella, A.; Heenan, R.K.; Radulescu, A.; Paduano, L.; d'Ischia, M. Eumelanin buildup on the nanoscale: Aggregate growth/assembly and visible absorption development in biomimetic 5,6-dihydroxyindole polymerization. *Biomacromolecules* **2012**, *13*, 2379–2390. [CrossRef]
19. d'Ischia, M.; Wakamatsu, K.; Cicoira, F.; Di Mauro, E.; Garcia-Borron, J.C.; Commo, S.; Galvan, I.; Ghanem, G.; Kenzo, K.; Meredith, P.; et al. Melanins and melanogenesis: From pigment cells to human health and technological applications. *Pigment. Cell Melanoma Res.* **2015**, *28*, 520–544. [CrossRef]
20. Wakamatsu, K.; Fujikawa, K.; Zucca, F.A.; Zecca, L.; Ito, S. The structure of neuromelanin as studied by chemical degradative methods. *J. Neurochem.* **2003**, *86*, 1015–1023. [CrossRef]
21. Bush, W.D.; Garguilo, J.; Zucca, F.A.; Albertini, A.; Zecca, L.; Edwards, G.S.; Nemanich, R.J.; Simon, J.D. The surface oxidation potential of human neuromelanin reveals a spherical architecture with a pheomelanin core and a eumelanin surface. *Proc. Natl. Acad. Sci. USA* **2006**, *103*, 14785–14789. [CrossRef] [PubMed]
22. Monzani, E.; Nicolis, S.; Dell'Acqua, S.; Capucciati, A.; Bacchella, C.; Zucca, F.A.; Mosharov, E.V.; Sulzer, D.; Zecca, L.; Casella, L. Dopamine, Oxidative Stress and Protein-Quinone Modifications in Parkinson's and Other Neurodegenerative Diseases. *Angew. Chem. Int. Ed. Engl.* **2019**, *58*, 6512–6527. [CrossRef] [PubMed]
23. Trist, B.G.; Hare, D.J.; Double, K.L. Oxidative stress in the aging substantia nigra and the etiology of Parkinson's disease. *Aging Cell* **2019**, *18*, e13031. [CrossRef] [PubMed]
24. Burbulla, L.F.; Song, P.; Mazzulli, J.R.; Zampese, E.; Wong, Y.C.; Jeon, S.; Santos, D.P.; Blanz, J.; Obermaier, C.D.; Strojny, C.; et al. Dopamine oxidation mediates mitochondrial and lysosomal dysfunction in Parkinson's disease. *Science* **2017**, *357*, 1255–1261. [CrossRef]
25. Tribl, F.; Arzberger, T.; Riederer, P.; Gerlach, M. Tyrosinase is not detected in human catecholaminergic neurons by immunohistochemistry and Western blot analysis. *J. Neural Transm. Suppl.* **2007**, 51–55. [CrossRef]
26. Masato, A.; Plotegher, N.; Boassa, D.; Bubacco, L. Impaired dopamine metabolism in Parkinson's disease pathogenesis. *Mol. Neurodegener.* **2019**, *14*, 35. [CrossRef]
27. Ferrari, E.; Capucciati, A.; Prada, I.; Zucca, F.A.; D'Arrigo, G.; Pontiroli, D.; Bridelli, M.G.; Sturini, M.; Bubacco, L.; Monzani, E.; et al. Synthesis, Structure Characterization, and Evaluation in Microglia Cultures of Neuromelanin Analogues Suitable for Modeling Parkinson's Disease. *ACS Chem. Neurosci.* **2017**, *8*, 501–512. [CrossRef]
28. Makin, O.S.; Serpell, L.C. Structures for amyloid fibrils. *FEBS J.* **2005**, *272*, 5950–5961. [CrossRef]
29. Meredith, P.; Sarna, T. The physical and chemical properties of eumelanin. *Pigment. Cell Res.* **2006**, *19*, 572–594. [CrossRef]
30. d'Ischia, M.; Napolitano, A.; Ball, V.; Chen, C.T.; Buehler, M.J. Polydopamine and eumelanin: From structure-property relationships to a unified tailoring strategy. *Acc. Chem. Res.* **2014**, *47*, 3541–3550. [CrossRef]
31. McGeer, P.L.; Itagaki, S.; Boyes, B.E.; McGeer, E.G. Reactive microglia are positive for HLA-DR in the substantia nigra of Parkinson's and Alzheimer's disease brains. *Neurology* **1988**, *38*, 1285–1291. [CrossRef]
32. Hirsch, E.; Graybiel, A.M.; Agid, Y.A. Melanized dopaminergic neurons are differentially susceptible to degeneration in Parkinson's disease. *Nature* **1988**, *334*, 345–348. [CrossRef]
33. Imamura, K.; Hishikawa, N.; Sawada, M.; Nagatsu, T.; Yoshida, M.; Hashizume, Y. Distribution of major histocompatibility complex class II-positive microglia and cytokine profile of Parkinson's disease brains. *Acta Neuropathol.* **2003**, *106*, 518–526. [CrossRef]
34. Langston, J.W.; Forno, L.S.; Tetrud, J.; Reeves, A.G.; Kaplan, J.A.; Karluk, D. Evidence of active nerve cell degeneration in the substantia nigra of humans years after 1-methyl-4-phenyl-1,2,3,6-tetrahydropyridine exposure. *Ann. Neurol.* **1999**, *46*, 598–605. [CrossRef]
35. Wilms, H.; Rosenstiel, P.; Sievers, J.; Deuschl, G.; Zecca, L.; Lucius, R. Activation of microglia by human neuromelanin is NF-kappaB dependent and involves p38 mitogen-activated protein kinase: Implications for Parkinson's disease. *FASEB J.* **2003**, *17*, 500–502. [CrossRef]
36. Block, M.L.; Zecca, L.; Hong, J.S. Microglia-mediated neurotoxicity: Uncovering the molecular mechanisms. *Nat. Rev. Neurosci.* **2007**, *8*, 57–69. [CrossRef]
37. Zecca, L.; Wilms, H.; Geick, S.; Claasen, J.H.; Brandenburg, L.O.; Holzknecht, C.; Panizza, M.L.; Zucca, F.A.; Deuschl, G.; Sievers, J.; et al. Human neuromelanin induces neuroinflammation and neurodegeneration in the rat substantia nigra: Implications for Parkinson's disease. *Acta Neuropathol.* **2008**, *116*, 47–55. [CrossRef]
38. Tousi, N.S.; Buck, D.J.; Zecca, L.; Davis, R.L. Neuromelanin inhibits CXCL10 expression in human astroglial cells. *Neurosci. Lett.* **2010**, *486*, 47–50. [CrossRef]
39. Cebrian, C.; Zucca, F.A.; Mauri, P.; Steinbeck, J.A.; Studer, L.; Scherzer, C.R.; Kanter, E.; Budhu, S.; Mandelbaum, J.; Vonsattel, J.P.; et al. MHC-I expression renders catecholaminergic neurons susceptible to T-cell-mediated degeneration. *Nat. Commun.* **2014**, *5*, 3633. [CrossRef]
40. Betts, M.J.; Kirilina, E.; Otaduy, M.C.G.; Ivanov, D.; Acosta-Cabronero, J.; Callaghan, M.F.; Lambert, C.; Cardenas-Blanco, A.; Pine, K.; Passamonti, L.; et al. Locus coeruleus imaging as a biomarker for noradrenergic dysfunction in neurodegenerative diseases. *Brain* **2019**, *142*, 2558–2571. [CrossRef]

41. Theofilas, P.; Ehrenberg, A.J.; Dunlop, S.; Di Lorenzo Alho, A.T.; Nguy, A.; Leite, R.E.P.; Rodriguez, R.D.; Mejia, M.B.; Suemoto, C.K.; Ferretti-Rebustini, R.E.L.; et al. Locus coeruleus volume and cell population changes during Alzheimer's disease progression: A stereological study in human postmortem brains with potential implication for early-stage biomarker discovery. *Alzheimers Dement.* **2017**, *13*, 236–246. [CrossRef]
42. Bjorklund, G.; Dadar, M.; Chirumbolo, S.; Aaseth, J. The Role of Xenobiotics and Trace Metals in Parkinson's Disease. *Mol. Neurobiol* **2020**, *57*, 1405–1417. [CrossRef]
43. Kalia, L.V.; Lang, A.E. Parkinson disease in 2015: Evolving basic, pathological and clinical concepts in PD. *Nat. Rev. Neurol.* **2016**, *12*, 65–66. [CrossRef]
44. Gorell, J.M.; Johnson, C.C.; Rybicki, B.A.; Peterson, E.L.; Kortsha, G.X.; Brown, G.G.; Richardson, R.J. Occupational exposure to manganese, copper, lead, iron, mercury and zinc and the risk of Parkinson's disease. *Neurotoxicology* **1999**, *20*, 239–247.
45. Zecca, L.; Pietra, R.; Goj, C.; Mecacci, C.; Radice, D.; Sabbioni, E. Iron and other metals in neuromelanin, substantia nigra, and putamen of human brain. *J. Neurochem.* **1994**, *62*, 1097–1101. [CrossRef]
46. Wilczok, T.; Stepien, K.; Dzierzega-Lecznar, A.; Zajdel, A.; Wilczok, A. Model neuromelanins as antioxidative agents during lipid peroxidation. *Neurotox. Res.* **1999**, *1*, 141–147. [CrossRef]
47. Lan, A.P.; Chen, J.; Chai, Z.F.; Hu, Y. The neurotoxicity of iron, copper and cobalt in Parkinson's disease through ROS-mediated mechanisms. *Biometals* **2016**, *29*, 665–678. [CrossRef]
48. Hofer, T. Oxidation of 2′-deoxyguanosine by H_2O_2-ascorbate: Evidence against free $OH^{\bullet}$ and thermodynamic support for two-electron reduction of H_2O_2. *J. Chem. Soc. Perkin Trans. 2* **2001**, 210–213. [CrossRef]
49. Keeney, P.M.; Xie, J.; Capaldi, R.A.; Bennett, J.P., Jr. Parkinson's disease brain mitochondrial complex I has oxidatively damaged subunits and is functionally impaired and misassembled. *J. Neurosci.* **2006**, *26*, 5256–5264. [CrossRef]
50. Trist, B.G.; Hilton, J.B.; Hare, D.J.; Crouch, P.J.; Double, K.L. Superoxide Dismutase 1 in Health and Disease: How a Frontline Antioxidant Becomes Neurotoxic. *Angew. Chem. Int. Ed. Engl.* **2021**, *60*, 9215–9246. [CrossRef]
51. Segura, A.; Sulzer, D.; Zucca, F.A.; Zecca, L. Overexpression of Vesicular Monoamine Transporter-2 may Block Neurotoxic Metabolites from Cytosolic Dopamine: A Potential Neuroprotective Therapy for Parkinson's Disease. *Clin. Pharmacol. Transl. Med.* **2019**, *3*, 143–148. [PubMed]
52. Fink, J.; Pathak, H.; Smith, J.; Achat-Mendes, C.; Haining, R.L. Development of a Competition-Binding Assay to Determine Binding Affinity of Molecules to Neuromelanin via Fluorescence Spectroscopy. *Biomolecules* **2019**, *9*, 175. [CrossRef] [PubMed]
53. Glinka, Y.; Gassen, M.; Youdim, M.B. Mechanism of 6-hydroxydopamine neurotoxicity. *J. Neural Transm. Suppl.* **1997**, *50*, 55–66. [CrossRef] [PubMed]
54. Kin, K.; Yasuhara, T.; Kameda, M.; Date, I. Animal Models for Parkinson's Disease Research: Trends in the 2000s. *Int. J. Mol. Sci.* **2019**, *20*, 5402. [CrossRef]
55. Marsden, C.D. Pigmentation in the nucleus substantiae nigrae of mammals. *J. Anat.* **1961**, *95*, 256–261.
56. DeMattei, M.; Levi, A.C.; Fariello, R.G. Neuromelanic pigment in substantia nigra neurons of rats and dogs. *Neurosci. Lett.* **1986**, *72*, 37–42. [CrossRef]
57. Uchida, K.; Kihara, N.; Hashimoto, K.; Nakayama, H.; Yamaguchi, R.; Tateyama, S. Age-related histological changes in the canine substantia nigra. *J. Vet. Med. Sci.* **2003**, *65*, 179–185. [CrossRef]
58. Merighi, A.; Peirone, S.M. Histochemical and ultrastructural features of neuronal pigment in some encephalic nuclei of ruminants. *Exp. Biol.* **1985**, *44*, 109–121.
59. Bianchi, M.; Merighi, A. The relationship between lipofuscin and neuromelanin in some sites of the nervous system of the horse. *Exp. Biol.* **1986**, *46*, 89–99.
60. Cozzi, B.; Pellegrini, M.; Droghi, A. Neuromelanin in the substantia nigra of adult horses. *Anat. Anz.* **1988**, *166*, 53–61.
61. Nielsen, M.S.; Sorensen, J.C.; Bjarkam, C.R. The substantia nigra pars compacta of the Gottingen minipig: An anatomical and stereological study. *Brain Struct. Funct.* **2009**, *213*, 481–488. [CrossRef]
62. Marsden, C.D. Neuromelanin and Parkinson's disease. *J. Neural Transm. Suppl.* **1983**, *19*, 121–141.
63. D'Amato, R.J.; Lipman, Z.P.; Snyder, S.H. Selectivity of the parkinsonian neurotoxin MPTP: Toxic metabolite MPP+ binds to neuromelanin. *Science* **1986**, *231*, 987–989. [CrossRef]
64. Levi, A.C.; DeMattei, M.; Ravazzani, R.; Corvetti, G.; Golden, G.T.; Fariello, R.G. Effects of 1-methyl-4-phenyl-1,2,3,6-tetrahydropyridine (MPTP) on ultrastructure of nigral neuromelanin in Macaca fascicularis. *Neurosci. Lett.* **1989**, *96*, 271–276. [CrossRef]
65. Beale, A.M.; Higgins, R.J.; Work, T.M.; Bailey, C.S.; Smith, M.O.; Shinka, T.; Hammock, B.D. MPTP-induced Parkinson-like disease in sheep: Clinical and pathologic findings. *J. Environ. Pathol. Toxicol. Oncol.* **1989**, *9*, 417–428.
66. Mikkelsen, M.; Moller, A.; Jensen, L.H.; Pedersen, A.; Harajehi, J.B.; Pakkenberg, H. MPTP-induced Parkinsonism in minipigs: A behavioral, biochemical, and histological study. *Neurotoxicol. Teratol.* **1999**, *21*, 169–175. [CrossRef]
67. Schneider, J.S. MPTP-induced parkinsonism: Acceleration of biochemical and behavioral recovery by GM1 ganglioside treatment. *J. Neurosci. Res.* **1992**, *31*, 112–119. [CrossRef]
68. Neafsey, E.J.; Albores, R.; Gearhart, D.; Kindel, G.; Raikoff, K.; Tamayo, F.; Collins, M.A. Methyl-beta-carbolinium analogs of MPP+ cause nigrostriatal toxicity after substantia nigra injections in rats. *Brain Res.* **1995**, *675*, 279–288. [CrossRef]
69. Boyce, S.; Kelly, E.; Reavill, C.; Jenner, P.; Marsden, C.D. Repeated administration of N-methyl-4-phenyl 1,2,5,6-tetrahydropyridine to rats is not toxic striatal dopamine neurones. *Biochem. Pharmacol.* **1984**, *33*, 1747–1752. [CrossRef]

70. Chiueh, C.C.; Markey, S.P.; Burns, R.S.; Johannessen, J.N.; Pert, A.; Kopin, I.J. Neurochemical and behavioral effects of systemic and intranigral administration of N-methyl-4-phenyl-1,2,3,6-tetrahydropyridine in the rat. *Eur. J. Pharmacol.* **1984**, *100*, 189–194. [CrossRef]
71. Munoz-Manchado, A.B.; Villadiego, J.; Romo-Madero, S.; Suarez-Luna, N.; Bermejo-Navas, A.; Rodriguez-Gomez, J.A.; Garrido-Gil, P.; Labandeira-Garcia, J.L.; Echevarria, M.; Lopez-Barneo, J.; et al. Chronic and progressive Parkinson's disease MPTP model in adult and aged mice. *J. Neurochem.* **2016**, *136*, 373–387. [CrossRef] [PubMed]
72. Date, I.; Felten, D.L.; Felten, S.Y. Long-term effect of MPTP in the mouse brain in relation to aging: Neurochemical and immunocytochemical analysis. *Brain Res.* **1990**, *519*, 266–276. [CrossRef]
73. Huang, B.; Wu, S.; Wang, Z.; Ge, L.; Rizak, J.D.; Wu, J.; Li, J.; Xu, L.; Lv, L.; Yin, Y.; et al. Phosphorylated alpha-Synuclein Accumulations and Lewy Body-like Pathology Distributed in Parkinson's Disease-Related Brain Areas of Aged Rhesus Monkeys Treated with MPTP. *Neuroscience* **2018**, *379*, 302–315. [CrossRef] [PubMed]
74. Kwak, T.H.; Kang, J.H.; Hali, S.; Kim, J.; Kim, K.P.; Park, C.; Lee, J.H.; Ryu, H.K.; Na, J.E.; Jo, J.; et al. Generation of homogeneous midbrain organoids with in vivo-like cellular composition facilitates neurotoxin-based Parkinson's disease modeling. *Stem Cells* **2020**. [CrossRef] [PubMed]
75. Bjorklund, G.; Hofer, T.; Nurchi, V.M.; Aaseth, J. Iron and other metals in the pathogenesis of Parkinson's disease: Toxic effects and possible detoxification. *J. Inorg. Biochem.* **2019**, *199*, 110717. [CrossRef]
76. Lindquist, N.G.; Larsson, B.S.; Lyden-Sokolowski, A. Autoradiography of [14C]paraquat or [14C]diquat in frogs and mice: Accumulation in neuromelanin. *Neurosci. Lett.* **1988**, *93*, 1–6. [CrossRef]
77. D'Amato, R.J.; Benham, D.F.; Snyder, S.H. Characterization of the binding of N-methyl-4-phenylpyridine, the toxic metabolite of the parkinsonian neurotoxin N-methyl-4-phenyl-1,2,3,6-tetrahydropyridine, to neuromelanin. *J. Neurochem.* **1987**, *48*, 653–658. [CrossRef]
78. Inden, M.; Kitamura, Y.; Takeuchi, H.; Yanagida, T.; Takata, K.; Kobayashi, Y.; Taniguchi, T.; Yoshimoto, K.; Kaneko, M.; Okuma, Y.; et al. Neurodegeneration of mouse nigrostriatal dopaminergic system induced by repeated oral administration of rotenone is prevented by 4-phenylbutyrate, a chemical chaperone. *J. Neurochem.* **2007**, *101*, 1491–1504. [CrossRef]
79. Nasuti, C.; Brunori, G.; Eusepi, P.; Marinelli, L.; Ciccocioppo, R.; Gabbianelli, R. Early life exposure to permethrin: A progressive animal model of Parkinson's disease. *J. Pharmacol. Toxicol. Methods* **2017**, *83*, 80–86. [CrossRef]
80. Herrera, A.; Munoz, P.; Paris, I.; Diaz-Veliz, G.; Mora, S.; Inzunza, J.; Hultenby, K.; Cardenas, C.; Jana, F.; Raisman-Vozari, R.; et al. Aminochrome induces dopaminergic neuronal dysfunction: A new animal model for Parkinson's disease. *Cell Mol. Life Sci.* **2016**, *73*, 3583–3597. [CrossRef]
81. Biesemeier, A.; Eibl, O.; Eswara, S.; Audinot, J.N.; Wirtz, T.; Pezzoli, G.; Zucca, F.A.; Zecca, L.; Schraermeyer, U. Elemental mapping of Neuromelanin organelles of human Substantia Nigra: Correlative ultrastructural and chemical analysis by analytical transmission electron microscopy and nano-secondary ion mass spectrometry. *J. Neurochem.* **2016**, *138*, 339–353. [CrossRef]
82. Brooks, J.; Everett, J.; Lermyte, F.; Tjhin, V.T.; Banerjee, S.; O'Connor, P.B.; Morris, C.M.; Sadler, P.J.; Telling, N.D.; Collingwood, J.F. Label-Free Nanoimaging of Neuromelanin in the Brain by Soft X-ray Spectromicroscopy. *Angew. Chem. Int. Ed. Engl.* **2020**, *59*, 11984–11991. [CrossRef]
83. Zecca, L.; Gallorini, M.; Schunemann, V.; Trautwein, A.X.; Gerlach, M.; Riederer, P.; Vezzoni, P.; Tampellini, D. Iron, neuromelanin and ferritin content in the substantia nigra of normal subjects at different ages: Consequences for iron storage and neurodegenerative processes. *J. Neurochem.* **2001**, *76*, 1766–1773. [CrossRef]
84. Aime, S.; Bergamasco, B.; Biglino, D.; Digilio, G.; Fasano, M.; Giamello, E.; Lopiano, L. EPR investigations of the iron domain in neuromelanin. *Biochim. Biophys. Acta (BBA) Mol. Basis Dis.* **1997**, *1361*, 49–58. [CrossRef]
85. Kropf, A.J.; Bunker, B.A.; Eisner, M.; Moss, S.C.; Zecca, L.; Stroppolo, A.; Crippa, P.R. X-ray absorption fine-structure spectroscopy studies of Fe sites in natural human neuromelanin and synthetic analogues. *Biophys. J.* **1998**, *75*, 3135–3142. [CrossRef]
86. Zecca, L.; Shima, T.; Stroppolo, A.; Goj, C.; Battiston, G.A.; Gerbasi, R.; Sarna, T.; Swartz, H.M. Interaction of neuromelanin and iron in substantia nigra and other areas of human brain. *Neuroscience* **1996**, *73*, 407–415. [CrossRef]
87. Shima, T.; Sarna, T.; Swartz, H.M.; Stroppolo, A.; Gerbasi, R.; Zecca, L. Binding of iron to neuromelanin of human substantia nigra and synthetic melanin: An electron paramagnetic resonance spectroscopy study. *Free Radic. Biol. Med.* **1997**, *23*, 110–119. [CrossRef]
88. Zecca, L.; Casella, L.; Albertini, A.; Bellei, C.; Zucca, F.A.; Engelen, M.; Zadlo, A.; Szewczyk, G.; Zareba, M.; Sarna, T. Neuromelanin can protect against iron-mediated oxidative damage in system modeling iron overload of brain aging and Parkinson's disease. *J. Neurochem.* **2008**, *106*, 1866–1875. [CrossRef]
89. Zucca, F.A.; Bellei, C.; Giannelli, S.; Terreni, M.R.; Gallorini, M.; Rizzio, E.; Pezzoli, G.; Albertini, A.; Zecca, L. Neuromelanin and iron in human locus coeruleus and substantia nigra during aging: Consequences for neuronal vulnerability. *J. Neural Transm.* **2006**, *113*, 757–767. [CrossRef]
90. Dexter, D.T.; Wells, F.R.; Lees, A.J.; Agid, F.; Agid, Y.; Jenner, P.; Marsden, C.D. Increased nigral iron content and alterations in other metal ions occurring in brain in Parkinson's disease. *J. Neurochem.* **1989**, *52*, 1830–1836. [CrossRef]
91. Earle, K.M. Studies on Parkinson's disease including x-ray fluorescent spectroscopy of formalin fixed brain tissue. *J. Neuropathol. Exp. Neurol.* **1968**, *27*, 1–14. [CrossRef] [PubMed]
92. Sofic, E.; Riederer, P.; Heinsen, H.; Beckmann, H.; Reynolds, G.P.; Hebenstreit, G.; Youdim, M.B. Increased iron (III) and total iron content in post mortem substantia nigra of parkinsonian brain. *J. Neural Transm.* **1988**, *74*, 199–205. [CrossRef] [PubMed]

93. Faucheux, B.A.; Martin, M.E.; Beaumont, C.; Hauw, J.J.; Agid, Y.; Hirsch, E.C. Neuromelanin associated redox-active iron is increased in the substantia nigra of patients with Parkinson's disease. *J. Neurochem.* **2003**, *86*, 1142–1148. [CrossRef] [PubMed]
94. Jellinger, K.; Kienzl, E.; Rumpelmair, G.; Riederer, P.; Stachelberger, H.; Ben-Shachar, D.; Youdim, M.B. Iron-melanin complex in substantia nigra of parkinsonian brains: An x-ray microanalysis. *J. Neurochem.* **1992**, *59*, 1168–1171. [CrossRef]
95. Bohic, S.; Murphy, K.; Paulus, W.; Cloetens, P.; Salome, M.; Susini, J.; Double, K. Intracellular chemical imaging of the developmental phases of human neuromelanin using synchrotron X-ray microspectroscopy. *Anal. Chem.* **2008**, *80*, 9557–9566. [CrossRef]
96. Good, P.F.; Olanow, C.W.; Perl, D.P. Neuromelanin-containing neurons of the substantia nigra accumulate iron and aluminum in Parkinson's disease: A LAMMA study. *Brain Res.* **1992**, *593*, 343–346. [CrossRef]
97. Crippa, P.R.; Eisner, M.; Morante, S.; Stellato, F.; Vicentin, F.C.; Zecca, L. An XAS study of the sulfur environment in human neuromelanin and its synthetic analogs. *Eur. Biophys. J.* **2010**, *39*, 959–970. [CrossRef]
98. Zecca, L.; Costi, P.; Mecacci, C.; Ito, S.; Terreni, M.; Sonnino, S. Interaction of human substantia nigra neuromelanin with lipids and peptides. *J. Neurochem.* **2000**, *74*, 1758–1765. [CrossRef]
99. Coon, S.; Stark, A.; Peterson, E.; Gloi, A.; Kortsha, G.; Pounds, J.; Chettle, D.; Gorell, J. Whole-body lifetime occupational lead exposure and risk of Parkinson's disease. *Environ. Health Perspect.* **2006**, *114*, 1872–1876. [CrossRef]
100. Kawashima, Y.; Horiguchi, A.; Taguchi, M.; Tuyuki, Y.; Karasawa, Y.; Araki, H.; Hatayama, K. Synthesis and Pharmacological Evaluation of 1,2,3,4-Tetrahydro-Beta-Carboline Derivatives. *Chem. Pharm. Bull.* **1995**, *43*, 783–787. [CrossRef]
101. Ostergren, A.; Annas, A.; Skog, K.; Lindquist, N.G.; Brittebo, E.B. Long-term retention of neurotoxic beta-carbolines in brain neuromelanin. *J. Neural Transm.* **2004**, *111*, 141–157. [CrossRef]
102. Matsubara, K.; Collins, M.A.; Akane, A.; Ikebuchi, J.; Neafsey, E.J.; Kagawa, M.; Shiono, H. Potential bioactivated neurotoxicants, N-methylated beta-carbolinium ions, are present in human brain. *Brain Res.* **1993**, *610*, 90–96. [CrossRef]
103. Abe, K.; Saitoh, T.; Horiguchi, Y.; Utsunomiya, I.; Taguchi, K. Synthesis and neurotoxicity of tetrahydroisoquinoline derivatives for studying Parkinson's disease. *Biol. Pharm. Bull.* **2005**, *28*, 1355–1362. [CrossRef]
104. D'Amato, R.J.; Alexander, G.M.; Schwartzman, R.J.; Kitt, C.A.; Price, D.L.; Snyder, S.H. Evidence for neuromelanin involvement in MPTP-induced neurotoxicity. *Nature* **1987**, *327*, 324–326. [CrossRef]
105. Lindquist, N.G. Accumulation in vitro of 35 S-chlorpromazine in the neuromelanin of human substantia nigra and locus coeruleus. *Arch. Int. Pharmacodyn. Ther.* **1972**, *200*, 190–195.
106. Salazar, M.; Sokoloski, T.D.; Patil, P.N. Binding of dopaminergic drugs by the neuromelanin of the substantia nigra, synthetic melanins and melanin granules. *Fed. Proc.* **1978**, *37*, 2403–2407.
107. Hansen, A.K.; Knudsen, K.; Lillethorup, T.P.; Landau, A.M.; Parbo, P.; Fedorova, T.; Audrain, H.; Bender, D.; Ostergaard, K.; Brooks, D.J.; et al. In vivo imaging of neuromelanin in Parkinson's disease using 18F-AV-1451 PET. *Brain* **2016**, *139*, 2039–2049. [CrossRef]
108. Marquie, M.; Normandin, M.D.; Vanderburg, C.R.; Costantino, I.M.; Bien, E.A.; Rycyna, L.G.; Klunk, W.E.; Mathis, C.A.; Ikonomovic, M.D.; Debnath, M.L.; et al. Validating novel tau positron emission tomography tracer [F-18]-AV-1451 (T807) on postmortem brain tissue. *Ann. Neurol.* **2015**, *78*, 787–800. [CrossRef]
109. Tago, T.; Toyohara, J.; Harada, R.; Furumoto, S.; Okamura, N.; Kudo, Y.; Takahashi-Fujigasaki, J.; Murayama, S.; Ishii, K. Characterization of the binding of tau imaging ligands to melanin-containing cells: Putative off-target-binding site. *Ann. Nucl. Med.* **2019**, *33*, 375–382. [CrossRef]
110. Lyden, A.; Larsson, B.; Lindquist, N.G. Studies on the melanin affinity of haloperidol. *Arch. Int. Pharmacodyn. Ther.* **1982**, *259*, 230–243. [PubMed]
111. Kawashima, H.; Iida, Y.; Kitamura, Y.; Saji, H. Binding of 4-(4-chlorophenyl)-1-[4-(4-fluorophenyl)-4-oxobutyl]pyridinium ion (HPP+), a metabolite of haloperidol, to synthetic melanin: Implications for the dopaminergic neurotoxicity of HPP+. *Neurotox. Res.* **2004**, *6*, 535–542. [CrossRef] [PubMed]
112. Kawashima, H.; Iida, Y.; Kitamura, Y.; Kiyono, Y.; Magata, Y.; Saji, H. Brain extraction of 4-(4-chlorophenyl)-1-[4-(4-fluorophenyl)-4-oxobutyl]pyridinium ion (HPP+), a neurotoxic metabolite of haloperidol: Studies using [3H]HPP+. *Jpn. J. Pharmacol.* **2002**, *89*, 426–428. [CrossRef]
113. Haining, R.L.; Jones, T.M.; Hernandez, A. Saturation Binding of Nicotine to Synthetic Neuromelanin Demonstrated by Fluorescence Spectroscopy. *Neurochem. Res.* **2016**, *41*, 3356–3363. [CrossRef] [PubMed]
114. Huang, L.Z.; Parameswaran, N.; Bordia, T.; Michael McIntosh, J.; Quik, M. Nicotine is neuroprotective when administered before but not after nigrostriatal damage in rats and monkeys. *J. Neurochem.* **2009**, *109*, 826–837. [CrossRef] [PubMed]
115. Myhre, O.; Utkilen, H.; Duale, N.; Brunborg, G.; Hofer, T. Metal dyshomeostasis and inflammation in Alzheimer's and Parkinson's diseases: Possible impact of environmental exposures. *Oxidative Med. Cell. Longev.* **2013**, *2013*, 726954. [CrossRef]
116. Cox, P.A.; Banack, S.A.; Murch, S.J.; Rasmussen, U.; Tien, G.; Bidigare, R.R.; Metcalf, J.S.; Morrison, L.F.; Codd, G.A.; Bergman, B. Diverse taxa of cyanobacteria produce beta-N-methylamino-L-alanine, a neurotoxic amino acid. *Proc. Natl. Acad. Sci. USA* **2005**, *102*, 5074–5078. [CrossRef] [PubMed]
117. Dunlop, R.A.; Cox, P.A.; Banack, S.A.; Rodgers, K.J. The non-protein amino acid BMAA is misincorporated into human proteins in place of L-serine causing protein misfolding and aggregation. *PLoS ONE* **2013**, *8*, e75376. [CrossRef]
118. Xie, X.; Basile, M.; Mash, D.C. Cerebral uptake and protein incorporation of cyanobacterial toxin beta-N-methylamino-L-alanine. *Neuroreport* **2013**, *24*, 779–784. [CrossRef]

119. Glover, W.B.; Mash, D.C.; Murch, S.J. The natural non-protein amino acid N-beta-methylamino-L-alanine (BMAA) is incorporated into protein during synthesis. *Amino Acids* **2014**, *46*, 2553–2559. [CrossRef]
120. Scott, L.; Downing, T. Dose-Dependent Adult Neurodegeneration in a Rat Model After Neonatal Exposure to beta-N-Methylamino-L-Alanine. *Neurotox. Res.* **2019**, *35*, 711–723. [CrossRef]
121. Cox, P.A.; Davis, D.A.; Mash, D.C.; Metcalf, J.S.; Banack, S.A. Dietary exposure to an environmental toxin triggers neurofibrillary tangles and amyloid deposits in the brain. *Proc. Biol. Sci.* **2016**, *283*. [CrossRef]
122. Myhre, O.; Eide, D.M.; Kleiven, S.; Utkilen, H.C.; Hofer, T. Repeated five-day administration of L-BMAA, microcystin-LR, or as mixture, in adult C57BL/6 mice-lack of adverse cognitive effects. *Sci. Rep.* **2018**, *8*, 2308. [CrossRef]
123. Delcourt, N.; Claudepierre, T.; Maignien, T.; Arnich, N.; Mattei, C. Cellular and Molecular Aspects of the beta-N-Methylamino-l-alanine (BMAA) Mode of Action within the Neurodegenerative Pathway: Facts and Controversy. *Toxins* **2017**, *10*, 6. [CrossRef]
124. Karlsson, O.; Berg, C.; Brittebo, E.B.; Lindquist, N.G. Retention of the cyanobacterial neurotoxin beta-N-methylamino-l-alanine in melanin and neuromelanin-containing cells–a possible link between Parkinson-dementia complex and pigmentary retinopathy. *Pigment. Cell Melanoma Res.* **2009**, *22*, 120–130. [CrossRef]
125. Karlsson, O.; Lindquist, N.G. Melanin and neuromelanin binding of drugs and chemicals: Toxicological implications. *Arch. Toxicol.* **2016**, *90*, 1883–1891. [CrossRef]
126. Diaz-Parga, P.; Goto, J.J.; Krishnan, V.V. Chemistry and Chemical Equilibrium Dynamics of BMAA and Its Carbamate Adducts. *Neurotox. Res.* **2018**, *33*, 76–86. [CrossRef]
127. Hofer, T.; Myhre, O.; Peltola-Thies, J.; Hirmann, D. Analysis of elimination half-lives in MamTKDB 1.0 related to bioaccumulation: Requirement of repeated administration and blood plasma values underrepresent tissues. *Environ. Int.* **2021**, in press. [CrossRef]
128. Hofer, T.; Peltola-Thies, J.; Karamertzanis, P.; Myhre, O.; Lantto, T.; Hirmann, D. *Mammalian Toxicokinetic Database (MamTKDB) 1.0*; Publications Office (EU Open Data Portal): Brussels, Belgium, 2021. [CrossRef]
129. Zadlo, A.; Mokrzynski, K.; Ito, S.; Wakamatsu, K.; Sarna, T. The influence of iron on selected properties of synthetic pheomelanin. *Cell Biochem. Biophys.* **2020**, *78*, 181–189. [CrossRef]
130. Ferrari, E.; Engelen, M.; Monzani, E.; Sturini, M.; Girotto, S.; Bubacco, L.; Zecca, L.; Casella, L. Synthesis and structural characterization of soluble neuromelanin analogs provides important clues to its biosynthesis. *J. Biol. Inorg. Chem.* **2013**, *18*, 81–93. [CrossRef]
131. Marszall, M.P.; Bucinski, A.; Gorynski, K.; Proszowska, A.; Kaliszan, R. Magnetic beads method for determination of binding of drugs to melanin. *J. Chromatogr. A* **2011**, *1218*, 229–236. [CrossRef] [PubMed]
132. Aubry, A.F. Applications of affinity chromatography to the study of drug-melanin binding interactions. *J. Chromatogr. B Anal. Technol. Biomed. Life Sci.* **2002**, *768*, 67–74. [CrossRef]
133. Marszall, M.P.; Proszowska, A.; Bucinski, A.; Kaliszan, R. Affinity Chromatography Method for Determination of Binding of Drugs to Melanin and Evaluation of Side Effect Potential of Antipsychotic Agents. *Curr. Pharm. Anal.* **2013**, *9*, 131–138. [CrossRef]
134. Jo, J.; Xiao, Y.; Sun, A.X.; Cukuroglu, E.; Tran, H.D.; Goke, J.; Tan, Z.Y.; Saw, T.Y.; Tan, C.P.; Lokman, H.; et al. Midbrain-like Organoids from Human Pluripotent Stem Cells Contain Functional Dopaminergic and Neuromelanin-Producing Neurons. *Cell Stem Cell* **2016**, *19*, 248–257. [CrossRef] [PubMed]
135. Marton, R.M.; Pasca, S.P. Neural Differentiation in the Third Dimension: Generating a Human Midbrain. *Cell Stem Cell* **2016**, *19*, 145–146. [CrossRef]
136. Tieng, V.; Stoppini, L.; Villy, S.; Fathi, M.; Dubois-Dauphin, M.; Krause, K.H. Engineering of midbrain organoids containing long-lived dopaminergic neurons. *Stem Cells Dev.* **2014**, *23*, 1535–1547. [CrossRef]
137. Nelson, P.T.; Greenberg, S.G.; Saper, C.B. Neurofibrillary Tangles in the Cerebral-Cortex of Sheep. *Neurosci. Lett.* **1994**, *170*, 187–190. [CrossRef]
138. Reid, S.J.; Mckean, N.E.; Henty, K.; Portelius, E.; Blennow, K.; Rudiger, S.R.; Bawden, C.S.; Handley, R.R.; Verma, P.J.; Faull, R.L.M.; et al. Alzheimer's disease markers in the aged sheep (Ovis aries). *Neurobiol. Aging* **2017**, *58*, 112–119. [CrossRef]
139. Kendrick, K.M.; da Costa, A.P.; Leigh, A.E.; Hinton, M.R.; Peirce, J.W. Sheep don't forget a face. *Nature* **2001**, *414*, 165–166. [CrossRef]
140. Knolle, F.; Goncalves, R.P.; Morton, A.J. Sheep recognize familiar and unfamiliar human faces from two-dimensional images. *R. Soc. Open Sci.* **2017**, *4*. [CrossRef]
141. Scheerlinck, J.P.; Snibson, K.J.; Bowles, V.M.; Sutton, P. Biomedical applications of sheep models: From asthma to vaccines. *Trends Biotechnol.* **2008**, *26*, 259–266. [CrossRef]
142. Palmer, D.N.; Neverman, N.J.; Chen, J.Z.; Chang, C.T.; Houweling, P.J.; Barry, L.A.; Tammen, I.; Hughes, S.M.; Mitchell, N.L. Recent studies of ovine neuronal ceroid lipofuscinoses from BARN, the Batten Animal Research Network. *Biochim. Biophys. Acta (BBA) Mol. Basis Dis.* **2015**, *1852*, 2279–2286. [CrossRef]
143. Braak, H.; Braak, E.; Strothjohann, M. Abnormally phosphorylated tau protein related to the formation of neurofibrillary tangles and neuropil threads in the cerebral cortex of sheep and goat. *Neurosci. Lett.* **1994**, *171*, 1–4. [CrossRef]

antioxidants

MDPI

Review

High-Content Screening for the Detection of Drug-Induced Oxidative Stress in Liver Cells

María Teresa Donato [1,2,*] and Laia Tolosa [1,*]

1 Unidad de Hepatología Experimental, Instituto de Investigación Sanitaria La Fe, 46026 Valencia, Spain
2 Departamento de Bioquímica y Biología Molecular, Facultad de Medicina, Universidad de Valencia, 46010 Valencia, Spain
* Correspondence: m.teresa.donato@uv.es (M.T.D.); laia_tolosa@iislafe.es (L.T.); Tel.: +34-961-246-649 (M.D.); +34-961-246-619 (L.T.)

Abstract: Drug-induced liver injury (DILI) remains a major cause of drug development failure, post-marketing warnings and restriction of use. An improved understanding of the mechanisms underlying DILI is required for better drug design and development. Enhanced reactive oxygen species (ROS) levels may cause a wide spectrum of oxidative damage, which has been described as a major mechanism implicated in DILI. Several cell-based assays have been developed as in vitro tools for early safety risk assessments. Among them, high-content screening technology has been used for the identification of modes of action, the determination of the level of injury and the discovery of predictive biomarkers for the safety assessment of compounds. In this paper, we review the value of in vitro high-content screening studies and evaluate how to assess oxidative stress induced by drugs in hepatic cells, demonstrating the detection of pre-lethal mechanisms of DILI as a powerful tool in human toxicology.

Keywords: oxidative stress; mitochondria; high-content screening; cell models; hepatotoxicity

Citation: Donato, M.T.; Tolosa, L. High-Content Screening for the Detection of Drug-Induced Oxidative Stress in Liver Cells. *Antioxidants* **2021**, *10*, 106. https://doi.org/10.3390/antiox10010106

Received: 21 December 2020
Accepted: 10 January 2021
Published: 13 January 2021

1. Introduction

Drug-induced liver injury (DILI) is triggered by prescription and non-prescription drugs, as well as herbal products or dietary supplements, leading to liver impairment or damage, and in the worst case liver failure [1,2]. DILI is one of the most serious adverse drug reactions that challenges pharmaceutical companies, regulatory agencies and health care professionals. It is a relevant entity in medical practice, which carries significant morbidity and mortality, constitutes the major cause of hepatic dysfunction and acute liver failure in Western countries [3] and has been responsible for 32% of drug withdrawals between 1975 and 2007 [4].

Several hundreds of drugs have been associated with liver damage and the list is ever-growing [5]. For a few drugs (e.g., acetaminophen) liver injury is dose-dependent, predictable and experimentally reproducible. However, most DILI episodes are idiosyncratic, that is, unexpected adverse reactions occurring in a minority of patients at doses that are safe for the general population [6]. Although ideally drug toxicity should be discovered during preclinical testing, in practice, due to its rarity and unpredictability, hepatotoxicity is seldom detected in in vitro assays or animal models. Most DILI cases appear post-approval when the drug is administered to several thousand patients.

It has been reported that DILI can be a multistep and multicellular disease process with a wide range of chemical etiologies. Thus, a better understanding of the mechanisms underlying DILI is essential in order to evaluate current strategies for the detection of hepatotoxicity [7]. Several specific signaling pathways that are activated during DILI and that could be predictive for hepatotoxicity have been identified [8]. Among them, oxidative stress, which is the imbalance between the production of reactive oxygen species (ROS) and the ability of cells to confront them through antioxidant enzymes, has been described to play an important role.

During the past years, numerous in vitro systems have been described in the literature in an attempt to advance the understanding of the underlying mechanisms of human DILI [1,9]. These liver model systems include conventionally cultured hepatic cell lines, primary human hepatocytes (PHHs), co-cultures and engineered liver platforms such as bioprinted hepatic models or perfusion systems [10].

Additionally, readout technologies have also evolved from single mechanistic endpoints toward approaches that consider global changes produced by drugs. In this sense, toxicologists have begun to apply newer technologies including toxicogenomics, metabolomics and high-content screening (HCS) assays [1]. HCS assays are multiplexed cell staining tests developed to gain a better understanding of complex biological functions and mechanisms of damage to the liver or other tissues. They have become an important tool for the safety evaluation of drug candidates [11]. Moreover, HCS has been used not only for toxicity screening but also in order to elucidate the mode of action of drugs, which makes it a powerful tool in drug discovery and development.

In this paper, we review the use of in vitro HCS to assess oxidative stress induced by drugs in hepatic cells, demonstrating the detection of pre-lethal mechanisms of DILI as a powerful tool in human toxicology.

2. Mechanisms of Drug-Induced Hepatotoxicity

DILI is a complex multistep phenomenon which encompasses a wide spectrum of clinical presentations and may mimic any form of liver disease (acute and chronic hepatitis, steatohepatitis, phospholipidosis, cholestasis, cirrhosis, etc.) [5]. Mechanisms underlying drug-induced hepatotoxicity remain poorly understood, which represents a major obstacle for the reliable prediction of DILI cases, in particular the idiosyncratic ones. Advances in the mechanistic understanding of DILI reveal that drugs can induce liver damage through multiple mechanisms [12]. Thus, the application of mechanism-based integrated approaches contributes to more reliable predictions of potential hepatotoxic effects induced by drugs [13]. Impairment of mitochondrial function, induction of oxidative stress, depletion of the glutathione (GSH) pool, covalent binding to macromolecules, inhibition of apical hepatic transporters (i.e., the bile salt export pump) and activation by cytochrome P450 (CYP) enzymes have been proposed as mechanistic indicators of drug-induced hepatotoxicity [14,15].

The liver is recognized as a frequent target of toxic damage due to its anatomic location and its high content of drug metabolizing enzymes. Most orally administered drugs are metabolized in the liver before they reach other tissues. Although for many drugs metabolism renders non-toxic metabolites (detoxication), some drugs may undergo metabolic bioactivation, with the generation of reactive metabolites able to induce liver injury. Thus, the parent drug itself, or more often any of its metabolites, may directly induce a chemical insult to hepatocytes or may eventually initiate a series of processes (activation of immune response, release of inflammatory mediators, mitochondrial dysfunction, endoplasmic reticulum stress, etc.) which contribute to the progression of liver damage [6].

Drug metabolism reactions, in particular oxidations by CYP enzymes, play an important role in hepatotoxicity. Reactive metabolites generated during drug bioactivation may covalently bind to macromolecules to form adducts with DNA or proteins (Figure 1). Covalent modification of enzymes, transporters and other key proteins may result in metabolic dysfunction, altered signaling pathways, loss of cell homeostasis and cell death. Some modified proteins may also serve as potential haptens or new antigens that trigger immunologically mediated liver injury. Covalent binding is dependent on the proportion of the drug converted into a reactive metabolite, the half-life of the reactive intermediate and its ability to react with cell macromolecules [6].

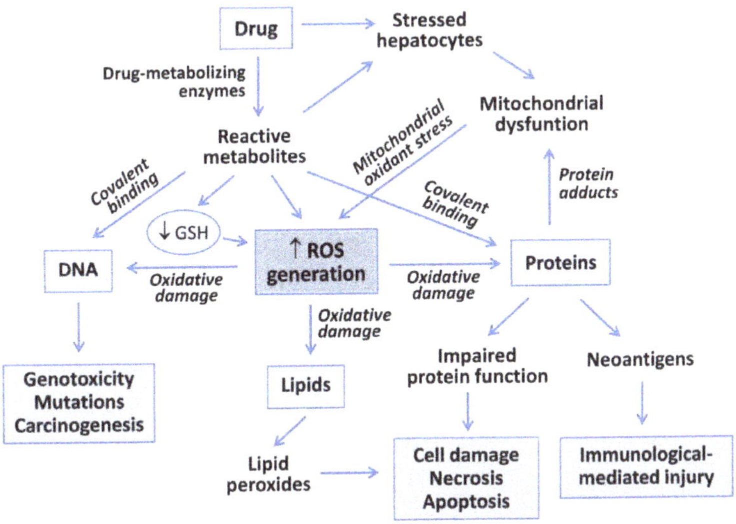

Figure 1. Mechanisms of drug-induced hepatotoxicity. Drug metabolism can render reactive metabolites that are able to induce hepatotoxicity via covalent binding to macromolecules or via oxidative damage induced by increased reactive oxygen species (ROS) generation. Reactive metabolites can interact with DNA to produce genotoxicity, induce lipid peroxidation or form adducts with proteins, which can lead to functional impairment, cell death or to the generation of neoantigens that trigger immune-based toxicity. Mitochondrial dysfunction induced by drugs, or more frequently by their reactive metabolites, contributes to increased ROS formation and oxidative damage to liver cells.

Mitochondria play a pivotal role in the maintenance of energy metabolism, the integration of cell signaling pathways and the mechanisms of cell death. Thus, the alteration of mitochondrial homeostasis is a common cause of cell damage induced by drugs or their metabolites [16]. Inhibition of the mitochondrial respiratory chain (MRC), impairment of β-oxidation of fatty acids, induction of mitochondrial permeability transition pore opening and depletion of mitochondrial DNA are among the mechanisms involved in drug-induced mitochondrial dysfunction. Direct consequences of these disturbances of liver mitochondria are ATP depletion, lipid overaccumulation, the release of cytochrome c into the cytosol and increased ROS formation, which can result in steatosis/steatohepatitis, apoptosis or necrosis [17].

ROS are highly reactive molecules formed as a by-product of aerobic metabolism. They participate in diverse cell signaling pathways and modulate cell growth and metabolism [18]. Disturbances in redox homeostasis of the cell can result in increased ROS levels with deleterious effects on diverse cell components [19]. The mitochondrial respiratory chain (MRC) is the main cellular source of ROS and, under physiological and homeostatic conditions, low levels of ROS generated by electron transport activity can be easily neutralized by antioxidant defenses. However, rates of ROS production can increase in the mitochondria or in other cell compartments under some stressful conditions, such as drug-induced mitochondrial injury. An excess of ROS that are not adequately neutralized can lead to many harmful effects including oxidative damage to DNA, proteins and lipids [19].

3. Drug-Induced ROS Generation

Oxidative stress results from an imbalance between the generation of oxidant species and the antioxidant capacity of the cells. This situation is produced under different pathological conditions, including exposure to drugs that induce an excessive formation of ROS or a depletion of antioxidant defense systems (Table 1).

Table 1. Examples of toxic compounds able to induce oxidative stress and mechanisms involved.

Mechanism	Compounds	References
Direct or indirect impairment of the mitochondrial respiratory chain (MRC)	Acetylsalicylic, amiodarone, azathioprine, buprenorphine, chloroquine, lovastatin, tamoxifen, nefazodone, troglitazone	[12,16,20,21]
Depletion of GSH pool and/or antioxidant enzymes	Azathioprine, doxorubicin, flutamide, isoniazid, valproic acid, paracetamol (acetaminophen)	[16,19,22–25]
Generation of electrophilic metabolites	Amitriptyline, benoxaprofen, diclofenac flutamide, paracetamol, ticlopidine, trogitazone	[12,13,16,25,26]
Redox cycling-induction	Diquat, paraquat, menadione, doxorubicin, flutamide	[22,27,28]

Although most electrons provided to the MRC are safely transferred to molecular oxygen to form water, direct or indirect impairment of the MRC by chemicals increase the occurrence of single-electron reductions of oxygen in the mitochondria to form ROS, mainly in the form of superoxide anions. The activity of superoxides dismutase and glutathione peroxidase, enzymes known to play important roles in the detoxification of ROS in the mitochondria, can be overtaken in some stressing circumstances. For instance, acetaminophen overdose leads to the formation of *N*-acetyl-p-benzoquinone imine (NAPQI), a reactive metabolite which depletes GSH, especially in mitochondria [19]. Thus, mitochondrial oxidative stress and peroxynitrite formation are induced, which trigger mitochondrial dysfunction, membrane permeability transition, mitochondrial depolarization, loss of calcium homeostasis and decreases in ATP synthesis [17]. An excess of oxidant species can cause a wide spectrum of oxidative damage to mitochondrial proteins, mitochondrial DNA and membrane lipids, which alters their structure and function. This oxidative damage aggravates mitochondrial dysfunction and further enhances ROS production, thus leading to important disturbances of cell bioenergetics and even cell death [17,19].

In addition to drug-induced disruption of the MRC, ROS can also be generated during metabolism of drugs by CYPs and other oxidative enzymes or by the presence of compounds that undergo repeated redox cycles. The formation of reactive intermediates by drug-metabolizing enzymes is involved in the hepatotoxicity of many compounds (Figure 1). Most reactive metabolites are electron-deficient species (electrophilic metabolites) that may react with nucleophilic sites in critical proteins or nucleic acids to form covalent adducts, leading to toxicity [6]. Alternatively, electrophilic metabolites form conjugates with GSH (via glutathione S-transferase activities) and deplete the GSH pool in the cell. There are many examples of CYP-mediated bioactivation of drugs into toxic metabolites in the endoplasmic reticulum of cells, such as the formation of electrophilic NAPQI from acetaminophen by CYP2E1, CYP1A2 and CYP3A4 [29]; the conversion of diclofenac into reactive quinone imines catalysed by CYP3A4 and CYP2C9 [12]; the formation of quinone and quinone methides from troglitazone by CYP3A4 [30]; and the oxidation of nitroaromatic moiety of flutamide by CYP1A2, CYP2C19 and CYP3A4 [26]. These bioactivation pathways are responsible for drug-induced hepatotoxicity via covalent binding and the formation of protein adducts, GSH depletion and oxidative stress [12].

Drug metabolism may also induce the formation of ROS or other radical species derived from oxidative or reductive metabolism of the drug. Oxidation of azathioprine (AZA) and other thiopurine drugs by xanthine oxidase has the potential to generate ROS that may result in liver damage via GSH depletion and mitochondrial injury [31]. Certain compounds increase intracellular levels of ROS through a cycling redox mechanism (Table 1). One-electron reduction of doxorubicin, an anthracycline anticancer drug, is carried out by several oxidoreductases to form a semiquinone radical that is re-oxidized back to doxorubicin with the formation of ROS [32]. Similarly, paraquat, a well-known toxicant to the liver and other tissues, undergoes one-electron reduction by diverse NADPH- or NADH-dependent oxidoreductases and is converted into a paraquat radical monocation. These radical species are then rapidly re-oxidized in the presence of molecular oxygen with the formation of a superoxide anion [33]. The result of these reactions is a futile redox cycle with the net regeneration of ROS leading to oxidative cell damage.

4. HCS Assays for the Detection of Oxidative Stress Induced by Drugs

4.1. HCS Technology

Multiparametric measurements of HCS assays are very useful in early toxicity and safety assessment in drug development. HCS technology, which allows automated image acquisition and analysis and provides information on multiple properties of individual cells loaded simultaneously with fluorescent dyes or transfected with fluorescent reporters, is used for drug safety evaluations. One of the main advantages of HCS technology is that distinct endpoints can be measured at a single-cell level simultaneously. A typical HCS assay for the detection of toxicity includes: (i) the selection of a relevant cell model; (ii) incubation with model test compounds; (iii) staining with a combination of fluorescent probes indicative of cell damage, and (iv) automated image acquisition and analysis using HCS equipment. Figure 2 summarizes the main steps in an HCS assay.

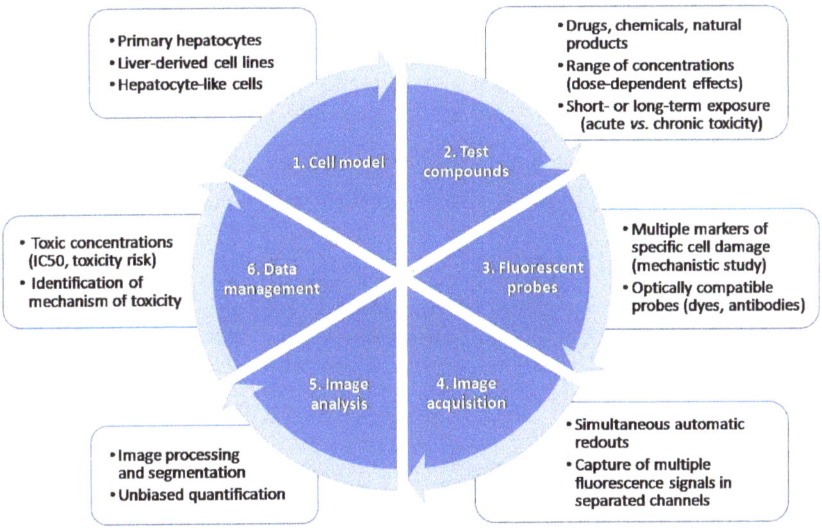

Figure 2. Major steps in a high-content screening (HCS) assay. An HCS assay includes the selection of an appropriate test system, incubation with test compounds at a range of concentrations, incubation with a combination of fluorescent probes, automated image acquisition, analysis and data management.

Different ex vivo and in vitro models have been used in HCS assays, from isolated primary hepatocytes to complex 3D models. Immortalized human cell lines such as HepG2 or HepaRG cells have been widely used for toxicological assessments as substitutes for PHHs in drug screening [13,34]. PHHs remain the gold standard in liver cell models, but they have many drawbacks like the scarcity of organs to isolate them and the loss of important enzymes and transporters when they are in culture [10]. To overcome these limitations, other models such as hepatocyte-like cells (HLCs) derived from induced-pluripotent stem cells (iPSCs) or more physiologically relevant 3D multicellular in vitro systems have been proposed [35]. An important technical issue in the first step of an HCS assay is to select an optimal cell density that allows an appropriate segmentation and the selection of plates that permit the best imaging.

The second step of an HCS assay for toxicology studies includes incubation with test compounds. In this phase, appropriate positive and negative controls may be properly selected to ensure that the assay allows the detection of the desired endpoint. These controls should be run in every set of experiments. Moreover, it should also be considered that some compounds can be autofluorescent and could interfere with some of the selected fluorescent probes.

HCS enables the multiplexing of the information that is obtained in a single assay, by using a combination of fluorescent probes. The major requirement for the use of multiple probes is their optical compatibility and the limitations of the instrument used for the measurements [36]. Section 4.2 describes in detail the most commonly used probes for oxidative stress detection in toxicological studies. Additionally, it should be considered that to get the brightest signal it is important to use illumination wavelengths that will optimally excite the fluorophore and to get the maximum of emission photons [37]. A comparison of the spectra of the fluorophores used and the fluorescent filter sets and/or laser line may be performed to ensure the optimal conditions [37]. It should be considered that some fluorescent dyes can be toxic to live cells, and therefore they can be only used with fixed cells.

After incubation with fluorescent probes, the automated acquisition of fluorescent images in separated channels is performed. Microscopy performance is usually assessed based on three criteria: sensitivity, resolution and speed [38]. It is impossible to have an optimal performance in each criterion, so a balance between them is pursued based on the compromise triangle of microscopy performance [38]. The selection of the objective affects the resolution, field of view and sensitivity, and should also be considered when setting up an HCS assay. Furthermore, considering that imaging a large quantity of cells to achieve high throughput is time-consuming [39], a correct focus is crucial for the correct development and use of HCS assays. Both image-based and laser-based autofocus methods are used in HCS. Image-based autofocus methods are slower, although they are better than laser autofocus for 3D structures [40]. Alternatively, machine learning methods are being developed to recognize and track only a subset of cells [40].

Due to the large numbers of images that are generated in HCS assays, automated image processing and analysis are decisive for large-scale image-based toxicity screenings. As a first step, in order to reduce noise and correct background or illumination, image-processing algorithms are used [37]. Secondly, cell segmentation is used to identify cellular or subcellular regions [41] and finally feature extraction allows the quantification of the changes in the identified regions [39], which in the case of toxicological assessment lead to the identification of significant changes produced by drugs compared to untreated cells. Thresholding is vulnerable to intensity variations from day to day and should be adjusted for day-to-day and user-to-user effects. On the other hand, accurate cell segmentation is a requisite to extract cell-by-cell information. Thus, the segmentation of touching cells is critical when analyzing HCS images [42]. Different algorithms, such as the evolving generalized Voronoi diagram [42], have been proposed in order to segment touching cells, which commonly have a variable morphology to improve the image analysis process.

Data analysis and management in HCS has advanced significantly in the past years, since the large volume of data produced in HCS assays is commonly a bottleneck in many projects. It should be considered that processing hundreds of images by applying image algorithms requires a powerful infrastructure [43]. Moreover, the data output of a 3D image-based screen is extremely high, which also poses a challenge in terms of data handling and image analysis [38].

4.2. HCS Probes for the Detection of Oxidative Stress

As ROS production is a major mechanism implicated in DILI, several fluorescent probes have been described for their use in toxicological studies. Fluorescent probes can simply detect ROS production in general or in specific radicals or locations. Table 2 summarizes the fluorescent probes and reporters for oxidative stress detection used in HCS.

Table 2. Probes and reporters used in HCS assays for the detection of oxidative stress.

Probe/Reporter	Indicator	Excitation	Emission	Reference
BODIPY 665/676	Peroxyl radicals (lipid peroxidation)	665	676	[13]
CellROX Green	ROS production (nuclei)	485	520	[44]
CellROX Deep Red	ROS production (cytoplasm)	644	665	[45]
CM-H_2DCFDA		495	527	[46–48]
Dihydroethidium	Superoxide anion	518	605	[49]
mBCl	Glutathione	390	478	[46–48]
MitoSOX Red	Mitochondrial superoxide	510	580	[50,51]
Srnx1-GFP reporter	Nrf2 oxidative stress response	488	510	[52]
CHOP-GFP reporter	Endoplasmic reticulum-stress/unfolded protein response	488	510	[52]
p21-GFP reporter	P53 dependent DNA damage-related signalling	488	510	[52]
ICAM1-GFP reporter	NF-κB-mediated pro-inflammatory cytokine signalling	488	510	[52]

For instance, the accumulation of the fluorescent compound 2′,7′-dichlorofluorescin (DCF), generated by intracellular oxidation of DHCF-DA, was used as an indicator of ROS generation, mainly H_2O_2 [53]. Once the membrane-permeant H_2DCFDA enters a cell, its acetate moieties are cleaved by intracellular esterases, resulting in an impermeable H_2DCF form. Subsequent oxidation of H_2DCF produces the fluorescent 2′,7′-dichlorofluorescein, which can be detected in the green spectrum [53].

Monochlorobimane (mBCl) is a membrane-permeant probe that fluoresces in the UV spectrum upon reacting with GSH in a reaction catalyzed by the enzyme glutathione-S-transferase [54]. Due to the enzymatic catalysis of mBCl–GSH adduct formation, mBCl has greater specificity for GSH compared with other thiol-specific probes such as monobromobimane, which reacts freely with both GSH and intracellular protein thiols [54].

CellROX® Oxidative Stress Reagents are fluorogenic probes designed to detect ROS in live cells. The CellROX Deep Red signal is localized in the cytoplasm of the cells and it has been used in several HCS assays for the detection of acute [45] or repeated-dose exposure to hepatotoxicants [51].

On the other hand, dihydroethidium (DHE), also called hydroethidine, dye is oxidized to fluorescent ethidium, which intercalates into DNA; the fluorescent signal is used as a measure of oxidative stress [49]. DHE exhibits blue fluorescence in the cytosol until it is oxidized, at which point it intercalates within the cell's DNA, staining its nucleus a bright fluorescent red.

Lipid peroxidation, as a measure of oxidative stress damage, can be detected with the lipophilic probe, BODIPY 665/676 dye. This probe exhibits a change in fluorescence after interaction with peroxyl radicals [55].

Mitochondrial superoxide production can be identified by measuring MitoSOX Red fluorescent intensity in the mitochondrial compartment [50]. It has been used in different hepatic models for the study of the specific mechanisms implicated in DILI [50,51].

Other researchers have used HCS in reporter-cell lines. In this case, it is based on the use of a bacterial artificial chromosome containing a genomic copy of a particular gene and a fluorescent luciferase reporter construct introduced by homologous recombination [56]. Major cell stress signaling routes activated in response to toxicants include antioxidant response element activation; the heat shock response; the unfolded protein response; the metal stress response; the DNA damage response and the induction of phase I, II and III enzymes/transporters by nuclear receptors [56]. Cell-based reporters for the study of the oxidative stress induced by drugs Keap1, Nfr2 and Srxn1 have been used [57].

4.3. Examples of HCS Assays for the Detection of Oxidative Stress Induced by Drugs in Hepatic Cell Models

Multiple HCS assays have included fluorescent probes for the detection of oxidative stress damage induced by drugs, since oxidative stress is one of the major mechanisms implicated in DILI. For instance, we have demonstrated that HCS can be used for the detection of oxidative stress damage. Figure 3 exemplifies the utility of this technique using different fluorescent probes. AZA is a widely used immunosuppressive drug that is generally well tolerated but that may produce severe hepatitis in a small number of patients [58]. Oxidative stress and the subsequently activated immune- and inflammation-related factors have been suggested as major mechanisms in AZA-induced hepatotoxicity [59]. Here, we demonstrate that the oxidative response induced by AZA can be measured by HCS (Figure 3). Three different probes (MitoSOX Red, DCF and CellROX) were used and a dose-response was detected when HepG2 cells were incubated with a range of concentrations for 1 h. No significant effects were seen in viability, whereas a significant increase in ROS production was detected. DCF fluorescence seemed to be the most sensitive parameter, although significant dose-response changes were also detected with CellROX and MitoSOX Red. These representative results indicate the suitability of this technique to detect subtle changes even at subcytotoxic concentrations.

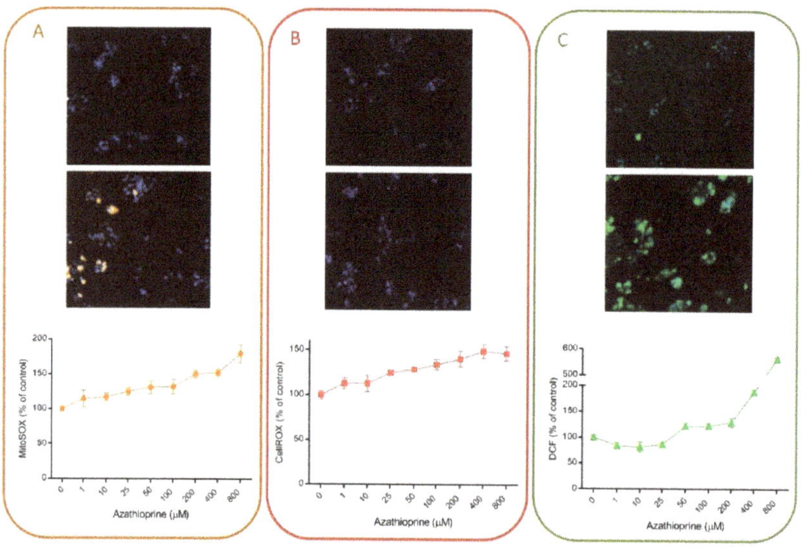

Figure 3. Azathioprine induces oxidative stress in liver cells. HepG2 cells were exposed to a range of concentrations of azathioprine (1–800 μM) for 1 h and then, incubated with MitoSOX Red (**A**), CellROX Deep Red (**B**) or DCF (**C**) and were analyzed using HCS. On the top of each panel, non-treated cells are shown, in contrast to cells treated with 400 μM of azathioprine, which are shown in the bottom part. HCS technology allows the quantification of changes induced by drugs and shows its utility for toxicological studies and to determine the major mechanisms implicated in their toxicity.

Many groups have used HCS for the detection of oxidative stress induced by drugs. Table 3 summarizes different studies that have used HCS for the detection of oxidative stress induced by drugs in different liver cell systems. In general, ROS production or GSH depletion are included in the HCS assays, as oxidative stress has been defined as a major mechanism implicated in DILI.

Table 3. HCS assays for the detection of oxidative stress-induced liver injury. DILI: drug-induced liver injury; HLCs: hepatocyte-like cells; iPSCs: induced pluripotent stem cells; PHH: primary human hepatocytes; UHH: Upcyte human hepatocytes.

Test Model	Probes/Reporters	Drugs	Reference
PHH	DRAQ5; TMRM; CM-H_2DCFDA; mBCl	300 DILI and non-DILI compounds	[48]
HepG2	Hoechst 33342; PI; TMRM; Fluo-4 AM; BODIPY 665/676	78 DILI and non-DILI compounds	[13]
HepG2 transfected with CYP adenovirus	Hoechst 33342; PI; TMRM; Fluo-4 AM; CellROX	15 DILI and non-DILI compounds	[45]
HepG2, HepG2 + S9, PHH	CM-H2DFFDA, TMRE Activated caspase-3, phosphorylated histone-H3 and HSP 70/72 assays LipidTox (Phospholipids + neutral lipids)	144 DILI and non-DILI compounds	[60]
Male and female PHHs	DHE, TMRE, TOTO3, Fluo4	6 chemicals	[49]
iPSC-HLCs	MitoTracker orange, carboxy-H2DCFDA, TOTO-3, Hoechst33342	8 toxicants	[61]
HepG2 and HepaRG cells	Hoechst 33342, CM-H2DCFDA, TMRM, TOTO-3 DRAQ5, mBCl, YOYO-1	28 DILI and non-DILI compounds	[62]
HepG2	Cellomics HCS reagent; CellROX Deep Red; Hoechst 33342	Complex mixtures of perfluorinated, brominated, and chlorinated compounds (persistent organic pollutants)	[63]
HepG2 reporter lines	11 BAC-GFP reporters containing target genes, representing the oxidative stress response pathway (KEAP1/NFR2/SRXN1), the unfolded protein response and DNA damage	30 hepatotoxicants	[57]
HepG2 reporter lines	SRXN1-GFP, CHOP-GFP, p21-GFP and ICAM-GFP	118 FDA-labelled drugs	[52]
HepG2 reporter lines in 3D spheroids	Srxn1-GFP, NQO1-GFP, BiP-GFP, Chop-GFP, p21-GFP and Btg2-GFP	33 compounds	[64]
UHH	BODIPY 493/503; LipidTOX Red Phspholipidosis, CellROX, Fluo-4 AM, Hoechst 33342; PI; MitoSOX Red, TMRM and mBCl	15 DILI and non-DILI compounds	[51]

One of the first descriptions of the potential of HCS for the evaluation of DILI was the measurement of mitochondrial damage, oxidative stress and intracellular GSH levels in PHHs using TMRM, CM-H_2DCFDA and mBCl, respectively, as fluorescent probes [48]. PHHs were incubated with 300 drugs at 100-fold of the therapeutic peak plasmatic concentration when possible. After 24 h incubation, they assessed the toxicity using HCS. The study showed a positive rate around 50–60% and a low false-positive rate (0–5%), which led many groups and companies to use it as a powerful tool in drug screening.

We have also included oxidative stress in our panels covering different mechanisms implicated in DILI [13,45]. We first described the use of HCS for developing and validating a cell-based protocol to assess the hepatotoxic potential of drugs and defining their mechanisms of action in HepG2 cells. In this assay, five different fluorescent probes to measure changes in mitochondrial membrane potential (TMRM), viability (PI), nuclear changes (Hoechst 33342), intracellular calcium concentration (Fluo-4 AM) and oxidative stress (BODIPY 665) were included. The assay turned out to be a simple screening that indicated the mechanisms implicated in both the toxicity and the degree of injury in a single incubation with high specificity and sensitivity [13]. Then, we applied the same strategy to a more complex test system, metabolically competent HepG2 cells, using CellROX as a measure of oxidative stress [45]. The assay allowed for the identification of bioactivable hepatotoxins and the mechanism(s) involved in their toxicity [45]. Finally, in order to assess the repeated-dose long term exposure to drugs, Upcyte human hepatocytes (UHH) were incubated with 15 toxic and non-toxic compounds for up to 21 days and the mechanisms implicated in their toxicity were assessed using HCS. In this case, three different probes for the study of oxidative stress-induced damage were used: mBCl, CellROX and MitoSOX Red [51].

Mennecozzi et al. (2015) investigated sex-specific differences in PHHs from different donors exposed to hepatotoxic drugs using HCS. They used a combination of probes which included DHE for ROS production. Differential effects in oxidative stress damage were detected depending on the compound assessed, although a general pattern could not be defined for female or male hepatocytes [49].

A comparative study of HepG2 and HepaRG cells for the prediction of DILI demonstrated that DILI compounds, which deplete GSH via reactive metabolites, showed a more significant decrease in GSH or increase in ROS in HepaRG than in HepG2 cells [62]. This study used two different panels for the detection of toxicity and included the oxidative stress probes such as CM-H_2DCFDA for ROS production and mBCl for GSH depletion. Additionally, they pointed out differences in drug metabolism due to the significant differences between both cell models in GSH depletion, indicating the value of HCS assays for DILI screening and also for studying the contribution of metabolism to hepatocyte toxicity [62].

Since current in vitro test systems have different limitations for toxicological studies, other cell-based models, including stem cells differentiated into a hepatic phenotype, are being explored as a functional source of human hepatocytes. HLC-derived iPSCs have been used to evaluate the hepatotoxic potential of well-known drugs at subcytotoxic concentrations by means of HCS [61]. The authors used carboxy-H_2DCFDA for the measurement of ROS and showed that ROS production was one of the most sensitive parameters for the detection of drug-induced toxicity, pointing out the suitability of this probe for the study of DILI in vitro [61].

On the other hand, the monitoring of adaptative stress response pathways has also been used as a predictive tool in toxicology [52,57]. Wink et al. (2016) established a panel of fluorescent protein reporter HepG2 cell lines using bacterial artificial chromosome (BAC) technology, reflecting three adaptative stress responses, and combined it with HCS technology. HCS allowed for the establishment of the temporal order of the activation of the distinct adaptative responses and the identification of the main ones induced by a selection of compounds that had previously been described to produce DILI [57]. The application of the HepG2 stress response reporters in 3D spheroid systems in combination with HCS has also been proposed [64]. HepG2 spheroids increase the expression of phase I and II metabolizing enzymes and show a high stability, which allows repeated-dose treatment. In this model, the authors demonstrated that a six-day repeated treatment with hepatotoxic drugs resulted in an increased sensitivity in the detection of cytotoxicity when compared to single-dose exposure, demonstrating their suitability for safety assessment [64].

Finally, HCS has been also used for the evaluation of the effects of natural products in hepatic cells, demonstrating the suitability of this tool for assessing both oxidative stress induction [65] and the antioxidant capacity of compounds [66].

HCS may be used in the pharmaceutical industry to screen candidate drugs for potential human toxicity, to rank and select lead compounds and to demonstrate the major mechanisms of toxicity [67]. However, despite their great sensitivity and specificity in identifying human DILI, HCS assays fail to provide translational biomarkers. Thus, HCS multiparametric assays could be used in combination with other highly sensitive approaches, such as expression analyses [68] or toxicometabonomics [69], that could contribute as complementary assays to better predictivity and sensitivity of the human hepatotoxicity potential for new compounds.

5. Conclusions and Future Outlook

The high financial investment required to bring a drug to the market, along with the high rates of attrition of new molecules, make necessary the optimization of screening procedures for more efficient evaluations of drug safety. Human cell models are increasingly used in preclinical safety testing, as they provide quick and relatively inexpensive information in the early phases of drug development. As previously stated, the mechanisms responsible for DILI are poorly understood, which complicate the establishment and validation of predictive screening tools and limit the reliability of in vitro toxicity testing. Traditional tests rely on the determination of unspecific endpoints indicative of overt loss of cell integrity and have been shown to be insufficient to predict the hepatotoxicity of many drugs. Multiparametric HCS or omics-based approaches have gained popularity for hepatotoxicity studies as they enable the simultaneous determination of multiple informative parameters that may contribute to more accurate DILI predictions. In particular, HCS allows for the simultaneous analysis of several markers of cell damage and enables the detection of subtle toxicity-related changes with greater sensitivity than conventional single-endpoint cytotoxicity assays.

The flexibility of HCS technology allows researchers to easily adapt assays to multiple applications by selecting the most appropriate cell model and fluorescent probes. HCS has been proven as a valuable screening tool for the identification of hepatotoxic drugs in diverse cell models (PHHs, hepatoma cell lines, UHHs, adenoviral-transfected cells, HLCs). Cell-based HCS assays have been evaluated and validated in the toxicological field not only for the study of hepatotoxicity but also for neurotoxicity, cardiotoxicity and nephrotoxicity applications [70]. A careful selection of the most appropriated probes allows the optimization of specific assay panels covering a wide spectrum of mechanistic parameters. In this line, many fluorescent labeling probes are available for oxidative stress detection and have been successfully applied to HCS assays in several cell systems. Oxidative stress is one of the main mechanisms implicated in the toxicity of many drugs, and ROS generation and GSH depletion are considered among the most sensitive parameters of drug-induced hepatotoxicity [61,62]. HCS-based testing allows multiplexed measurements in cells cultured in multi-well plate formats. Thus, there are considerable reductions in time, cells, reagents and tested compounds needed for hepatotoxicity evaluations, rendering the assays amenable to medium/high-throughput screenings in early drug development steps.

One potential limitation of traditional monolayer cell systems used in hepatotoxicity evaluations is the loss of the native 3D microenvironment of liver cells. Recent technological advances have allowed the development of more complex in vitro models that better reproduce the spatial organization and inter-cellular interactions in the liver [71]. 3D culture systems stabilize many hepatocyte functions and seem to be a valuable alternative for long-term toxicity studies. The application of readout technologies to cellular models that allow repeated exposure assessments to mimic the prolonged clinical exposure in patients, and not only acute liver damage, is a promising strategy to improve DILI predictions. Thus, the combination of mechanistic multiparametric HCS measurements and physiologically relevant culture models will provide potent screening platforms for human toxicity.

The widespread application of HCS assays to the study of oxidative stress induced by drugs could provide fast, sensitive and accurate safety evaluations in early drug development. Cell-based HCS tools could also enhance the understanding of mechanisms and pathways involved in oxidative cell damage and could contribute to the identification of drugs, natural products or foods with antioxidant capacity.

Author Contributions: Writing—original draft preparation, M.T.D., L.T.; writing—review and editing, M.T.D., L.T. All authors have read and agreed to the published version of the manuscript.

Funding: This work has been supported by the Institute of Health Carlos III (Plan Estatal de I+D+i 2013-2016) and co-financed by the European Regional Development Fund "A way to achieve Europe" (FEDER) through grants PI16/00333, PI18/00993 and CP16/00097, by the Spanish Ministry of Science through the PID2019-106000RB-C22 Project (including FEDER funds), and by the Generalitat Valenciana (PROMETEO/2019/060). L.T. was supported by ISCIII CP16/00097.

Acknowledgments: The authors acknowledge M.A. Herrero for his support in linguistic concerns.

Conflicts of Interest: The authors declare no conflict of interest. The funders had no role in the design of the study; in the collection, analyses or interpretation of data; in the writing of the manuscript or in the decision to publish the results.

Abbreviations

AZA	azathioprine
CYP	cytochrome P450
DCF	2′,7′-dichlorofluorescin
DILI	drug-induced liver injury
GSH	glutathione
HCS	high-content screening
HLCs	hepatocyte-like cells
iPSCs	induced pluripotent stem cells
mBCl	monochlorobimane
MRC	mitochondrial respiratory chain
NAPQI	*N*-acetyl-p-benzoquinone imine
PHH	primary human hepatocytes
ROS	reactive oxygen species
UHH	Upcyte human hepatocytes

References

1. Funk, C.; Roth, A. Current limitations and future opportunities for prediction of DILI from in vitro. *Arch. Toxicol.* **2017**, *91*, 131–142. [CrossRef] [PubMed]
2. Garcia-Cortes, M.; Robles-Diaz, M.; Stephens, C.; Ortega-Alonso, A.; Lucena, M.I.; Andrade, R.J. Drug induced liver injury: An update. *Arch. Toxicol.* **2020**, *94*, 3381–3407. [CrossRef]
3. Lee, W.M. Drug-induced hepatotoxicity. *N. Engl. J. Med.* **2003**, *349*, 474–485. [CrossRef] [PubMed]
4. Stevens, J.L.; Baker, T.K. The future of drug safety testing: Expanding the view and narrowing the focus. *Drug Discov. Today* **2009**, *14*, 162–167. [CrossRef] [PubMed]
5. Fontana, R.J. Pathogenesis of idiosyncratic drug-induced liver injury and clinical perspectives. *Gastroenterology* **2014**, *146*, 914–928. [CrossRef] [PubMed]
6. Gomez-Lechon, M.J.; Tolosa, L.; Donato, M.T. Metabolic activation and drug-induced liver injury: In vitro approaches for the safety risk assessment of new drugs. *J. Appl. Toxicol.* **2016**, *36*, 752–768. [CrossRef]
7. Weaver, R.J.; Blomme, E.A.; Chadwick, A.E.; Copple, I.M.; Gerets, H.H.J.; Goldring, C.E.; Guillouzo, A.; Hewitt, P.G.; Ingelman-Sundberg, M.; Jensen, K.G.; et al. Managing the challenge of drug-induced liver injury: A roadmap for the development and deployment of preclinical predictive models. *Nat. Rev. Drug Discov.* **2020**, *19*, 131–148. [CrossRef]

8. Kuijper, I.A.; Yang, H.; Van De Water, B.; Beltman, J.B. Unraveling cellular pathways contributing to drug-induced liver injury by dynamical modeling. *Expert Opin. Drug Metab. Toxicol.* **2017**, *13*, 5–17. [CrossRef]
9. Jiang, J.; Pieterman, C.D.; Ertaylan, G.; Peeters, R.L.M.; de Kok, T. The application of omics-based human liver platforms for investigating the mechanism of drug-induced hepatotoxicity in vitro. *Arch. Toxicol.* **2019**, *93*, 3067–3098. [CrossRef]
10. Gomez-Lechon, M.J.; Tolosa, L.; Conde, I.; Donato, M.T. Competency of different cell models to predict human hepatotoxic drugs. *Expert Opin. Drug Metab. Toxicol.* **2014**, *10*, 1553–1568. [CrossRef]
11. Joshi, P.; Lee, M.Y. High Content Imaging (HCI) on Miniaturized Three-Dimensional (3D) Cell Cultures. *Biosensors* **2015**, *5*, 768–790. [CrossRef] [PubMed]
12. Dragovic, S.; Vermeulen, N.P.; Gerets, H.H.; Hewitt, P.G.; Ingelman-Sundberg, M.; Park, B.K.; Juhila, S.; Snoeys, J.; Weaver, R.J. Evidence-based selection of training compounds for use in the mechanism-based integrated prediction of drug-induced liver injury in man. *Arch. Toxicol.* **2016**, *90*, 2979–3003. [CrossRef] [PubMed]
13. Tolosa, L.; Pinto, S.; Donato, M.T.; Lahoz, A.; Castell, J.V.; O'Connor, J.E.; Gomez-Lechon, M.J. Development of a multiparametric cell-based protocol to screen and classify the hepatotoxicity potential of drugs. *Toxicol. Sci.* **2012**, *127*, 187–198. [CrossRef]
14. Hewitt, M.; Enoch, S.J.; Madden, J.C.; Przybylak, K.R.; Cronin, M.T. Hepatotoxicity: A scheme for generating chemical categories for read-across, structural alerts and insights into mechanism(s) of action. *Crit. Rev. Toxicol.* **2013**, *43*, 537–558. [CrossRef] [PubMed]
15. Thompson, R.A.; Isin, E.M.; Li, Y.; Weidolf, L.; Page, K.; Wilson, I.; Swallow, S.; Middleton, B.; Stahl, S.; Foster, A.J.; et al. In vitro approach to assess the potential for risk of idiosyncratic adverse reactions caused by candidate drugs. *Chem. Res. Toxicol.* **2012**, *25*, 1616–1632. [CrossRef] [PubMed]
16. Pessayre, D.; Fromenty, B.; Berson, A.; Robin, M.A.; Letteron, P.; Moreau, R.; Mansouri, A. Central role of mitochondria in drug-induced liver injury. *Drug Metab. Rev.* **2012**, *44*, 34–87. [CrossRef]
17. Labbe, G.; Pessayre, D.; Fromenty, B. Drug-induced liver injury through mitochondrial dysfunction: Mechanisms and detection during preclinical safety studies. *Fundam. Clin. Pharmacol.* **2008**, *22*, 335–353. [CrossRef] [PubMed]
18. Glasauer, A.; Chandel, N.S. Targeting antioxidants for cancer therapy. *Biochem. Pharmacol.* **2014**, *92*, 90–101. [CrossRef]
19. Pereira, C.V.; Nadanaciva, S.; Oliveira, P.J.; Will, Y. The contribution of oxidative stress to drug-induced organ toxicity and its detection in vitro and in vivo. *Expert Opin. Drug Metab. Toxicol.* **2012**, *8*, 219–237. [CrossRef]
20. Grattagliano, I.; Bonfrate, L.; Diogo, C.V.; Wang, H.H.; Wang, D.Q.; Portincasa, P. Biochemical mechanisms in drug-induced liver injury: Certainties and doubts. *World J. Gastroenterol.* **2009**, *15*, 4865–4876. [CrossRef]
21. Thompson, R.A.; Isin, E.M.; Ogese, M.O.; Mettetal, J.T.; Williams, D.P. Reactive Metabolites: Current and Emerging Risk and Hazard Assessments. *Chem. Res. Toxicol.* **2016**, *29*, 505–533. [CrossRef] [PubMed]
22. Giampieri, F.; Alvarez-Suarez, J.M.; Gasparrini, M.; Forbes-Hernandez, T.Y.; Afrin, S.; Bompadre, S.; Rubini, C.; Zizzi, A.; Astolfi, P.; Santos-Buelga, C.; et al. Strawberry consumption alleviates doxorubicin-induced toxicity by suppressing oxidative stress. *Food Chem. Toxicol.* **2016**, *94*, 128–137. [CrossRef] [PubMed]
23. Simeonova, R.; Kondeva-Burdina, M.; Vitcheva, V.; Mitcheva, M. Some in vitro/in vivo chemically-induced experimental models of liver oxidative stress in rats. *Biomed. Res. Int.* **2014**, *2014*, 706302. [CrossRef] [PubMed]
24. Singh, D.; Cho, W.C.; Upadhyay, G. Drug-Induced Liver Toxicity and Prevention by Herbal Antioxidants: An Overview. *Front. Physiol.* **2015**, *6*, 363. [CrossRef]
25. Tong, V.; Teng, X.W.; Chang, T.K.; Abbott, F.S. Valproic acid II: Effects on oxidative stress, mitochondrial membrane potential, and cytotoxicity in glutathione-depleted rat hepatocytes. *Toxicol. Sci.* **2005**, *86*, 436–443. [CrossRef]
26. Kang, P.; Dalvie, D.; Smith, E.; Zhou, S.; Deese, A.; Nieman, J.A. Bioactivation of flutamide metabolites by human liver microsomes. *Drug Metab. Dispos.* **2008**, *36*, 1425–1437. [CrossRef]
27. Jaeschke, H.; McGill, M.R.; Ramachandran, A. Oxidant stress, mitochondria, and cell death mechanisms in drug-induced liver injury: Lessons learned from acetaminophen hepatotoxicity. *Drug Metab. Rev.* **2012**, *44*, 88–106. [CrossRef]
28. Teppner, M.; Boess, F.; Ernst, B.; Pahler, A. Biomarkers of Flutamide-Bioactivation and Oxidative Stress in Vitro and in Vivo. *Drug Metab. Dispos.* **2016**, *44*, 560–569. [CrossRef]
29. Park, B.K.; Kitteringham, N.R.; Maggs, J.L.; Pirmohamed, M.; Williams, D.P. The role of metabolic activation in drug-induced hepatotoxicity. *Annu. Rev. Pharmacol. Toxicol.* **2005**, *45*, 177–202. [CrossRef]
30. Masubuchi, Y.; Suda, C.; Horie, T. Involvement of mitochondrial permeability transition in acetaminophen-induced liver injury in mice. *J. Hepatol.* **2005**, *42*, 110–116. [CrossRef]
31. Petit, E.; Langouet, S.; Akhdar, H.; Nicolas-Nicolaz, C.; Guillouzo, A.; Morel, F. Differential toxic effects of azathioprine, 6-mercaptopurine and 6-thioguanine on human hepatocytes. *Toxicol. In Vitro* **2008**, *22*, 632–642. [CrossRef]
32. Ingawale, D.K.; Mandlik, S.K.; Naik, S.R. Models of hepatotoxicity and the underlying cellular, biochemical and immunological mechanism(s): A critical discussion. *Environ. Toxicol. Pharmacol.* **2014**, *37*, 118–133. [CrossRef] [PubMed]
33. Castell, J.V.; Donato, M.T.; Gomez-Lechon, M.J. Metabolism and bioactivation of toxicants in the lung. The in vitro cellular approach. *Exp. Toxicol. Pathol.* **2005**, *57* (Suppl. S1), 189–204. [CrossRef] [PubMed]
34. O'Brien, P.J.; Irwin, W.; Diaz, D.; Howard-Cofield, E.; Krejsa, C.M.; Slaughter, M.R.; Gao, B.; Kaludercic, N.; Angeline, A.; Bernardi, P.; et al. High concordance of drug-induced human hepatotoxicity with in vitro cytotoxicity measured in a novel cell-based model using high content screening. *Arch. Toxicol.* **2006**, *80*, 580–604. [CrossRef]

35. Donato, M.T.; Tolosa, L. Stem-cell derived hepatocyte-like cells for the assessment of drug-induced liver injury. *Differentiation* **2019**, *106*, 15–22. [CrossRef]
36. Persson, M.; Loye, A.F.; Jacquet, M.; Mow, N.S.; Thougaard, A.V.; Mow, T.; Hornberg, J.J. High-content analysis/screening for predictive toxicology: Application to hepatotoxicity and genotoxicity. *Basic Clin. Pharmacol. Toxicol.* **2014**, *115*, 18–23. [CrossRef] [PubMed]
37. Waters, J.C. Accuracy and precision in quantitative fluorescence microscopy. *J. Cell Biol.* **2009**, *185*, 1135–1148. [CrossRef] [PubMed]
38. Wright, G.D.; Ward, A.M.; Bard, F.; Calvert, M.E. Pushing the boundaries of high content imaging. *Cytom. A* **2017**, *91*, 113–114. [CrossRef]
39. Bougen-Zhukov, N.; Loh, S.Y.; Lee, H.K.; Loo, L.H. Large-scale image-based screening and profiling of cellular phenotypes. *Cytom. A* **2017**, *91*, 115–125. [CrossRef]
40. Liron, Y.; Paran, Y.; Zatorsky, N.G.; Geiger, B.; Kam, Z. Laser autofocusing system for high-resolution cell biological imaging. *J. Microsc* **2006**, *221*, 145–151. [CrossRef]
41. Meijering, E.; Dzyubachyk, O.; Smal, I. Methods for cell and particle tracking. *Methods Enzymol.* **2012**, *504*, 183–200. [CrossRef] [PubMed]
42. Yu, W.; Lee, H.K.; Hariharan, S.; Bu, W.; Ahmed, S. Evolving generalized Voronoi diagrams for accurate cellular image segmentation. *Cytom. A* **2010**, *77*, 379–386. [CrossRef] [PubMed]
43. Kozak, K.; Rinn, B.; Leven, O.; Emmenlauer, M. Strategies and Solutions to Maintain and Retain Data from High Content Imaging, Analysis, and Screening Assays. *Methods Mol. Biol.* **2018**, *1683*, 131–148. [CrossRef] [PubMed]
44. Berliocchi, L.; Chiappini, C.; Adornetto, A.; Gentile, D.; Cerri, S.; Russo, R.; Bagetta, G.; Corasaniti, M.T. Early LC3 lipidation induced by d-limonene does not rely on mTOR inhibition, ERK activation and ROS production and it is associated with reduced clonogenic capacity of SH-SY5Y neuroblastoma cells. *Phytomedicine* **2018**, *40*, 98–105. [CrossRef] [PubMed]
45. Tolosa, L.; Gomez-Lechon, M.J.; Perez-Cataldo, G.; Castell, J.V.; Donato, M.T. HepG2 cells simultaneously expressing five P450 enzymes for the screening of hepatotoxicity: Identification of bioactivable drugs and the potential mechanism of toxicity involved. *Arch. Toxicol.* **2013**, *87*, 1115–1127. [CrossRef]
46. Cosgrove, B.D.; King, B.M.; Hasan, M.A.; Alexopoulos, L.G.; Farazi, P.A.; Hendriks, B.S.; Griffith, L.G.; Sorger, P.K.; Tidor, B.; Xu, J.J.; et al. Synergistic drug-cytokine induction of hepatocellular death as an in vitro approach for the study of inflammation-associated idiosyncratic drug hepatotoxicity. *Toxicol. Appl. Pharmacol.* **2009**, *237*, 317–330. [CrossRef] [PubMed]
47. Dykens, J.A.; Jamieson, J.D.; Marroquin, L.D.; Nadanaciva, S.; Xu, J.J.; Dunn, M.C.; Smith, A.R.; Will, Y. In vitro assessment of mitochondrial dysfunction and cytotoxicity of nefazodone, trazodone, and buspirone. *Toxicol. Sci.* **2008**, *103*, 335–345. [CrossRef]
48. Xu, J.J.; Henstock, P.V.; Dunn, M.C.; Smith, A.R.; Chabot, J.R.; de Graaf, D. Cellular imaging predictions of clinical drug-induced liver injury. *Toxicol. Sci.* **2008**, *105*, 97–105. [CrossRef]
49. Mennecozzi, M.; Landesmann, B.; Palosaari, T.; Harris, G.; Whelan, M. Sex differences in liver toxicity-do female and male human primary hepatocytes react differently to toxicants in vitro? *PLoS ONE* **2015**, *10*, e0122786. [CrossRef]
50. Tolosa, L.; Carmona, A.; Castell, J.V.; Gomez-Lechon, M.J.; Donato, M.T. High-content screening of drug-induced mitochondrial impairment in hepatic cells: Effects of statins. *Arch. Toxicol.* **2015**, *89*, 1847–1860. [CrossRef]
51. Tolosa, L.; Jimenez, N.; Pelecha, M.; Castell, J.V.; Gomez-Lechon, M.J.; Donato, M.T. Long-term and mechanistic evaluation of drug-induced liver injury in Upcyte human hepatocytes. *Arch. Toxicol.* **2019**, *93*, 519–532. [CrossRef] [PubMed]
52. Wink, S.; Hiemstra, S.W.; Huppelschoten, S.; Klip, J.E.; van de Water, B. Dynamic imaging of adaptive stress response pathway activation for prediction of drug induced liver injury. *Arch. Toxicol.* **2018**, *92*, 1797–1814. [CrossRef] [PubMed]
53. Donato, M.T.; Martinez-Romero, A.; Jimenez, N.; Negro, A.; Herrera, G.; Castell, J.V.; O'Connor, J.E.; Gomez-Lechon, M.J. Cytometric analysis for drug-induced steatosis in HepG2 cells. *Chem. Biol. Interact.* **2009**, *181*, 417–423. [CrossRef] [PubMed]
54. Hedley, D.W.; Chow, S. Evaluation of methods for measuring cellular glutathione content using flow cytometry. *Cytometry* **1994**, *15*, 349–358. [CrossRef] [PubMed]
55. Naguib, Y.M. Antioxidant activities of astaxanthin and related carotenoids. *J. Agric. Food Chem.* **2000**, *48*, 1150–1154. [CrossRef] [PubMed]
56. Wink, S.; Hiemstra, S.; Huppelschoten, S.; Danen, E.; Niemeijer, M.; Hendriks, G.; Vrieling, H.; Herpers, B.; van de Water, B. Quantitative high content imaging of cellular adaptive stress response pathways in toxicity for chemical safety assessment. *Chem. Res. Toxicol.* **2014**, *27*, 338–355. [CrossRef]
57. Wink, S.; Hiemstra, S.; Herpers, B.; van de Water, B. High-content imaging-based BAC-GFP toxicity pathway reporters to assess chemical adversity liabilities. *Arch. Toxicol.* **2017**, *91*, 1367–1383. [CrossRef]
58. Aithal, G.P. Hepatotoxicity related to antirheumatic drugs. *Nat. Rev. Rheumatol.* **2011**, *7*, 139–150. [CrossRef]
59. Matsuo, K.; Sasaki, E.; Higuchi, S.; Takai, S.; Tsuneyama, K.; Fukami, T.; Nakajima, M.; Yokoi, T. Involvement of oxidative stress and immune- and inflammation-related factors in azathioprine-induced liver injury. *Toxicol. Lett.* **2014**, *224*, 215–224. [CrossRef]
60. Garside, H.; Marcoe, K.F.; Chesnut-Speelman, J.; Foster, A.J.; Muthas, D.; Kenna, J.G.; Warrior, U.; Bowes, J.; Baumgartner, J. Evaluation of the use of imaging parameters for the detection of compound-induced hepatotoxicity in 384-well cultures of HepG2 cells and cryopreserved primary human hepatocytes. *Toxicol. In Vitro* **2014**, *28*, 171–181. [CrossRef]

61. Pradip, A.; Steel, D.; Jacobsson, S.; Holmgren, G.; Ingelman-Sundberg, M.; Sartipy, P.; Bjorquist, P.; Johansson, I.; Edsbagge, J. High Content Analysis of Human Pluripotent Stem Cell Derived Hepatocytes Reveals Drug Induced Steatosis and Phospholipidosis. *Stem Cells Int.* **2016**, *2016*, 2475631. [CrossRef]
62. Saito, J.; Okamura, A.; Takeuchi, K.; Hanioka, K.; Okada, A.; Ohata, T. High content analysis assay for prediction of human hepatotoxicity in HepaRG and HepG2 cells. *Toxicol. In Vitro* **2016**, *33*, 63–70. [CrossRef] [PubMed]
63. Wilson, J.; Berntsen, H.F.; Zimmer, K.E.; Frizzell, C.; Verhaegen, S.; Ropstad, E.; Connolly, L. Effects of defined mixtures of persistent organic pollutants (POPs) on multiple cellular responses in the human hepatocarcinoma cell line, HepG2, using high content analysis screening. *Toxicol. Appl. Pharmacol.* **2016**, *294*, 21–31. [CrossRef]
64. Hiemstra, S.; Ramaiahgari, S.C.; Wink, S.; Callegaro, G.; Coonen, M.; Meerman, J.; Jennen, D.; van den Nieuwendijk, K.; Dankers, A.; Snoeys, J.; et al. High-throughput confocal imaging of differentiated 3D liver-like spheroid cellular stress response reporters for identification of drug-induced liver injury liability. *Arch. Toxicol.* **2019**, *93*, 2895–2911. [CrossRef]
65. Rodeiro, I.; Hernandez, I.; Herrera, J.A.; Riera, M.; Donato, M.T.; Tolosa, L.; Gonzalez, K.; Ansoar, Y.; Gomez-Lechon, M.J.; Vanden Berghe, W.; et al. Assessment of the cytotoxic potential of an aqueous-ethanolic extract from Thalassia testudinum angiosperm marine grown in the Caribbean Sea. *J. Pharm. Pharmacol.* **2018**, *70*, 1553–1560. [CrossRef] [PubMed]
66. Tolosa, L.; Rodeiro, I.; Donato, M.T.; Herrera, J.A.; Delgado, R.; Castell, J.V.; Gomez-Lechon, M.J. Multiparametric evaluation of the cytoprotective effect of the Mangifera indica L. stem bark extract and mangiferin in HepG2 cells. *J. Pharm. Pharmacol.* **2013**, *65*, 1073–1082. [CrossRef] [PubMed]
67. O'Brien, P.J. High-content analysis in toxicology: Screening substances for human toxicity potential, elucidating subcellular mechanisms and in vivo use as translational safety biomarkers. *Basic Clin. Pharmacol. Toxicol.* **2014**, *115*, 4–17. [CrossRef] [PubMed]
68. Carbonell, P.; Lopez, O.; Amberg, A.; Pastor, M.; Sanz, F. Hepatotoxicity prediction by systems biology modeling of disturbed metabolic pathways using gene expression data. *ALTEX* **2017**, *34*, 219–234. [CrossRef]
69. Garcia-Canaveras, J.C.; Castell, J.V.; Donato, M.T.; Lahoz, A. A metabolomics cell-based approach for anticipating and investigating drug-induced liver injury. *Sci. Rep.* **2016**, *6*, 27239. [CrossRef]
70. Li, S.; Xia, M. Review of high-content screening applications in toxicology. *Arch. Toxicol.* **2019**, *93*, 3387–3396. [CrossRef]
71. Godoy, P.; Hewitt, N.J.; Albrecht, U.; Andersen, M.E.; Ansari, N.; Bhattacharya, S.; Bode, J.G.; Bolleyn, J.; Borner, C.; Bottger, J.; et al. Recent advances in 2D and 3D in vitro systems using primary hepatocytes, alternative hepatocyte sources and non-parenchymal liver cells and their use in investigating mechanisms of hepatotoxicity, cell signaling and ADME. *Arch. Toxicol.* **2013**, *87*, 1315–1530. [CrossRef] [PubMed]

antioxidants

MDPI

Review

Methylmercury-Mediated Oxidative Stress and Activation of the Cellular Protective System

Masatake Fujimura [1,†] **and Fusako Usuki** [2,*,†]

1 Department of Basic Medical Sciences, National Institute for Minamata Disease, Kumamoto 867-0008, Japan; fujimura@nimd.go.jp
2 Division of Neuroimmunology, Joint Research Center for Human Retrovirus Infection, Kagoshima University, Kagoshima 890-8544, Japan
* Correspondence: fusuki@kufm.kagoshima-u.ac.jp; Tel.: +81-99-275-6246; Fax: +81-99-275-5942
† These authors contributed equally to this work.

Received: 30 September 2020; Accepted: 14 October 2020; Published: 16 October 2020

Abstract: Methylmercury (MeHg) is a well-known neurotoxicant that causes severe intoxication in humans. In Japan, it is referred to as Minamata disease, which involves two characteristic clinical forms: fetal type and adult type depending on the exposed age. In addition to MeHg burden level, individual susceptibility to MeHg plays a role in the manifestation of MeHg toxicity. Research progress has pointed out the importance of oxidative stress in the pathogenesis of MeHg toxicity. MeHg has a high affinity for selenohydryl groups, sulfhydryl groups, and selenides. It has been clarified that such affinity characteristics cause the impairment of antioxidant enzymes and proteins, resulting in the disruption of antioxidant systems. Furthermore, MeHg-induced intracellular selenium deficiency due to the greater affinity of MeHg for selenohydryl groups and selenides leads to failure in the recoding of a UGA codon for selenocysteine and results in the degradation of antioxidant selenoenzyme mRNA by nonsense-mediated mRNA decay. The defect of antioxidant selenoenzyme replenishment exacerbates MeHg-mediated oxidative stress. On the other hand, it has also been revealed that MeHg can directly activate the antioxidant Keap1/Nrf2 signaling pathway. This review summarizes the incidence of MeHg-mediated oxidative stress from the viewpoint of the individual intracellular redox system interactions and the MeHg-mediated aforementioned intracellular events. In addition, the mechanisms of cellular stress pathways and neuronal cell death triggered by MeHg-mediated oxidative stress and direct interactions of MeHg with reactive residues of proteins are mentioned.

Keywords: methylmercury; oxidative stress; binding affinity; redox signaling; selenoenzyme; nonsense-mediated mRNA decay; posttranscriptional defect; thiol antioxidant capacity; Keap1/Nrf2 pathway

1. Introduction

Methylmercury (MeHg) is a well-established neurotoxicant that affects various cellular functions depending on the cellular context and developmental phase. Severe MeHg-intoxication episodes in humans have been recognized in a number of countries, including Japan (Minamata disease) [1], Iraq [2], and the USA [3]. Minamata disease was named for the first case of MeHg poisoning due to ingestion of seafood contaminated by MeHg discharged from a chemical plant. The disease involves two characteristic clinical forms: fetal type and adult type depending on the exposed age. Fetal-type Minamata disease, which is caused by exposure to MeHg in utero, shows cerebral palsy-like clinical features with delayed psychomotor development [4]. In contrast, adult-type Minamata disease, which is caused by MeHg intoxication in adulthood, shows Hunter-Russell syndrome-like features [5]. The main lesions are found in the central nervous and peripheral sensory nervous systems [6,7]. The patients show neurological signs associated with such pathological lesions. MeHg toxicity is an environmental concern for human health, especially in susceptible populations who frequently consume substantial amounts of fish or fish predators [8].

To date, many studies on the pathogenetic processes caused by MeHg exposure have been reported using various cells and animal models. The pathological changes in MeHg toxicity do not correspond directly to the accumulation of Hg in tissues. In a MeHg-exposed subacute rat model (20 ppm MeHg in drinking water every day for 28 days), for example, the cerebellum showed the most severe pathology with a lower Hg concentration than that in the liver or kidneys, which displayed fewer pathological changes from high Hg content [9]. These findings suggest that cellular responsive conditions play a role in the development of pathogenetic changes due to MeHg exposure. Research progress has noted the importance of oxidative stress in the pathogenesis of MeHg toxicity. The early incidence of reactive oxygen species (ROS) has been reported in many MeHg intoxication experiments [10–13]. The elimination of ROS by co-treatment with radical scavengers can prevent MeHg-cytotoxicity in vitro and in vivo [9,11,14], but failure to protect cells against such early oxidative stress leads to further progression of MeHg-toxicity, such as the occurrence of subsequent endoplasmic reticulum (ER) stress and apoptosis [12].

It has been revealed that high affinity for selenohydryl groups, sulfhydryl groups, and selenides [15] plays a critical role in the incidence of MeHg toxicity [16–18]. Many antioxidant enzymes and proteins have thiol and selenol residues. The high affinity of MeHg for selenohydryl groups, sulfhydryl groups, and selenides results in the impairment of antioxidant enzymes and proteins as well as the subsequent disruption of antioxidant systems, which leads to MeHg-mediated oxidative stress. On the other hand, it has been clarified that MeHg can also directly activate an antioxidant-signaling pathway. The high affinity of MeHg for sulfhydryl groups can activate the cellular antioxidant transcription factor Nrf2 through the interaction of MeHg and Nrf2 regulator Keap1 [19]. Nrf2 activation and translocation to the nucleus induce downstream antioxidant proteins and enzymes [20–22].

Furthermore, MeHg has been shown to induce posttranscriptional defects in antioxidant selenoenzymes [23]. The high affinity of MeHg for selenohydryl groups and selenides causes intracellular relative active selenium (Se) deficiency. Such MeHg-induced Se deficiency leads to failure in the recoding of a UGA codon for selenocysteine and results in the degradation of antioxidant selenoenzyme mRNA by nonsense-mediated mRNA decay (NMD). NMD is a cellular mechanism that detects the premature termination codon (PTC) located 5′-upstream of the last exon-exon junction and degrades PTC-containing mRNAs [24]. Generally, PTC is recognized when it is located upstream of an exon–exon junction with a distance of at least 55 nucleotides [25]. Targets for NMD can include mutationally-induced nonsense or frameshift codons, upstream open reading frames, alternatively spliced or mis-spliced mRNA [26]. NMD has been considered an mRNA quality surveillance mechanism to protect an organism against deleterious dominant-negative or gain-of-function effects of truncated proteins that arise from PTCs.

The incidence of MeHg-mediated oxidative stress should depend on the individual capacities of the intracellular redox systems to respond to the results of the interactions among the aforementioned MeHg-induced events. Once MeHg-mediated oxidative stress occurs, it may trigger activation of various cellular signaling pathways leading to cellular damage. Recent works clarified the molecular mechanism of MeHg-induced apoptosis and cortical neuronal cell death caused by MeHg-mediated oxidative stress.

2. Disruption of the Cellular Redox Systems by MeHg Exposure

The critical role of oxidative stress in the pathogenesis of MeHg toxicity has been demonstrated both in vitro [11,12,27–30] and in vivo [9,31,32]. Many studies have reported an increase in ROS after MeHg exposure [10,13], and suppression of such increases by co-treatment with the antioxidant Trolox, sodium selenite, and ebselen have also been reported [12,23]. Oxidative stress is a condition in which the normal function of the redox network is disrupted. MeHg can disrupt many antioxidant proteins and enzymes in the cellular redox system because of the high affinity of MeHg for selenohydryl groups, sulfhydryl groups, and selenides [15,33]. In particular, the high affinity of MeHg for selenohydryl groups and selenides leads to intracellular relative Se-deficient conditions, resulting in enhanced degradation of antioxidant selenoenzyme mRNA targeted by NMD [23]. Defects in antioxidant selenoenzyme replenishment due to mRNA degradation should exacerbate MeHg-induced oxidative stress.

2.1. Suppression of Antioxidant Protein and Enzyme Activity by MeHg Exposure

Glutathione (GSH) and thioredoxin (Trx) systems, which are the central regulators of cellular redox status, include antioxidant proteins and enzymes with the selenohydryl group or sulfhydryl group at the redox-active centers.

GSH is the most abundant thiol compound and the major antioxidant that functions as a redox buffer. MeHg interacts with GSH and forms a GSH–MeHg complex, which is excreted from the ATP-binding cassette sub-family C member 4 (ABCC4) transporter. GSH consumption under MeHg exposure elicited GSH synthesis through the upregulation of glutamate–cysteine ligase (GCL), a rate-limiting enzyme for GSH synthesis [34,35]. It has been reported that GSH content shows no change in the brain, liver, kidneys, and muscles [9,35] or an increase in the brain [34] in rat or mice MeHg-exposed models, suggesting the replenishment of GSH occurs during the consumption of GSH under MeHg exposure. Glutathione peroxidase (GPx), the most abundant selenoenzyme, plays a role in preventing the production of ROS by reducing hydrogen peroxide (H_2O_2) and free fatty acid hydroperoxides. Exposure to MeHg decreases GPx1 activity [9,36,37] and GPx1 mRNA in vitro and in vivo [23]. It has been reported that overexpression of GPx1 prevents MeHg-induced neurotoxicity in cultured cerebellar granule cells, suggesting that GPx1 plays a critical role in MeHg-mediated disruption of the cellular redox systems [37].

In contrast, another selenoenzyme, thioredoxin reductase (TrxR) and its substrate Trx, are known to be involved in the regulation of a large network of redox reactions, including metallothionein, ribonucleotide reductase, and redox-factor-1 (REF-1) [38]. REF-1 is known to maintain the cysteine residues of transcription factors, such as nuclear factor kappa B (NF-κB), Nrf2, and p53, in the reduced form required for DNA binding. The active site of TrxR has a redox-active selenothiol/selenylsulfide [39] and is known to be sensitive to MeHg [40]. Since selenols have a lower pK_a than thiols and are fully ionized to selenolates under physiological conditions [39,41], selenols are more reactive toward Hg [40]. The decrease in TrxR1 activity caused by MeHg exposure has been shown in vitro and in vivo [23,40,42]. The reduction in activity of the cellular Trx system under MeHg exposure has also been demonstrated [40]. Such damage to activity in the Trx system should lead to the disruption of a large network of redox reactions [38]. It is generally known that males are more vulnerable to MeHg toxicity than females. The recent study on litters from dam mice exposed to MeHg (5 ppm MeHg in drinking water from early gestational period until postnatal day 21) showed that basal levels of GPx1 and TrxR1 mRNAs in cerebrum were lower in males than in females and that the activities of TrxR, GPx1, and Gpx1 mRNA after MeHg exposure decreased at a greater rate in males than that in females [43]. The results suggest that sex differences in the antioxidant system may affect sex differences in the susceptibility to MeHg.

Manganese-superoxide dismutase (Mn-SOD) localized in mitochondria functions in preventing the production of ROS by reducing superoxide radicals. The study of the effect of MeHg on Mn-SOD in mouse brains revealed that MeHg reduced Mn-SOD activity whereas Mn-SOD mRNA levels and protein synthesis were not affected by MeHg administration [44]. It has been reported that MeHg interacts with Mn-SOD through a reactive sulfhydryl group [19]. A decrease in Mn-SOD activity caused by MeHg contributes to an increase in ROS. It has been reported that overexpression of Mn-SOD prevents MeHg toxicity in HeLa cells [45].

2.2. Basal Level of Antioxidant Enzymes Associated with MeHg Toxicity

Fragile cells exposed to oxidative stress should be more damaged by MeHg. The main target organ of MeHg toxicity is the central nervous system. Adult-type Minamata disease caused the site-specific brain lesions in the cerebellum and cerebrum. Neurons in the cerebellar cortex are arranged in three layers. The innermost layer is densely packed with numerous, small neurons called granule cells (cerebellar granule cells). The middle layer contains Purkinje cells, and the outmost layer, or the molecular layer, contains stellate and basket cells. Among these cerebellar neurons, cerebellar granule cells are the most vulnerable to MeHg. Furthermore, autopsy studies of human cerebrum revealed that the lesions are localized in the deeper layers cerebrocortical neurons, especially layer IV, compared to the shallow layers cerebrocortical neurons [46,47]. Such site-specific cerebral lesions were also observed in MeHg intoxicated animal models of rats and mice [48,49].

We previously demonstrated that cerebellar granule cells susceptible to MeHg have lower in situ expression of Mn-SOD, GPx1, and TRxR1 mRNAs than cerebellar molecular layers and Purkinje cells, which are known to be resistant to MeHg [49]. Furthermore, our in situ analyses of antioxidative enzymes expression using quantitative reverse transcription polymerase chain reaction from laser micro-dissected mouse cerebral cortex samples and immunohistochemistry revealed that lower basal expression levels of Mn-SOD and GPx1 mRNAs and proteins in the cerebrocortical neurons of deeper layers than those in shallow layers. In addition, an increase in Mn-SOD mRNA expression induced by MeHg exposure is lower in deeper layers than that in shallow layers [50]. These findings suggest that the different antioxidative systems in situ, including basal levels of antioxidant enzymes, play a role in the site-specific neurotoxicity of MeHg in the brain.

2.3. Posttranscriptional Defects of Selenoenzymes

MeHg-mediated increases in intracellular ROS cause changes in antioxidant gene expression. Our previous study demonstrated that MeHg exposure upregulated Mn-SOD, copper, zinc (Cu, Zn)-SOD, catalase, and TrxR1 mRNAs [23]. The upregulation of these mRNAs was mediated by ROS because treatment with the antioxidant Trolox suppressed the increase in these mRNAs. In contrast, selenoenzyme GPx1 mRNA was downregulated despite its decreased activity in vitro and in vivo [23]. In addition, Trolox failed to rescue such GPx1 mRNA decrease. Our in situ antioxidative enzymes expression analyses using laser micro-dissected mouse cerebrocortical neuron samples also revealed downregulation of GPx1 mRNA [50]. This is intriguing because oxidative stress due to the general burden of H_2O_2 caused upregulation of GPx1 mRNA, indicating that the MeHg-induced GPx1 mRNA decrease is specific to the burden of MeHg [23].

GPx1 has a single selenocysteine (Sec), in which Se is co-translationally inserted. Sec is encoded by a UGA codon, which shares a common codon to function as a terminator for protein synthesis. The biosynthesis of Sec occurs on its tRNA (Sec tRNA $^{[Ser]Sec}$), unlike the other 20 amino acids. Once activated Se is donated to the structure, Sec is completed [51]. The insertion of Sec into protein requires the Sec insertion sequence (SECIS) [52], SECIS binding protein (SBP2) [53], and Sec-specific elongation factor [54,55]. Under Se deficiency, however, the UGA codon for Sec may be recognized as a nonsense codon, known as a PTC, due to the incomplete biosynthesis of Sec. mRNAs harboring PTCs are known to be deleted by NMD, an mRNA quality control mechanism that is executed when PTC is located sufficiently upstream of the exon–exon junction [24,56–58]. A previous report demonstrated that Sec on GPx1 mRNA that resides 105 nucleotides upstream of the sole exon–exon junction was recognized as a PTC and degraded by NMD under active Se-deficient conditions [59].

It is known that the selenohydryl group has a high affinity for Hg compared to those of the sulfhydryl and amino groups. The order of binding affinity of the coordination groups toward MeHg is as follows: SeH > SH ≥ Se-Se > NH_2 > S-S [15]. The high affinity of MeHg for the selenohydryl group and selenide should cause relative intracellular Se-deficient conditions under MeHg exposure. Our previous study demonstrated that the MeHg-induced decrease in GPx1 mRNA is a post-transcriptional event by NMD, enhanced degradation of mRNA that is most likely mediated by cellular Se deficiency [23]. This finding was confirmed by two studies: (1) MeHg-induced decrease in GPx1 mRNA was rescued by pretreatment with sodium selenite, and (2) MeHg-induced decrease in GPx1 mRNA was inhibited in siRNA-mediated NMD component knockdown cells. In contrast to GPx1, mRNA of another antioxidant selenoenzyme, TrxR1, was not downregulated by MeHg exposure. The Sec codon UGA-498 on TrxR1 that resides in the last exon cannot be a substrate for NMD because at least one downstream intron is required to trigger NMD [25,60,61]. In theory, the TrxR1 protein synthesized by NMD-skipped TrxR1 mRNA should be truncated because the Sec codon is recognized as a nonsense codon under Se-deficient conditions. The different pathways involved in the synthesis of GPx1 and TrxR1 under the sufficient or MeHg-induced deficient active form of Se are summarized in Figure 1.

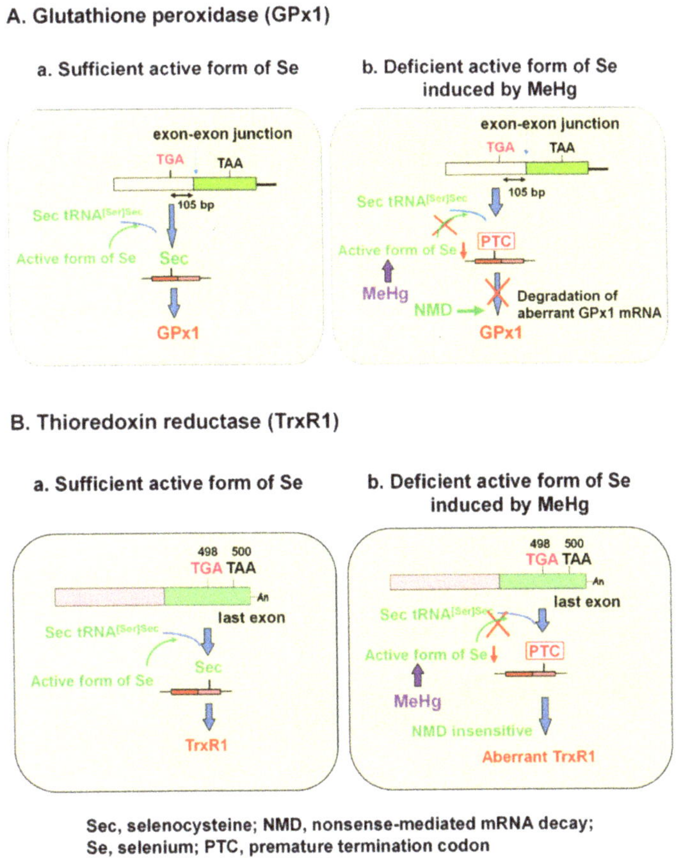

Figure 1. Posttranscriptional effect of methylmercury (MeHg) on antioxidant selenoenzymes. (**A**) Glutathione peroxidase 1 (GPx1). The encoded UGA codon for selenocysteine (Sec) resides 105 nucleotides upstream of the sole exon–exon junction. When a UGA codon is recognized as a Sec codon under sufficient active form of selenium (Se) (left panel), GPx1 is synthesized. However, since UGA codon is recognized as a nonsense codon under MeHg-induced active Se deficiency (right), GPx1 mRNA should be a natural substrate for nonsense-mediated mRNA decay (NMD; right panel). (**B**) Thioredoxin reductase 1 (TrxR1). The Sec codon UGA-498 resides in the last exon on TrxR1 mRNA; thus, TrxR1 mRNA cannot be a substrate for NMD even when a UGA codon is recognized as a nonsense codon under MeHg-induced Se deficiency and aberrant Trx1 is synthesized (right panel).

There are 25 known Se-containing proteins in mammals [62]. Under Se deficiency, it has been noted that selenoprotein expression has a hierarchy regarding the level of individual selenoproteins and Se in different organs [51]. The different distributions of MeHg accumulation after MeHg exposure [9] should also cause MeHg-induced intracellular Se-deficient conditions to be responsive to Se status in various tissues. In addition, the sensitivity of selenoenzyme activity to Se deficiency and mRNA turnover may affect selenoprotein expression. Many eukaryotic selenoproteins serve as antioxidant and redox proteins. Among them, the greater sensitivity of GPx1 activity to Se deficiency is attributed to an increased turnover in mRNA [63], resulting in enhanced degradation by NMD. Defects in the synthesis of GPx1 protein should exacerbate MeHg-mediated oxidative stress.

3. MeHg-Induced Mitochondrial Damage

Mitochondria is the most important sites of ROS generation. Imbalance of cellular ROS elimination system and RNA leakage by the disruption of mitochondrial function contributes to the occurrence of intracellular oxidative stress. It has been reported that isolated rat mitochondria show higher oxygen consumption levels and ROS production rates in the cerebrum and the cerebellum, which are the main lesion in MeHg toxicity, than in the liver [64]. In addition, it is known that GPx and SOD activities are lower in cerebrum and cerebellum than those in liver [64]. These findings may explain the difference in the distribution of MeHg-induced pathological changes.

3.1. Inhibition of Electron Transport Chain

The mitochondrial electron transport chain (ETC), composed of four respiratory enzyme complexes (I–IV) on the inner membrane, has been recognized as one of the most important sites of ROS generation [65,66]. Under physiological conditions, electrons are transferred fluently through the chains via adequate levels of ETC activity, resulting in the release of a minimum of mitochondrial ROS. When the ETC is damaged, however, electrons flowing through the chain will be disrupted, leading to the elevation of mitochondrial ROS leakage. If MeHg directly attacks the respiratory enzyme complexes or MeHg-mediated oxidative stress damages them, the function of ETC may be disrupted to produce excess ROS. The elevation of mitochondrial ROS leakage should cause oxidative stress and cell damage under the insufficient ROS elimination system. In the previous study, we demonstrated that MeHg treatment (orally administered at 5 mg/kg/day for 12 days) decreased mitochondrial enzymes (cytochrome c oxidase (CCO, complex IV) and succinate dehydrogenase (SDH, complex II)) activities in mitochondria-rich soleus muscle of rats [67]. Co-treatment with antioxidant Trolox clearly prevented such decrease in CCO and SDH activities despite the retention of MeHg, suggesting that MeHg-mediated oxidative stress caused a decrease in CCO and SDH activities [9]. MeHg-induced increase in the generation of H_2O_2 and O_2^- in the mitochondria has been shown using extracted mitochondria from MeHg-exposed rat cerebellum [64]. Furthermore, the subsequent study showed that the complex II activity decreased in the cerebellum mitochondria and released cytochrome c in MeHg-exposed rats [68].

3.2. Mitochondria-Dependent Apoptotic Pathway

Many studies have shown that MeHg induces apoptosis due to both mitochondria- and ER-generated processes. Mitochondrial apoptotic pathway is governed by members of B-cell lymphoma-2 (Bcl2) protein family, which regulates mitochondrial outer membrane permeabilization (MOMP) [69]. Once MOMP occurs, cytochrome c is released from the mitochondria to the cytosol. Cytosolic cytochrome c interacts with apoptotic protease activating factor 1 (Apaf-1) and procaspase-9 to form a complex called apoptosome, which facilitates caspase-9 activation [70]. Caspase-9 is one of the initiators that cleave and activate the effector caspase-3, a central component of the apoptotic response.

It has been reported that MeHg induces activation of caspase-9 and caspase-3, high levels of cytoplasm cytochrome C, and apoptosis in cultured cortical neurons. Pre-treatment with Trolox significantly inhibited such neuronal apoptosis as well as mitochondrial dysfunction, suggesting these events were caused by MeHg-mediated oxidative stress [71]. MeHg-induced activation of mitochondria-dependent apoptotic pathway was also shown in the mouse cerebrum [72] and in developing rat hippocampus [73]. MeHg induced an increase in executioner caspase-3 by both mitochondrial-dependent caspase-9 and mitochondrial-independent caspase-8 in developing rat hippocampus. Studies using cultured neuronal cells suggested that ROS-mediated activation of ERK1/2 and p38 pathways regulated mitochondria-dependent apoptotic pathways that were involved in MeHg-induced neurotoxicity [72].

4. Cellular Stress Pathways Triggered by MeHg-Mediated Oxidative Stress

Oxidative stress generates not only free radicals but also nonradical oxidants. Free radicals tend to be reactive, and their initiation leads to macromolecular damage. In contrast, nonradical oxidants (e.g., H_2O_2, peroxynitrite, lipid hydroperoxide, and disulfides) are generated more than free radicals during oxidative stress, resulting in the disruption of redox signaling and physiological regulation pathways. Two major thiol- and selenol-containing GSH and Trx systems regulate such redox pathways, including receptor signaling (e.g., estrogen and glucocorticoid receptors), transcriptional regulation (e.g., NF-κB, Nrf2, p53, and HIF-1α), and apoptosis [38].

4.1. Apoptosis Signaling Pathway

MeHg disturbs intracellular redox systems through direct attacks on the cysteine and/or selenocysteine residues of antioxidant enzymes and a post-transcriptional effect on the major antioxidant selenoenzymes GPx1 and TrxR1 [23,40]. GPx1 is the most abundant selenoprotein that plays a critical role in the reduction in cellular H_2O_2. In contrast, TrxR1 transfers electrons from NADPH to thioredoxin, which in turn reduces thioredoxin peroxidase and other redox proteins [74]. It is known that TrxR1 can reduce some substrates other than Trx, including selenite [75], lipid hydroperoxides [76], and H_2O_2 [77]. As such, MeHg disturbs the GSH and Trx cellular redox systems at the early stage of cytotoxicity, which is followed by cellular stress responses.

Trx is known to be a direct physiological inhibitor of apoptosis signal-regulating kinase-1 (ASK1) through binding of the active site dithiol with the N-terminus of ASK1 [78]. ASK1 activates c-Jun N-terminal kinase (JNK) and p38 MAP kinase pathways and is required for stress-induced apoptosis [79]. Oxidation of the Trx active site results in the release and activation of ASK1 followed by ASK1-dependent apoptosis. It has been demonstrated that MeHg exposure activates ASK1, subsequently resulting in activation of the stress-activated protein kinase (SAPK)/JNK pathways and apoptosis [12,29].

4.2. ER Stress

The ER is a membrane-bound organelle specialized for folding and post-translational maturation of secretory and membrane proteins. The ER redox state is linked to ER protein-folding homeostasis. Disulfide bond formation in the ER lumen is highly sensitive to altered redox balance, where both reducing and oxidizing reagents disrupt protein folding and cause ER stress [80]. In a stressed ER, dysregulated disulfide bond formation and breakage may result in ROS accumulation and cause oxidative stress. As such, ER stress can cause mitochondrial dysfunction and increase mitochondrial ROS production [81].

In a previous study, we showed that failure to protect cells against MeHg-mediated early oxidative stress triggers the subsequent ER stress and apoptosis [12]. A subsequent study showed that pretreatment with Trolox significantly blocked MeHg-induced ER stress, unfolded protein response activation, and apoptosis in neuronal cells [71], confirming that MeHg-mediated oxidative stress causes ER stress and apoptosis. In contrast, it has been reported that MeHg directly causes ER stress through interaction with protein disulfide isomerase (PDI) [18]. PDI localizes in the ER and catalyzes all of the reactions involved in native disulfide bond formation in the ER [82]. The oxidoreductase activity of PDI is derived from thiol groups of active site cysteines [83]. Furthermore, PDI has been identified as a multi-domain protein related to the cytoplasmic thioredoxin [84,85]. It has been demonstrated that the dysfunction of PDI enzymatic activity by the oxidative modification of active cysteine by nitric oxide (NO) induced by treatment with N-Methyl-D-aspartate causes the accumulation of newly synthesized unfolded protein in the ER lumen, resulting ER stress [86]. Similar to NO, the MeHg-modified C-terminal catalytic domain in PDI was detected using matrix-assisted laser desorption/ionization-time of flight mass spectrometry (MALDI-TOF/MS) analysis, suggesting that PDI is a target protein for MeHg in the ER [18]. In addition, treatment with MeHg significantly attenuated the enzymatic activity of PDI. These findings suggest that MeHg can also cause ER stress through the direct disruption of PDI function.

5. Activation of the Keap1/Nrf2 Pathway by MeHg Exposure

As a cellular compensatory function against oxidative stress, it has been known that MeHg upregulates antioxidant gene expression of GCL [34,35] as well as Mn-SOD, Cu, Zn-SOD, catalase, and TrxR1 mRNAs [23]. GCL is a rate-limiting enzyme for GSH synthesis and is known to be regulated by the transcription factor Nrf2 [87]. Nrf2 is an essential factor for the protective antioxidant response and detoxification against various environmental toxicants through the antioxidant responsive element (ARE)-mediated induction of enzyme genes [88]. Under physiological conditions, Nrf2 localizes in the cytoplasm and is inactivated through binding to Keap1 to repress its translocation to the nucleus [88]. Keap1 has highly reactive cysteine residues, which are the preferential targets of electrophiles and ROS. Electrophilic agents can functionally liberate Nrf2 from repression by Keap1, allowing Nrf2 to move to the nucleus and potentiate the ARE-mediated induction of enzyme genes [88].

It has been demonstrated by MALDI-TOF/MS analysis that MeHg reacts with Keap1 cysteine residues to cause a structural alteration [19]. Since pretreatment with the antioxidant Trolox could not activate Nrf2 in MeHg-exposed cells, MeHg may activate Nrf2 through direct interaction with the cysteine residues of Keap1 rather than MeHg-mediated oxidative stress [19]. Thus, MeHg exposure led to Keap1/Nrf2 dissociation, Nrf2 translocation to the nucleus, and ARE-mediated induction of oxidative stress enzyme genes, such as GCL and heme oxygenase-1 (HO-1) [20–22]. The cytoprotective role of Nrf2/Keap1 system to MeHg toxicity was confirmed by studies using Nrf2-overexpressed SH-SY5Y cells, and Nrf2- or Keap1-deficient mouse hepatocytes [20]. On the other hand, recent findings have shown that MeHg activates Nrf2 through the Keap1-independent pathway. It is known that Src subfamily kinase Fyn phosphorylates Nrf2, leading to nuclear export and degradation of Nrf2 [89]. It has been demonstrated that MeHg downregulates Fyn through the phosphorylation of Akt and glycogen synthase kinase 3 beta (GSK-3β), leading to sustained Nrf2 activity [90].

As such, MeHg can activate the antioxidant signaling pathway through direct interaction with the cysteine residues of the Keap1 and/or Akt/GSK-3β/Fyn pathway.

6. Neuronal Hyperactivity and Cell Death Triggered by MeHg-Mediated Oxidative Stress

Recently, the mechanism of neuronal cell death caused by MeHg-mediated oxidative stress has been clarified. Autopsy studies of human cerebrum revealed that the lesions were localized in the cerebrocortical neurons of deeper layers, especially layer IV, compared to shallow layers [46,47]. The neocortex is formed of six cortical layers, which are numbered I to VI from the outermost to the innermost. Layer IV, the internal granular layer, contains different types of stellate and pyramidal cells. The site-specific cerebral lesions were also observed in MeHg-intoxicated animal models. We previously reported that neuronal damage caused by MeHg is localized in layer IV of the cerebral cortex in adult mice, especially within the somatosensory cortex [48]. We further demonstrated that site-specific upregulation of c-fos and brain-derived neurotrophic factor (BDNF) preceded neuronal degeneration in layer IV of the cerebral somatosensory cortex of MeHg-intoxicated mice [91]. Layer IV has been shown to be mainly composed of excitable cells [92], suggesting that the characteristic of neuronal excitability may be related to susceptibility to MeHg neurotoxicity and subsequent damage.

c-fos and BDNF are known to be proper markers of neural activity. In particular, c-fos expression analysis is carried out extensively to identify the site-specific neural activity in the brain [93,94]. It has been reported that the induction of c-fos expression is observed in brain-derived tissue exposed to several pathological agents. Exposure of a mouse hippocampal cell line to the 25–35 peptide fragments of A beta caused a rapid and sustained increase in nuclear c-fos immunoreactivity with a decrease in viability [95]. c-fos expression was also observed in the neocortex and paleocortex in picrotoxin-treated rats [96]. The expressions of c-fos and c-jun were significantly up-regulated with apoptosis by T-2 toxin treatment in human neuroblastoma cells [97]. In addition, it has been demonstrated that MeHg exposure significantly induces the expression of c-fos protein in cortex and hippocampus of rats [98]. As such, since c-fos expression has been observed during neuronal degeneration such as neuronal cell loss, and astroglia and microglia accumulation, c-fos may play a role as a trigger of neuronal degeneration. On the other hand, BDNF expression bears on neural activity concerning physiological activity and social stress [99,100].

The molecular mechanism of the relation between MeHg-induced upregulation of c-fos and neuronal cell death was investigated using *all-trans*-retinoic acid (RA) differentiated SH-SY5Y cells, which show neuron-like morphological changes and express neuron/synapse markers for cerebrocortical neurons [101]. Time course studies revealed that MeHg-induced upregulation of c-fos preceded neuronal cell death in RA-differentiated SH-SY5Y cells. We demonstrated early expression of the oxidative stress marker thymidine glycerol followed by activation of p44/42 mitogen-activated protein kinase (MAPK) and p38 MAPK, and an increase in cAMP response element binding protein (CREB) pathways. The antioxidants Trolox and edaravone significantly suppressed such MeHg-induced thymidine glycerol expression, p38 MAPK-CREB pathway activation, and neurotoxicity. Furthermore, treatment with SB203580, a specific inhibitor of p38 MAPK, significantly blocked the upregulation of c-fos and neuronal cell death. These results suggest that MeHg-induced oxidative stress and subsequent activation of the p38 MAPK-CREB pathway contribute to cerebrocortical neuronal hyperactivity and subsequent site-specific neuronal cell death.

7. Biomarkers for Ongoing MeHg-Induced Oxidative Stress

The Hg level in the hair, blood, or nails is known to be a useful biomarker to assess the MeHg body burden for a period of several months prior to sample collection, and the fetal umbilical cord is useful to assess the MeHg burden during fetal development. However, neither of these biomarkers sufficiently reflects individual toxic effects because MeHg toxicity depends on individual susceptibility to MeHg in addition to the MeHg burden. An assessment of MeHg in preserved umbilical cords collected in the Minamata region of Japan demonstrated that some patients with adult-type Minamata disease or mental retardation had MeHg concentrations that were as high as fetal-type Minamata disease despite not showing delayed psychomotor development [4], which suggests that the individual protective capacity against the MeHg body burden plays an important role in the severity of MeHg toxicity.

As mentioned above, MeHg-induced oxidative stress is generated by disruption of the antioxidant system caused by the high affinity of MeHg for selenohydryl groups, sulfhydryl groups, or selenides. Such characteristics may allow plasma thiols and selenoproteins to be used as biomarkers for individual susceptibility to MeHg and thus predict the early effects of MeHg intoxication. Thiol-disulfide pools in plasma can be related to systemic oxidative stress through continuous interactions with tissues and organ systems [102]. Therefore, we assessed the plasma levels of three oxidative stress markers, dROMs (diacron reactive oxidant metabolites) as derivatives of reactive oxygen metabolites, -SHp (thiol antioxidant barrier) as an overall measure of thiol antioxidants, and BAP (biological antioxidant potential) as an overall measure of antioxidants. We also assessed plasma selenoproteins, glutathione peroxidase (GPx3), and selenoprotein P1 (SeP1). We used MeHg-intoxicated model rats (20 ppm MeHg in drinking water every day for 28 days) that showed neuropathological changes after 4 weeks of MeHg exposure and compared the results to those obtained from rats treated with lead (Pb) acetate or cadmium (Cd) chloride, which have been reported to cause oxidative stress in tissues [103–105]. Three plasma oxidative stress markers were measured using a free radical elective evaluator (FREE) (Diacron International srl, Grosseto, Italy) and dedicated reagents. In this measurement, dROM quantifies the metabolite ROOH to evaluate oxidative stress, -SHp was measured using Ellman's reagent (5,5′-dithiobis-2-nitrobenzoic acid), and BAP quantifies the reducing power of endogenous antioxidants to iron, which acts as a sensitive antioxidant. Such measurements of systemic oxidative stress markers using FREE have already been used clinically [106,107]. In this study, we identified the decreased capacity of -SHp, GPx3 activity, and SeP1 as useful potential plasma biomarkers of ongoing MeHg cytotoxicity [108]. Among them, the -SHp level significantly decreased 2 weeks after MeHg exposure, which is an early stage at which no systemic oxidative stress, histopathological changes, or clinical signs were detected. These findings suggest that the -SHp level is useful for assessing the early effects of MeHg exposure and subsequent changes in structure/function induced by MeHg intoxication. On the other hand, exposure to Pb or Cd did not alter oxidative stress markers in the plasma, although glial fibrillary acidic protein immunolabeled astrocytes were detected in the cerebellum of Pb-treated rats as previously reported [109,110], and the doses of Cd used in the study were reported to induce oxidative stress or neurotoxicity in previous studies [111–113]. In contrast to MeHg, an increase in SeP1 was observed in both Pb- and Cd-treated rats. The disruption of the antioxidant system caused by the high affinity of MeHg for the selenohydryl and sulfhydryl groups is a specific feature of MeHg-induced oxidative stress. Plasma biomarkers reflecting such features are useful for the evaluation of ongoing MeHg-mediated oxidative stress.

8. Prevention against MeHg-Induced Cytotoxicity

Based on the findings of the mechanism of MeHg toxicity, there have been many reports on effective treatments to prevent cytotoxicity triggered by MeHg-induced oxidative stress in vitro and in vivo. MeHg toxicity can be prevented by co-treatment or pretreatment with below-mentioned chemicals. Effective treatments to prevent in vivo toxicity triggered by MeHg-induced oxidative stress are summarized in Table 1.

Table 1. In vivo protection against methylmercury (MeHg) toxicity.

Treatment	Function	Animal	Effectiveness	Ref.
Trolox	radical scavenging	rat	cytotoxicity clinical feature	[9]
α-tocopherol	radical scavenging	rat	cytotoxicity clinical feature	[14]
sodium selenite	replenishment of selenium	rat	clinical feature	[114, 115]
selenomethionine	replenishment of selenium	postnatal rat	clinical feature	[116]
NAC	replenishment of GSH	rat	DNA synthesis hippocampal cell number	[117]
sulforaphane	Nrf2 activator	mice	cellular Hg content clinical feature	[118]

NAC, N-acetyl-L-cysteine; GSH, glutathione.

8.1. Radical Scavenging Chemicals

The early incidence of ROS leads to further progression of MeHg toxicity. Therefore, a number of radical-scavenging chemicals have been tried to prevent MeHg toxicity in vitro and in vivo. They include α-tocopherol [14,119], Trolox (a water-soluble derivative of vitamin E) [9,30], n-propyl gallate [30,120], and tocotrienol (unsaturated vitamin E) [119]. Inhibitory effect of α-tocopherol on MeHg-induced oxidative stress has been demonstrated in rats, which were treated with 5 mg/kg MeHg for 12 consecutive days and 150 mg/kg α-tocopherol for 20 consecutive days after initial MeHg administration. α-tocopherol-treated rats showed decreased lipid peroxidation and manifestation of clinical signs (hind limb crossing sign and ataxic gait) compared to rats with MeHg alone [14]. Co-treatment with Trolox protected MeHg-treated rat skeletal muscle (20 ppm MeHg in drinking water every day for 28 days) against the decrease in mitochondrial electron transport system enzyme activities (cytochrome c oxidase and succinate dehydrogenase) despite the retention of MeHg. In addition, Trolox was effective for protecting cerebellum from MeHg-induced apoptosis in rats [9]. Pre-treatment with n-propyl gallate protected cultured human cerebellar granular cells established from fetal brain tissue from MeHg cytotoxicity [120].

8.2. Replenishment of GSH, Se, or GPx1

GSH interacts with MeHg to form a GSH-MeHg complex which is excreted from the ABCC4 transporter. Further, GSH functions as a redox buffer, leading to GSH consumption under MeHg exposure. N-acetyl-L-cysteine (NAC), an amino acid derivative, provides cysteine for GSH production. In addition, it has been shown that NAC can scavenge oxidants directly, reducing hydroxyl radicals and hypochlorous acid [121]. Co-treatment with NAC protected cultured cells against MeHg-cytotoxicity [11,117]. It has been reported that co-treatment with NAC reduces MeHg-induced neurotoxicity in the developing rat hippocampus [117]. Furthermore, post-treatment with NAC accelerated urinary MeHg excretion in mice [122].

The high affinity of MeHg for the selenohydryl group and selenide should cause relative intracellular Se-deficient conditions, leading to the degradation of antioxidant selenoenzyme mRNA by NMD. Therefore, the replenishment of Se should improve such impairment. It has been reported that pretreatment with sodium selenite rescues MeHg-induced downregulation of GPx1 mRNA, an increase in intracellular ROS, and MeHg-induced decrease in TrxR1 activity in vitro [23]. Co-treatment with sodium selenite can suppress MeHg-mediated neurotoxicity in rats [114] and fetotoxicity in mice [115]. Co-treatment with selenomethionine, a food-based Se, prevented MeHg-induced neuronal degeneration and reactive astrocytosis in a postnatal rat model, suggesting that dietary Se is useful for the protection of neurons against MeHg cytotoxicity [116].

Probucol is a phenolic lipid-lowering drug with anti-inflammatory and antioxidant properties. Probucol has been shown to increase GPx1 activity in the rat heart and prevent adriamycin-induced myocardial toxicity [123]. Co-treatment with probucol protected mouse cerebellar granular cells against MeHg cytotoxicity, which was correlated with increased GPx1 activity and decreased lipid peroxidation [37].

8.3. Seleno-Organic Compound Ebselen

Ebselen, a seleno-organic compound, exhibits GPx1 mimic activity and can quench free radicals and singlet oxygen [124]. Furthermore, ebselen is an excellent direct substrate for mammalian TrxR and Trx [125,126]. Ebselen treatment does not increase the amount of bioavailable Se [124] but generates the selenol form of the compound [125]. A number of reports have demonstrated the effectiveness of ebselen. MeHg-induced glutamate release from rat brain synaptosomal preparations was inhibited by co-treatment with ebselen [127]. Co-treatment with ebselen could also protect MeHg-induced glutamate uptake inhibition in rat cerebral cortical slices [128]. Furthermore, pre-treatment with ebselen suppressed MeHg-induced inhibition of glutamine uptake in rat neonatal cortical astrocytes [129]. We also demonstrated that co-treatment with ebselen effectively suppressed the downregulation of GPx1 mRNA and the ROS increase, and finally cytotoxicity after MeHg exposure in vitro [23].

8.4. Nrf2 Activators

It has been reported that isothiocyanates causes release of Nrf2 from sequestration by Keap1, and its subsequent translocation into the nucleus, resulting in the ARE-mediated induction of oxidative stress enzyme genes [130]. ARE-mediated genes include GCL, glutathione S-transferase, and multidrug resistance-associated protein, which are associated with MeHg excretion. It has been reported that pretreatment with isothiocyanates 6-methylsulfinylhexyl isothiocyanate (6-HITC) or sulforaphane (SFN) before MeHg exposure suppresses Hg accumulation and cytotoxicity in mouse hepatocytes [118]. Furthermore, SFN treatment prior to administration of MeHg also suppressed Hg accumulation in the brains of mice and clinical sign hind-limb flaccidity [118].

8.5. ER Stress Preconditioning

Cells and tissues can be protected against potentially lethal stress by pre-exposure to the same or different milder stress. Preconditioning cytoprotection has been described in ischemic preconditioning against myocardial infarction [131,132] or delayed neuronal cell death [133] and ER stress preconditioning against renal epithelial cell oxidative injury [134] or cardiomyocyte oxidative injury [135]. We demonstrated that cells preconditioned with an inhibitor of ER Ca^{2+}-ATPase, thapsigargin, showed resistance to MeHg-induced cytotoxicity through favorable stress responses, which included phosphorylation of eukaryotic initiation factor 2 alpha, accumulation of activating transcription factor 4, upregulation of stress-related proteins, and activation of the extracellular signal-regulated kinase pathway [136].

9. Conclusions

High affinity for selenohydryl groups, sulfhydryl groups, and selenides plays a critical role in the incidence of MeHg-mediated oxidative stress. It causes the impairment of many antioxidant enzymes and proteins, resulting in the disruption of antioxidant systems. On the other hand, MeHg can also activate an antioxidant signaling pathway. Cellular antioxidant transcription factor Nrf2 can be activated through the direct interaction of MeHg and Nrf2 regulator Keap1. Furthermore, MeHg-induced Se deficiency due to the high affinity of MeHg for selenohydryl groups and selenides leads to failure in the recoding of a UGA codon for Sec and results in the degradation of antioxidant selenoenzyme mRNA by NMD. The incidence of MeHg-mediated oxidative stress arises from the interactions of individual intracellular redox systems and the aforementioned MeHg-mediated events. MeHg-mediated oxidative stress causes apoptosis through activation of ASK1 and the subsequent SAPK/JNK pathways. MeHg induces site-specific cerebrocortical neuronal cell death, which was observed in layer IV of the cerebral cortex mainly composed of excitable cells. It has been demonstrated that the mechanism of MeHg-induced site specific cortical neuronal damage is caused by cerebrocortical neuronal hyperactivity triggered by MeHg-mediated oxidative stress and subsequent activation of the p38 MAPK-CREB pathway.

A schematic overview of MeHg-mediated oxidative stress and activation of the antioxidant Nrf2 pathway is presented in Figure 2.

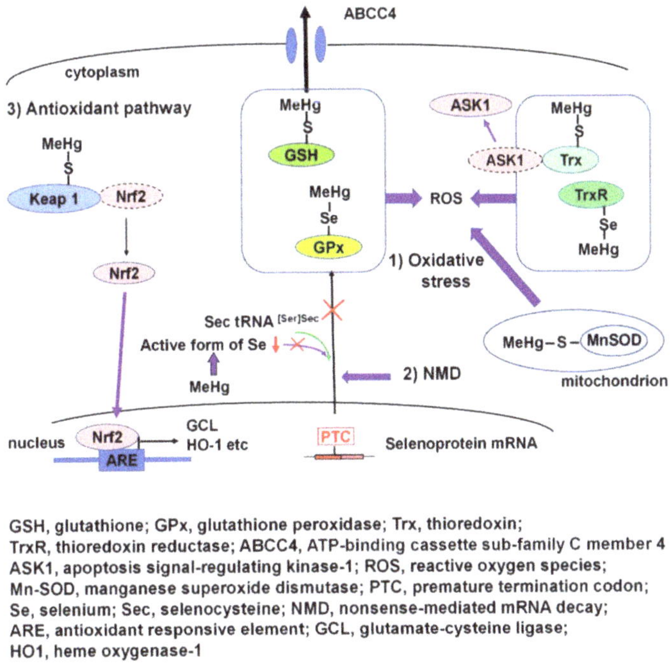

Figure 2. Schematic overview of MeHg-mediated oxidative stress and activation of the antioxidant Nrf2 pathway. (1) Thiols and selenols of many antioxidant proteins and enzymes of glutathione (GSH) and thioredoxin (Trx) cellular redox systems, such as GSH, GPx1, and TrxR, interact with MeHg, resulting in the disruption of the cellular redox system and the incidence of oxidative stress. The interaction of MeHg and Trx results in release and activation of apoptosis signal-regulating kinase-1 (ASK1) followed by ASK1-dependent apoptosis. In addition, interaction of mitochondrial Mn-SOD with MeHg may contribute to an increase in reactive oxygen species (ROS). (2) MeHg-induced relative intracellular Se deficiency causes failure in the recoding of a UGA codon for Sec because of active Se deficiency and results in the degradation of antioxidant selenoenzyme mRNA by nonsense-mediated mRNA decay (NMD). (3) The interaction of MeHg and Nrf2 regulator Keap1 leads to the activation of antioxidant transcription factor Nrf2, resulting in the antioxidant responsive element-mediated induction of oxidative stress enzyme genes, such as glutamate-cysteine ligase (GCL) and heme oxygenase-1 (HO-1).

Author Contributions: Conceptualization, M.F. and F.U.; writing—original draft preparation, F.U.; writing—review and editing, M.F. and F.U. All authors have read and agreed to the published version of the manuscript.

Funding: This research received no external funding.

Conflicts of Interest: The authors declare no conflict of interest.

References

1. Takeuchi, T. Pathology of Minamata disease. With special reference to its pathogenesis. *Acta Pathol. Jpn.* **1982**, *32* (Suppl. 1), 73–99.
2. Bakir, F.; Damluji, S.F.; Amin-Zaki, L.; Murtadha, M.; Khalidi, A.; Al-Rawi, N.Y.; Tikriti, S.; Dahahir, H.I.; Clarkson, T.W.; Smith, J.C.; et al. Methylmercury poisoning in Iraq. *Science* **1973**, *181*, 230–241. [CrossRef]
3. Davis, L.E.; Kornfeld, M.; Mooney, H.S.; Fiedler, K.J.; Haaland, K.Y.; Orrison, W.W.; Cernichiari, E.; Clarkson, T.W. Methylmercury poisoning: Long-term clinical, radiological, toxicological, and pathological studies of an affected family. *Ann. Neurol.* **1994**, *35*, 680–688. [CrossRef]

4. Harada, M.; Akagi, H.; Tsuda, T.; Kizaki, T.; Ohno, H. Methylmercury level in umbilical cords from patients with congenital Minamata disease. *Sci. Total Environ.* **1999**, *234*, 59–62. [CrossRef]
5. Hunter, D.; Russell, D.S. Focal cerebellar and cerebellar atrophy in a human subject due to organic mercury compounds. *J. Neurol. Neurosurg. Psychiatry* **1954**, *17*, 235–241. [CrossRef]
6. Eto, K. Pathology of Minamata disease. *Toxicol. Pathol.* **1997**, *25*, 614–623. [CrossRef]
7. Eto, K.; Tokunaga, H.; Nagashima, K.; Takeuchi, T. An autopsy case of minamata disease (methylmercury poisoning)—pathological viewpoints of peripheral nerves. *Toxicol. Pathol.* **2002**, *30*, 714–722. [CrossRef]
8. Nakamura, M.; Hachiya, N.; Murata, K.Y.; Nakanishi, I.; Kondo, T.; Yasutake, A.; Miyamoto, K.; Ser, P.H.; Omi, S.; Furusawa, H.; et al. Methylmercury exposure and neurological outcomes in Taiji residents accustomed to consuming whale meat. *Environ. Int.* **2014**, *68*, 25–32. [CrossRef]
9. Usuki, F.; Yasutake, A.; Umehara, F.; Tokunaga, H.; Matsumoto, M.; Eto, K.; Ishiura, S.; Higuchi, I. In vivo protection of a water-soluble derivative of vitamin E, Trolox, against methylmercury-intoxication in the rat. *Neurosci. Lett.* **2001**, *304*, 199–203. [CrossRef]
10. Ali, S.F.; LeBel, C.P.; Bondy, S.C. Reactive oxygen species formation as a biomarker of methylmercury and trimethyltin neurotoxicity. *Neurotoxicology* **1992**, *13*, 637–648.
11. Usuki, F.; Ishiura, S. Expanded CTG repeats in myotonin protein kinase increase susceptibility to oxidative stress. *Neuroreport* **1998**, *9*, 2291–2296. [CrossRef]
12. Usuki, F.; Fujita, E.; Sasagawa, N. Methylmercury activates ASK1/JNK signaling pathways, leading to apoptosis due to both mitochondria- and endoplasmic reticulum (ER)-generated processes in myogenic cell lines. *Neurotoxicology* **2008**, *29*, 22–30. [CrossRef]
13. Kaur, P.; Schulz, K.; Heggland, I.; Aschner, M.; Syversen, T. The use of fluorescence for detecting MeHg-induced ROS in cell cultures. *Toxicol. Vitr. Int. J. Publ. Assoc. Bibra* **2008**, *22*, 1392–1398. [CrossRef]
14. Yamashita, T.; Ando, Y.; Nakamura, M.; Obayashi, K.; Terazaki, H.; Haraoka, K.; Guo, S.X.; Ueda, M.; Uchino, M. Inhibitory effect of α-tocopherol on methylmercury-induced oxidative steress. *Environ. Health Prev. Med.* **2004**, *9*, 111–117. [CrossRef]
15. Sugiura, Y.; Tamai, Y.; Tanaka, H. Selenium protection against mercury toxicity: High binding affinity of methylmercury by selenium-containing ligands in comparison with sulfur-containing ligands. *Bioinorg. Chem.* **1978**, *9*, 167–180. [CrossRef]
16. Hughes, W.L. A physicochemical rationale for the biological activity of mercury and its compounds. *Ann. N. Y. Acad. Sci.* **1957**, *65*, 454–460. [CrossRef]
17. Kanda, H.; Shinkai, Y.; Kumagai, Y. S-mercuration of cellular proteins by methylmercury and its toxicological implications. *J. Toxicol. Sci.* **2014**, *39*, 687–700. [CrossRef]
18. Makino, K.; Okuda, K.; Sugino, E.; Nishiya, T.; Toyama, T.; Iwawaki, T.; Fujimura, M.; Kumagai, Y.; Uehara, T. Correlation between attenuation of protein disulfide isomerase activity through S-mercuration and neurotoxicity induced by methylmercury. *Neurotox. Res.* **2015**, *27*, 99–105. [CrossRef]
19. Kumagai, Y.; Kanda, H.; Shinkai, Y.; Toyama, T. The role of the Keap1/Nrf2 pathway in the cellular response to methylmercury. *Oxidative Med. Cell. Longev.* **2013**, *2013*, 848279. [CrossRef]
20. Toyama, T.; Sumi, D.; Shinkai, Y.; Yasutake, A.; Taguchi, K.; Tong, K.I.; Yamamoto, M.; Kumagai, Y. Cytoprotective role of Nrf2/Keap1 system in methylmercury toxicity. *Biochem. Biophys. Res. Commun.* **2007**, *363*, 645–650. [CrossRef]
21. Wang, L.; Jiang, H.; Yin, Z.; Aschner, M.; Cai, J. Methylmercury toxicity and Nrf2-dependent detoxification in astrocytes. *Toxicol. Sci. Off. J. Soc. Toxicol.* **2009**, *107*, 135–143. [CrossRef] [PubMed]
22. Ni, M.; Li, X.; Yin, Z.; Jiang, H.; Sidoryk-Wegrzynowicz, M.; Milatovic, D.; Cai, J.; Aschner, M. Methylmercury induces acute oxidative stress, altering Nrf2 protein level in primary microglial cells. *Toxicol. Sci. Off. J. Soc. Toxicol.* **2010**, *116*, 590–603. [CrossRef] [PubMed]
23. Usuki, F.; Yamashita, A.; Fujimura, M. Post-transcriptional defects of antioxidant selenoenzymes cause oxidative stress under methylmercury exposure. *J. Biol. Chem.* **2011**, *286*, 6641–6649. [CrossRef] [PubMed]
24. Maquat, L.E. Nonsense-mediated mRNA decay: Splicing, translation and mRNP dynamics. *Nat. Rev. Mol. Cell Biol.* **2004**, *5*, 89–99. [CrossRef]
25. Zhang, J.; Sun, X.; Qian, Y.; LaDuca, J.P.; Maquat, L.E. At least one intron is required for the nonsense-mediated decay of triosephosphate isomerase mRNA: A possible link between nuclear splicing and cytoplasmic translation. *Mol. Cell. Biol.* **1998**, *18*, 5272–5283. [CrossRef]

26. Mendell, J.T.; Sharifi, N.A.; Meyers, J.L.; Martinez-Murillo, F.; Dietz, H.C. Nonsense surveillance regulates expression of diverse classes of mammalian transcripts and mutes genomic noise. *Nat. Genet.* **2004**, *36*, 1073–1078. [CrossRef]
27. Park, S.T.; Lim, K.T.; Chung, Y.T.; Kim, S.U. Methylmercury-induced neurotoxicity in cerebral neuron culture is blocked by antioxidants and NMDA receptor antagonists. *Neurotoxicology* **1996**, *17*, 37–45.
28. Yee, S.; Choi, B.H. Oxidative stress in neurotoxic effects of methylmercury poisoning. *Neurotoxicology* **1996**, *17*, 17–26.
29. Usuki, F.; Takahashi, N.; Sasagawa, N.; Ishiura, S. Differential signaling pathways following oxidative stress in mutant myotonin protein kinase cDNA-transfected C2C12 cell lines. *Biochem. Biophys. Res. Commun.* **2000**, *267*, 739–743. [CrossRef]
30. Shanker, G.; Aschner, M. Methylmercury-induced reactive oxygen species formation in neonatal cerebral astrocytic cultures is attenuated by antioxidants. *Brain Res. Mol. Brain Res.* **2003**, *110*, 85–91. [CrossRef]
31. Yee, S.; Choi, B.H. Methylmercury poisoning induces oxidative stress in the mouse brain. *Exp. Mol. Pathol.* **1994**, *60*, 188–196. [CrossRef] [PubMed]
32. Usuki, F.; Yasutake, A.; Umehara, F.; Higuchi, I. Beneficial effects of mild lifelong dietary restriction on skeletal muscle: Prevention of age-related mitochondrial damage, morphological changes, and vulnerability to a chemical toxin. *Acta Neuropathol.* **2004**, *108*, 1–9. [CrossRef]
33. Ganther, H.E.; Goudie, C.; Sunde, M.L.; Kopecky, M.J.; Wagner, P. Selenium: Relation to decreased toxicity of methylmercury added to diets containing tuna. *Science* **1972**, *175*, 1122–1124. [CrossRef] [PubMed]
34. Li, S.; Thompson, S.A.; Woods, J.S. Localization of gamma-glutamylcysteine synthetase mRNA expression in mouse brain following methylmercury treatment using reverse transcription in situ PCR amplification. *Toxicol. Appl. Pharmacol.* **1996**, *140*, 180–187. [CrossRef] [PubMed]
35. Thompson, S.A.; White, C.C.; Krejsa, C.M.; Diaz, D.; Woods, J.S.; Eaton, D.L.; Kavanagh, T.J. Induction of glutamate-cysteine ligase (gamma-glutamylcysteine synthetase) in the brains of adult female mice subchronically exposed to methylmercury. *Toxicol. Lett.* **1999**, *110*, 1–9. [CrossRef]
36. Hirota, Y.; Yamaguchi, S.; Shimojoh, N.; Sano, K.I. Inhibitory effect of methylmercury on the activity of glutathione peroxidase. *Toxicol. Appl. Pharmacol.* **1980**, *53*, 174–176. [CrossRef]
37. Farina, M.; Campos, F.; Vendrell, I.; Berenguer, J.; Barzi, M.; Pons, S.; Suñol, C. Probucol increases glutathione peroxidase-1 activity and displays long-lasting protection against methylmercury toxicity in cerebellar granule cells. *Toxicol. Sci. Off. J. Soc. Toxicol.* **2009**, *112*, 416–426. [CrossRef]
38. Jones, D.P. Radical-free biology of oxidative stress. *Am. J. Physiol. Cell Physiol.* **2008**, *295*, C849–C868. [CrossRef]
39. Zhong, L.; Arnér, E.S.; Holmgren, A. Structure and mechanism of mammalian thioredoxin reductase: The active site is a redox-active selenolthiol/selenenylsulfide formed from the conserved cysteine-selenocysteine sequence. *Proc. Natl. Acad. Sci. USA* **2000**, *97*, 5854–5859. [CrossRef]
40. Carvalho, C.M.; Chew, E.H.; Hashemy, S.I.; Lu, J.; Holmgren, A. Inhibition of the human thioredoxin system. A molecular mechanism of mercury toxicity. *J. Biol. Chem.* **2008**, *283*, 11913–11923. [CrossRef]
41. Sandalova, T.; Zhong, L.; Lindqvist, Y.; Holmgren, A.; Schneider, G. Three-dimensional structure of a mammalian thioredoxin reductase: Implications for mechanism and evolution of a selenocysteine-dependent enzyme. *Proc. Natl. Acad. Sci. USA* **2001**, *98*, 9533–9538. [CrossRef] [PubMed]
42. Dalla Corte, C.L.; Wagner, C.; Sudati, J.H.; Comparsi, B.; Leite, G.O.; Busanello, A.; Soares, F.A.; Aschner, M.; Rocha, J.B. Effects of diphenyl diselenide on methylmercury toxicity in rats. *BioMed Res. Int.* **2013**, *2013*, 983821. [CrossRef] [PubMed]
43. Ruszkiewicz, J.A.; Bowman, A.B.; Farina, M.; Rocha, J.B.T.; Aschner, M. Sex-and structure-specific differences in antioxidant responses to methylmercury during early development. *Neurotoxicology* **2016**, *56*, 118–126. [CrossRef] [PubMed]
44. Shinyashiki, M.; Kumagai, Y.; Homma-Takeda, S.; Nagafune, J.; Takasawa, N.; Suzuki, J.; Matsuzaki, I.; Satoh, S.; Sagai, M.; Shimojo, N. Selective inhibition of the mouse brain Mn-SOD by methylmercury. *Environ. Toxicol. Pharmacol.* **1996**, *2*, 359–366. [CrossRef]
45. Naganuma, A.; Miura, K.; Tanaka-Kagawa, T.; Kitahara, J.; Seko, Y.; Toyoda, H.; Imura, N. Overexpression of manganese-superoxide dismutase prevents methylmercury toxicity in HeLa cells. *Life Sci.* **1998**, *62*, Pl157–Pl161. [CrossRef]

46. Eto, K.; Takeuchi, T. A pathological study of prolonged cases of Minamata disease. With particular reference to 83 autopsy cses. *Acta Pathol. Jpn.* **1978**, *28*, 565–584. [CrossRef]
47. Eto, K.; Takizawa, Y.; Akagi, H.; Haraguchi, K.; Asano, S.; Takahata, N.; Tokunaga, H. Differential diagnosis between organic and inorganic mercury poisoning in human cases–the pathologic point of view. *Toxicol. Pathol.* **1999**, *27*, 664–671. [CrossRef]
48. Fujimura, M.; Usuki, F.; Sawada, M.; Takashima, A. Methylmercury induces neuropathological changes with tau hyperphosphorylation mainly through the activation of the c-jun-N-terminal kinase pathway in the cerebral cortex, but not in the hippocampus of the mouse brain. *Neurotoxicology* **2009**, *30*, 1000–1007. [CrossRef]
49. Fujimura, M.; Usuki, F. Low in situ expression of antioxidative enzymes in rat cerebellar granular cells susceptible to methylmercury. *Arch. Toxicol.* **2014**, *88*, 109–113. [CrossRef]
50. Fujimura, M.; Usuki, F. In situ different antioxidative systems contribute to the site-specific methylmercury neurotoxicity in mice. *Toxicology* **2017**, *392*, 55–63. [CrossRef]
51. Hatfield, D.L.; Gladyshev, V.N. How selenium has altered our understanding of the genetic code. *Mol. Cell. Biol.* **2002**, *22*, 3565–3576. [CrossRef] [PubMed]
52. Low, S.C.; Berry, M.J. Knowing when not to stop: Selenocysteine incorporation in eukaryotes. *Trends Biochem. Sci.* **1996**, *21*, 203–208. [CrossRef]
53. Low, S.C.; Grundner-Culemann, E.; Harney, J.W.; Berry, M.J. SECIS-SBP2 interactions dictate selenocysteine incorporation efficiency and selenoprotein hierarchy. *EMBO J.* **2000**, *19*, 6882–6890. [CrossRef]
54. Fagegaltier, D.; Hubert, N.; Yamada, K.; Mizutani, T.; Carbon, P.; Krol, A. Characterization of mSelB, a novel mammalian elongation factor for selenoprotein translation. *EMBO J.* **2000**, *19*, 4796–4805. [CrossRef] [PubMed]
55. Tujebajeva, R.M.; Copeland, P.R.; Xu, X.M.; Carlson, B.A.; Harney, J.W.; Driscoll, D.M.; Hatfield, D.L.; Berry, M.J. Decoding apparatus for eukaryotic selenocysteine insertion. *EMBO Rep.* **2000**, *1*, 158–163. [CrossRef]
56. Frischmeyer, P.A.; Dietz, H.C. Nonsense-mediated mRNA decay in health and disease. *Hum. Mol. Genet.* **1999**, *8*, 1893–1900. [CrossRef]
57. Holbrook, J.A.; Neu-Yilik, G.; Hentze, M.W.; Kulozik, A.E. Nonsense-mediated decay approaches the clinic. *Nat. Genet.* **2004**, *36*, 801–808. [CrossRef]
58. Yamashita, A.; Kashima, I.; Ohno, S. The role of SMG-1 in nonsense-mediated mRNA decay. *Biochim. Et Biophys. Acta* **2005**, *1754*, 305–315. [CrossRef]
59. Moriarty, P.M.; Reddy, C.C.; Maquat, L.E. Selenium deficiency reduces the abundance of mRNA for Se-dependent glutathione peroxidase 1 by a UGA-dependent mechanism likely to be nonsense codon-mediated decay of cytoplasmic mRNA. *Mol. Cell. Biol.* **1998**, *18*, 2932–2939. [CrossRef]
60. Nagy, E.; Maquat, L.E. A rule for termination-codon position within intron-containing genes: When nonsense affects RNA abundance. *Trends Biochem. Sci.* **1998**, *23*, 198–199. [CrossRef]
61. Sun, X.; Moriarty, P.M.; Maquat, L.E. Nonsense-mediated decay of glutathione peroxidase 1 mRNA in the cytoplasm depends on intron position. *EMBO J.* **2000**, *19*, 4734–4744. [CrossRef]
62. Kryukov, G.V.; Castellano, S.; Novoselov, S.V.; Lobanov, A.V.; Zehtab, O.; Guigó, R.; Gladyshev, V.N. Characterization of mammalian selenoproteomes. *Science* **2003**, *300*, 1439–1443. [CrossRef] [PubMed]
63. Lei, X.G.; Evenson, J.K.; Thompson, K.M.; Sunde, R.A. Glutathione peroxidase and phospholipid hydroperoxide glutathione peroxidase are differentially regulated in rats by dietary selenium. *J. Nutr.* **1995**, *125*, 1438–1446. [CrossRef]
64. Mori, N.; Yasutake, A.; Hirayama, K. Comparative study of activities in reactive oxygen species production/defense system in mitochondria of rat brain and liver, and their susceptibility to methylmercury toxicity. *Arch. Toxicol.* **2007**, *81*, 769–776. [CrossRef]
65. Chance, B.; Sies, H.; Boveris, A. Hydroperoxide metabolism in mammalian organs. *Physiol. Rev.* **1979**, *59*, 527–605. [CrossRef]
66. Liu, Y.; Fiskum, G.; Schubert, D. Generation of reactive oxygen species by the mitochondrial electron transport chain. *J. Neurochem.* **2002**, *80*, 780–787. [CrossRef]
67. Usuki, F.; Yasutake, A.; Matsumoto, M.; Umehara, F.; Higuchi, I. The effect of methylmercury on skeletal muscle in the rat: A histopathological study. *Toxicol. Lett.* **1998**, *94*, 227–232. [CrossRef]

68. Mori, N.; Yasutake, A.; Marumoto, M.; Hirayama, K. Methylmercury inhibits electron transport chain activity and induces cytochrome c release in cerebellum mitochondria. *J. Toxicol. Sci.* **2011**, *36*, 253–259. [CrossRef] [PubMed]
69. Chipuk, J.E.; Bouchier-Hayes, L.; Green, D.R. Mitochondrial outer membrane permeabilization during apoptosis: The innocent bystander scenario. *Cell Death Differ.* **2006**, *13*, 1396–1402. [CrossRef]
70. Zou, H.; Li, Y.; Liu, X.; Wang, X. An APAF-1.cytochrome c multimeric complex is a functional apoptosome that activates procaspase-9. *J. Biol. Chem.* **1999**, *274*, 11549–11556. [CrossRef]
71. Liu, W.; Yang, T.; Xu, Z.; Xu, B.; Deng, Y. Methyl-mercury induces apoptosis through ROS-mediated endoplasmic reticulum stress and mitochondrial apoptosis pathways activation in rat cortical neurons. *Free Radic. Res.* **2019**, *53*, 26–44. [CrossRef]
72. Lu, T.H.; Hsieh, S.Y.; Yen, C.C.; Wu, H.C.; Chen, K.L.; Hung, D.Z.; Chen, C.H.; Wu, C.C.; Su, Y.C.; Chen, Y.W.; et al. Involvement of oxidative stress-mediated ERK1/2 and p38 activation regulated mitochondria-dependent apoptotic signals in methylmercury-induced neuronal cell injury. *Toxicol. Lett.* **2011**, *204*, 71–80. [CrossRef] [PubMed]
73. Sokolowski, K.; Falluel-Morel, A.; Zhou, X.; DiCicco-Bloom, E. Methylmercury (MeHg) elicits mitochondrial-dependent apoptosis in developing hippocampus and acts at low exposures. *Neurotoxicology* **2011**, *32*, 535–544. [CrossRef] [PubMed]
74. Mustacich, D.; Powis, G. Thioredoxin reductase. *Biochem. J.* **2000**, *346 Pt 1*, 1–8.
75. Kumar, S.; Björnstedt, M.; Holmgren, A. Selenite is a substrate for calf thymus thioredoxin reductase and thioredoxin and elicits a large non-stoichiometric oxidation of NADPH in the presence of oxygen. *Eur. J. Biochem.* **1992**, *207*, 435–439. [CrossRef]
76. Björnstedt, M.; Hamberg, M.; Kumar, S.; Xue, J.; Holmgren, A. Human thioredoxin reductase directly reduces lipid hydroperoxides by NADPH and selenocystine strongly stimulates the reaction via catalytically generated selenols. *J. Biol. Chem.* **1995**, *270*, 11761–11764. [CrossRef] [PubMed]
77. Björnstedt, M.; Xue, J.; Huang, W.; Akesson, B.; Holmgren, A. The thioredoxin and glutaredoxin systems are efficient electron donors to human plasma glutathione peroxidase. *J. Biol. Chem.* **1994**, *269*, 29382–29384. [PubMed]
78. Saitoh, M.; Nishitoh, H.; Fujii, M.; Takeda, K.; Tobiume, K.; Sawada, Y.; Kawabata, M.; Miyazono, K.; Ichijo, H. Mammalian thioredoxin is a direct inhibitor of apoptosis signal-regulating kinase (ASK) 1. *EMBO J.* **1998**, *17*, 2596–2606. [CrossRef]
79. Ichijo, H.; Nishida, E.; Irie, K.; ten Dijke, P.; Saitoh, M.; Moriguchi, T.; Takagi, M.; Matsumoto, K.; Miyazono, K.; Gotoh, Y. Induction of apoptosis by ASK1, a mammalian MAPKKK that activates SAPK/JNK and p38 signaling pathways. *Science* **1997**, *275*, 90–94. [CrossRef]
80. Malhotra, J.D.; Kaufman, R.J. Endoplasmic reticulum stress and oxidative stress: A vicious cycle or a double-edged sword? *Antioxid. Redox Signal.* **2007**, *9*, 2277–2293. [CrossRef]
81. Cao, S.S.; Kaufman, R.J. Endoplasmic reticulum stress and oxidative stress in cell fate decision and human disease. *Antioxid. Redox Signal.* **2014**, *21*, 396–413. [CrossRef] [PubMed]
82. Ellgaard, L.; Ruddock, L.W. The human protein disulphide isomerase family: Substrate interactions and functional properties. *EMBO Rep.* **2005**, *6*, 28–32. [CrossRef] [PubMed]
83. Xu, S.; Sankar, S.; Neamati, N. Protein disulfide isomerase: A promising target for cancer therapy. *Drug Discov. Today* **2014**, *19*, 222–240. [CrossRef] [PubMed]
84. Ferrari, D.M.; Söling, H.D. The protein disulphide-isomerase family: Unravelling a string of folds. *Biochem. J.* **1999**, *339 Pt 1*, 1–10. [CrossRef]
85. Freedman, R.B.; Klappa, P.; Ruddock, L.W. Protein disulfide isomerases exploit synergy between catalytic and specific binding domains. *EMBO Rep.* **2002**, *3*, 136–140. [CrossRef]
86. Uehara, T.; Nakamura, T.; Yao, D.; Shi, Z.Q.; Gu, Z.; Ma, Y.; Masliah, E.; Nomura, Y.; Lipton, S.A. S-nitrosylated protein-disulphide isomerase links protein misfolding to neurodegeneration. *Nature* **2006**, *441*, 513–517. [CrossRef]
87. Wild, A.C.; Moinova, H.R.; Mulcahy, R.T. Regulation of gamma-glutamylcysteine synthetase subunit gene expression by the transcription factor Nrf2. *J. Biol. Chem.* **1999**, *274*, 33627–33636. [CrossRef]
88. Itoh, K.; Wakabayashi, N.; Katoh, Y.; Ishii, T.; Igarashi, K.; Engel, J.D.; Yamamoto, M. Keap1 represses nuclear activation of antioxidant responsive elements by Nrf2 through binding to the amino-terminal Neh2 domain. *Genes Dev.* **1999**, *13*, 76–86. [CrossRef]

89. Niture, S.K.; Jain, A.K.; Shelton, P.M.; Jaiswal, A.K. Src subfamily kinases regulate nuclear export and degradation of transcription factor Nrf2 to switch off Nrf2-mediated antioxidant activation of cytoprotective gene expression. *J. Biol. Chem.* **2011**, *286*, 28821–28832. [CrossRef]
90. Culbreth, M.; Zhang, Z.; Aschner, M. Methylmercury augments Nrf2 activity by downregulation of the Src family kinase Fyn. *Neurotoxicology* **2017**, *62*, 200–206. [CrossRef]
91. Fujimura, M.; Usuki, F. Site-specific neural hyperactivity via the activation of MAPK and PKA/CREB pathways triggers neuronal degeneration in methylmercury-intoxicated mice. *Toxicol. Lett.* **2017**, *271*, 66–73. [CrossRef] [PubMed]
92. Jones, E.G. Varieties and distribution of non-pyramidal cells in the somatic sensory cortex of the squirrel monkey. *J. Comp. Neurol.* **1975**, *160*, 205–267. [CrossRef]
93. Morgan, J.I.; Curran, T. Stimulus-transcription coupling in the nervous system: Involvement of the inducible proto-oncogenes fos and jun. *Annu. Rev. Neurosci.* **1991**, *14*, 421–451. [CrossRef] [PubMed]
94. Herdegen, T.; Leah, J.D. Inducible and constitutive transcription factors in the mammalian nervous system: Control of gene expression by Jun, Fos and Krox, and CREB/ATF proteins. *Brain Res. Brain Res. Rev.* **1998**, *28*, 370–490. [CrossRef]
95. Gillardon, F.; Skutella, T.; Uhlmann, E.; Holsboer, F.; Zimmermann, M.; Behl, C. Activation of c-Fos contributes to amyloid beta-peptide-induced neurotoxicity. *Brain Res.* **1996**, *706*, 169–172. [CrossRef]
96. Carballo-Quintás, M.; Martínez-Silva, I.; Cadarso-Suárez, C.; Alvarez-Figueiras, M.; Ares-Pena, F.J.; López-Martín, E. A study of neurotoxic biomarkers, c-fos and GFAP after acute exposure to GSM radiation at 900 MHz in the picrotoxin model of rat brains. *Neurotoxicology* **2011**, *32*, 478–494. [CrossRef]
97. Agrawal, M.; Bhaskar, A.S.; Lakshmana Rao, P.V. Involvement of mitogen-activated protein kinase pathway in T-2 toxin-induced cell cycle alteration and apoptosis in human neuroblastoma cells. *Mol. Neurobiol.* **2015**, *51*, 1379–1394. [CrossRef]
98. Cheng, J.P.; Wang, W.H.; Jia, J.P.; Zheng, M.; Shi, W.; Lin, X.Y. Expression of c-fos in rat brain as a prelude marker of central nervous system injury in response to methylmercury-stimulation. *Biomed. Environ. Sci.* **2006**, *19*, 67–72.
99. Walsh, J.J.; Friedman, A.K.; Sun, H.; Heller, E.A.; Ku, S.M.; Juarez, B.; Burnham, V.L.; Mazei-Robison, M.S.; Ferguson, D.; Golden, S.A.; et al. Stress and CRF gate neural activation of BDNF in the mesolimbic reward pathway. *Nat. Neurosci.* **2014**, *17*, 27–29. [CrossRef]
100. Jasińska, K.K.; Molfese, P.J.; Kornilov, S.A.; Mencl, W.E.; Frost, S.J.; Lee, M.; Pugh, K.R.; Grigorenko, E.L.; Landi, N. The BDNF Val66Met Polymorphism Influences Reading Ability and Patterns of Neural Activation in Children. *PLoS ONE* **2016**, *11*, e0157449. [CrossRef]
101. Fujimura, M.; Usuki, F. Methylmercury induces oxidative stress and subsequent neural hyperactivity leading to cell death through the p38 MAPK-CREB pathway in differentiated SH-SY5Y cells. *Neurotoxicology* **2018**, *67*, 226–233. [CrossRef] [PubMed]
102. Jones, D.P.; Liang, Y. Measuring the poise of thiol/disulfide couples in vivo. *Free Radic. Biol. Med.* **2009**, *47*, 1329–1338. [CrossRef] [PubMed]
103. Chen, J.; Shaikh, Z.A. Activation of Nrf2 by cadmium and its role in protection against cadmium-induced apoptosis in rat kidney cells. *Toxicol. Appl. Pharmacol.* **2009**, *241*, 81–89. [CrossRef] [PubMed]
104. Tandon, S.K.; Singh, S.; Prasad, S.; Khandekar, K.; Dwivedi, V.K.; Chatterjee, M.; Mathur, N. Reversal of cadmium induced oxidative stress by chelating agent, antioxidant or their combination in rat. *Toxicol. Lett.* **2003**, *145*, 211–217. [CrossRef]
105. Soltaninejad, K.; Kebriaeezadeh, A.; Minaiee, B.; Ostad, S.N.; Hosseini, R.; Azizi, E.; Abdollahi, M. Biochemical and ultrastructural evidences for toxicity of lead through free radicals in rat brain. *Hum. Exp. Toxicol.* **2003**, *22*, 417–423. [CrossRef]
106. Komatsu, F.; Kudoh, H.; Kagawa, Y. Evaluation of oxidative stress and effectiveness of low-dose glucocorticoid therapy on exacerbation of chronic obstructive pulmonary disease. *J. Gerontol. Ser. A Biol. Sci. Med Sci.* **2007**, *62*, 459–464. [CrossRef]
107. Tanito, M.; Kaidzu, S.; Takai, Y.; Ohira, A. Status of systemic oxidative stresses in patients with primary open-angle glaucoma and pseudoexfoliation syndrome. *PLoS ONE* **2012**, *7*, e49680. [CrossRef]
108. Usuki, F.; Fujimura, M. Decreased plasma thiol antioxidant barrier and selenoproteins as potential biomarkers for ongoing methylmercury intoxication and an individual protective capacity. *Arch. Toxicol.* **2016**, *90*, 917–926. [CrossRef]

109. Selvín-Testa, A.; Loidl, C.F.; López-Costa, J.J.; López, E.M.; Pecci-Saavedra, J. Chronic lead exposure induces astrogliosis in hippocampus and cerebellum. *Neurotoxicology* **1994**, *15*, 389–401.
110. Partl, S.; Herbst, H.; Schaeper, F.; Mohnhaupt, A.; Stoltenburg-Didinger, G. GFAP gene expression is altered in young rats following developmental low level lead exposure. *Neurotoxicology* **1998**, *19*, 547–551.
111. Ferramola, M.L.; Antón, R.I.; Anzulovich, A.C.; Giménez, M.S. Myocardial oxidative stress following sub-chronic and chronic oral cadmium exposure in rats. *Environ. Toxicol. Pharmacol.* **2011**, *32*, 17–26. [CrossRef]
112. Abdalla, F.H.; Cardoso, A.M.; Pereira, L.B.; Schmatz, R.; Gonçalves, J.F.; Stefanello, N.; Fiorenza, A.M.; Gutierres, J.M.; Serres, J.D.; Zanini, D.; et al. Neuroprotective effect of quercetin in ectoenzymes and acetylcholinesterase activities in cerebral cortex synaptosomes of cadmium-exposed rats. *Mol. Cell. Biochem.* **2013**, *381*, 1–8. [CrossRef] [PubMed]
113. Shagirtha, K.; Muthumani, M.; Prabu, S.M. Melatonin abrogates cadmium induced oxidative stress related neurotoxicity in rats. *Eur. Rev. Med Pharmacol. Sci.* **2011**, *15*, 1039–1050. [PubMed]
114. Ohi, G.; Nishigaki, S.; Seki, H.; Tamura, Y.; Maki, T. Efficacy of selenium in tuna and selenite in modifying methylmercury intoxication. *Environ. Res.* **1976**, *12*, 49–58. [CrossRef]
115. Satoh, H.; Yasuda, N.; Shimai, S. Development of reflexes in neonatal mice prenatally exposed to methylmercury and selenite. *Toxicol. Lett.* **1985**, *25*, 199–203. [CrossRef] [PubMed]
116. Sakamoto, M.; Yasutake, A.; Kakita, A.; Ryufuku, M.; Chan, H.M.; Yamamoto, M.; Oumi, S.; Kobayashi, S.; Watanabe, C. Selenomethionine protects against neuronal degeneration by methylmercury in the developing rat cerebrum. *Environ. Sci. Technol.* **2013**, *47*, 2862–2868. [CrossRef]
117. Falluel-Morel, A.; Lin, L.; Sokolowski, K.; McCandlish, E.; Buckley, B.; DiCicco-Bloom, E. N-acetyl cysteine treatment reduces mercury-induced neurotoxicity in the developing rat hippocampus. *J. Neurosci. Res.* **2012**, *90*, 743–750. [CrossRef]
118. Toyama, T.; Shinkai, Y.; Yasutake, A.; Uchida, K.; Yamamoto, M.; Kumagai, Y. Isothiocyanates reduce mercury accumulation via an Nrf2-dependent mechanism during exposure of mice to methylmercury. *Environ. Health Perspect.* **2011**, *119*, 1117–1122. [CrossRef]
119. Shichiri, M.; Takanezawa, Y.; Uchida, K.; Tamai, H.; Arai, H. Protection of cerebellar granule cells by tocopherols and tocotrienols against methylmercury toxicity. *Brain Res.* **2007**, *1182*, 106–115. [CrossRef]
120. Gassó, S.; Cristòfol, R.M.; Selema, G.; Rosa, R.; Rodríguez-Farré, E.; Sanfeliu, C. Antioxidant compounds and Ca(2+) pathway blockers differentially protect against methylmercury and mercuric chloride neurotoxicity. *J. Neurosci. Res.* **2001**, *66*, 135–145. [CrossRef]
121. Aruoma, O.I.; Halliwell, B.; Hoey, B.M.; Butler, J. The antioxidant action of N-acetylcysteine: Its reaction with hydrogen peroxide, hydroxyl radical, superoxide, and hypochlorous acid. *Free Radic. Biol. Med.* **1989**, *6*, 593–597. [CrossRef]
122. Ballatori, N.; Lieberman, M.W.; Wang, W. N-acetylcysteine as an antidote in methylmercury poisoning. *Environ. Health Perspect.* **1998**, *106*, 267–271. [CrossRef]
123. Li, T.; Singal, P.K. Adriamycin-induced early changes in myocardial antioxidant enzymes and their modulation by probucol. *Circulation* **2000**, *102*, 2105–2110. [CrossRef]
124. Sies, H. Ebselen, a selenoorganic compound as glutathione peroxidase mimic. *Free Radic. Biol. Med.* **1993**, *14*, 313–323. [CrossRef]
125. Zhao, R.; Masayasu, H.; Holmgren, A. Ebselen: A substrate for human thioredoxin reductase strongly stimulating its hydroperoxide reductase activity and a superfast thioredoxin oxidant. *Proc. Natl. Acad. Sci. USA* **2002**, *99*, 8579–8584. [CrossRef] [PubMed]
126. Zhao, R.; Holmgren, A. A novel antioxidant mechanism of ebselen involving ebselen diselenide, a substrate of mammalian thioredoxin and thioredoxin reductase. *J. Biol. Chem.* **2002**, *277*, 39456–39462. [CrossRef] [PubMed]
127. Farina, M.; Dahm, K.C.; Schwalm, F.D.; Brusque, A.M.; Frizzo, M.E.; Zeni, G.; Souza, D.O.; Rocha, J.B. Methylmercury increases glutamate release from brain synaptosomes and glutamate uptake by cortical slices from suckling rat pups: Modulatory effect of ebselen. *Toxicol. Sci. Off. J. Soc. Toxicol.* **2003**, *73*, 135–140. [CrossRef] [PubMed]
128. Moretto, M.B.; Funchal, C.; Santos, A.Q.; Gottfried, C.; Boff, B.; Zeni, G.; Pureur, R.P.; Souza, D.O.; Wofchuk, S.; Rocha, J.B. Ebselen protects glutamate uptake inhibition caused by methyl mercury but does not by Hg2+. *Toxicology* **2005**, *214*, 57–66. [CrossRef] [PubMed]

129. Yin, Z.; Lee, E.; Ni, M.; Jiang, H.; Milatovic, D.; Rongzhu, L.; Farina, M.; Rocha, J.B.; Aschner, M. Methylmercury-induced alterations in astrocyte functions are attenuated by ebselen. *Neurotoxicology* **2011**, *32*, 291–299. [CrossRef]
130. Xu, C.; Yuan, X.; Pan, Z.; Shen, G.; Kim, J.H.; Yu, S.; Khor, T.O.; Li, W.; Ma, J.; Kong, A.N. Mechanism of action of isothiocyanates: The induction of ARE-regulated genes is associated with activation of ERK and JNK and the phosphorylation and nuclear translocation of Nrf2. *Mol. Cancer Ther.* **2006**, *5*, 1918–1926. [CrossRef]
131. Murry, C.E.; Jennings, R.B.; Reimer, K.A. Preconditioning with ischemia: A delay of lethal cell injury in ischemic myocardium. *Circulation* **1986**, *74*, 1124–1136. [CrossRef] [PubMed]
132. Yellon, D.M.; Downey, J.M. Spotlight on preconditioning. *Cardiovasc. Res.* **2002**, *55*, 425–428. [CrossRef]
133. Hayashi, T.; Saito, A.; Okuno, S.; Ferrand-Drake, M.; Chan, P.H. Induction of GRP78 by ischemic preconditioning reduces endoplasmic reticulum stress and prevents delayed neuronal cell death. *J. Cereb. Blood Flow Metab.* **2003**, *23*, 949–961. [CrossRef] [PubMed]
134. Hung, C.C.; Ichimura, T.; Stevens, J.L.; Bonventre, J.V. Protection of renal epithelial cells against oxidative injury by endoplasmic reticulum stress preconditioning is mediated by ERK1/2 activation. *J. Biol. Chem.* **2003**, *278*, 29317–29326. [CrossRef] [PubMed]
135. Zhang, P.L.; Lun, M.; Teng, J.; Huang, J.; Blasick, T.M.; Yin, L.; Herrera, G.A.; Cheung, J.Y. Preinduced molecular chaperones in the endoplasmic reticulum protect cardiomyocytes from lethal injury. *Ann. Clin. Lab. Sci.* **2004**, *34*, 449–457. [PubMed]
136. Usuki, F.; Fujimura, M.; Yamashita, A. Endoplasmic reticulum stress preconditioning attenuates methylmercury-induced cellular damage by inducing favorable stress responses. *Sci. Rep.* **2013**, *3*, 2346. [CrossRef]

MDPI
St. Alban-Anlage 66
4052 Basel
Switzerland
Tel. +41 61 683 77 34
Fax +41 61 302 89 18
www.mdpi.com

Antioxidants Editorial Office
E-mail: antioxidants@mdpi.com
www.mdpi.com/journal/antioxidants

www.ingramcontent.com/pod-product-compliance
Lightning Source LLC
Chambersburg PA
CBHW040752250926
51850CB00007BB/1482
9783036580548